ADVANCED TOPICS
IN SCIENCE AND TECHNOLOGY IN CHINA

ADVANCED TOPICS IN SCIENCE AND TECHNOLOGY IN CHINA

Zhejiang University is one of the leading universities in China. In Advanced Topics in Science and Technology in China, Zhejiang University Press and Springer jointly publish monographs by Chinese scholars and professors, as well as invited authors and editors from abroad who are outstanding experts and scholars in their fields. This series will be of interest to researchers, lecturers, and graduate students alike.

Advanced Topics in Science and Technology in China aims to present the latest and most cutting-edge theories, techniques, and methodologies in various research areas in China. It covers all disciplines in the fields of natural science and technology, including but not limited to, computer science, materials science, life sciences, engineering, environmental sciences, mathematics, and physics.

Zheng-Ming Huang
Ye-Xin Zhou

Strength of Fibrous Composites

With 109 figures

Authors
Prof. Zheng-Ming Huang
School of Aerospace Engineering
& Applied Mechanics
Tongji University
1239 Siping Road, Shanghai 200092
China
E-mail: huangzm@tongji.edu.cn

Mr. Ye-Xin Zhou
Department of Mechanical Engineering
The University of Hong Kong
Pokfulam, Hong Kong
China
E-mail: Caihui6871@163.com

Additional material to this book can be downloaded from http://extra.springer.com.

ISSN 1995-6819 e-ISSN 1995-6827
Advanced Topics in Science and Technology in China

ISBN 978-7-308-08268-6
Zhejiang University Press, Hangzhou

ISBN 978-3-642-22957-2 e-ISBN 978-3-642-22958-9
Springer Heidelberg Dordrecht London New York

Library of Congress Control Number: 2011933465

Printed on acid-free paper

Springer is a part of Springer Science+Business Media (www.springer.com)

Preface

Laminated composites made of continuous fibers and metal, ceramic, or polymer matrices have been used for structural applications for more than half a century. Many modern industries such as aerospace engineering or wind power energy engineering would not have advanced to their current levels if composites had not been used. Among all of the superior characteristics of composites in comparison with other more traditional, isotropic structural materials, three are the most well known. They are high-specific stiffness (stiffness to mass ratio), high-specific strength and the ability to tailor desired properties by choosing suitable fiber and matrix materials as well as the fiber architecture geometry.

Determination of the composite mechanical properties has attracted the attention of scientists, researchers and engineers. From an application point of view, it would be best if all of the mechanical properties of the composites can be estimated by using their constituent fiber and matrix properties and the fiber architecture parameters, i.e., by using a micromechanical approach. For the composite stiffness, this is feasible. There are many micromechanical models for efficiently estimating the effective elastic properties of laminated composites, which have been the focus of most of the available mechanics of composite materials textbooks and monographs. A very challenging problem, however, is to estimate the composite strength as well as other inelastic behaviors micromechanically. In the current literature, there is a lack of a book systematically addressing this problem. Almost all of the monographs dealing with laminate strength follow a phenomenological philosophy. Namely, the laminate strength is estimated based on the information of lamina strengths, which must be measured on composites themselves. However, predicting laminate strength micromechanically is very important, as one of the most critical issues in designing a composite structure is to know its load carrying capacity *in priori*. Only when this capacity has been explicitly related to the constituent properties and geometric parameters, can an optimal design choosing proper constituent materials, fiber content and architecture, and laminate layups for the structure before fabrication, be achieved.

Would it be possible to dream that any mechanical property, including the ultimate load carrying capacity of a composite made using any continuous fiber architecture subjected to arbitrary loads, would be simply available without any experiment on it but be based only on an established database containing the required constituent properties? Will this become a reality? More than a decade ago, the first author of this book established a unified micromechanical theory, the bridging model, to describe the constitutive relationship of a composite up to the point of failure. The unique feature of this theory is that the internal stresses in the constituent fiber and matrix materials of the composite under any arbitrary load conditions, including a temperature variation, can be evaluated using rigorous and explicit equations. By assuming that a composite failure is caused by either the fiber or the matrix failure, a micromechanical strength theory for the composite is established. The last decade has seen sound development of the bridging model as well as its applications to the analysis of mechanical properties, especially strengths of various fibrous composites. The assessment by the World Wide Failure Exercises (WWFE-I and WWFE-II, also known as "Failure Olympics" in the composite community) has confirmed the efficiency and accuracy of this model.

This book systematically deals with the bridging model development as well as applications to strength prediction of unidirectional (UD) laminas and multidirectional laminates. The model can be derived in terms of an Eshelby's tensor. Presented in Chapter 1 is the classical Eshelby's problem as well as other pre-requirements in mechanics and mathematics to understand the bridging model theory and applications. Chapter 2 addresses a general elastic-plastic constitutive theory, the Prandtl-Reuss theory, for isotropic materials. This theory is used to describe the matrix behavior in a composite. Chapter 3 is the key to this book, where the bridging model development is shown in detail. An interesting outcome is that by making use of a bridging matrix, any micromechanical model for predicting effective elastic moduli of a UD composite can be formulated into a unified expression. In Chapter 4, the strength of UD composites is dealt with. Closed-form formulae for strengths of a UD lamina under uniaxial loads are derived. Modified maximum normal stress failure criteria for both multiaxial tension and multiaxial compression of a constituent are set forth. Strengths at elevated temperatures or subjected to fatigue loads are analyzed. Application of the bridging model to predict the strength of multidirectional laminates subjected to various load conditions is a main focus of this book, and is addressed in Chapter 5. Either the classical or a pseudo 3D laminate theory is incorporated with the bridging model to determine the internal stresses in the fibers and matrix of the laminate subjected to 2D or 3D load conditions. Fatal and nonfatal failures are classified. In additional to a variety of strength prediction examples, the WWFE-I and WWFE-II problems are analyzed with detailed discussions. The chapter ends with the highlight of the simulation procedure for inelastic and strength properties of woven, braided and knitted fabric reinforced composite laminates. The analyzing formulae have been programmed into a computer routine in the FORTRAN language, which is shown in Chapter 6. Supplymentary materials to this book containing the original code of the computer routine can be found from http://extra.springer.com. Input data for running the routine to resolve several illustrated examples and to analyze the WWFE-I and WWFE-II problems are included in the supplymentary materials.

The book is intended for senior and postgraduate students in engineering. It can be regarded as an extension to Strength/Mechanics of Materials textbooks. Researchers and engineers who are working with composite materials will also find this book useful. Any comment on the book can be sent to huangzm@tongji.edu.cn or huangzm@email.com. The authors would like to express their heartiest gratitude for any comments, in advance.

Zheng-Ming Huang
Ye-Xin Zhou
July,2011

Contents

1

Background

1.1 Scope of This Book

It has been recognized that technological development depends on advances in the field of materials. Whatever the field may be, the final limitations will rest on the available materials. In some industries, conventional monolithic materials are currently operating at or near their limits and do not offer the potential for meeting the demands of further technical advancement (Lerch & Saltsman, 1993). In this regard, composites represent nothing less than a giant step in the ever-lasting endeavor to achieve optimisation of materials.

Composite materials are made on a macroscopic scale from two or more distinct phases of constituent materials. They are developed to achieve unique mechanical properties and other superior performance characteristics that would be impossible with any of the constituent materials alone. As most practical synthetic composites are essentially constructed from two-phase composite components, we thus only need to focus on those composites having two distinct constituent materials, a continuous phase and a reinforcement phase. The continuous phase is commonly referred to as a matrix, which may be metal, ceramic or polymer. The geometric form of the reinforcement phase can be powders, particles, short fibers, whiskers or continuous fibers. Only continuous fiber reinforced composites are considered in this book. Thus, the fiber reinforced or, simply, the fibrous composites referred to throughout this book are considered as those made from continuous fiber reinforcement. However, the continuous fibers can be arranged in an arbitrary form, such as uni-/multi-directional, woven, braided or knitted preforms.

Modern composites made using continuous fiber preforms and various types of matrices have generated a revolution in high-performance structures in a number of industries such as aerospace, shipbuilding, sports equipment, automobile construction, energy, and so on. Advanced fibrous composites offer significantly high stiffness and strength to weight ratios, compared to conventional monolithic materials such as metallic materials. This is mainly because a material in very thin fiber form has a much higher mechanical performance than in its bulk form (Griffith, 1920; Gordon, 1976). Another advantage of fibrous composites is that people can freely select different constituent materials, their contents and their arrangement for an optimum performance.

A fundamental issue in making use of a composite, the same as in the use of any other material, is to understand thoroughly its mechanical properties, especially its ultimate load-carrying capacity. Metals, polymers and ceramics are essentially isotropic and homogenous, and have predictable properties. Hence, material selection, component design and manufacturing are fairly straightforward. On the other hand, composites essentially display anisotropic behaviors and their mechanical responses are different if loaded in different directions. Not surprisingly, the use of composites presents a whole new array of challenges for a designer. The designer must deal with anisotropic materials in his component design and understand how the properties of raw constituent materials, together with the specifics of potential manufacturing methods (possible reinforcement form and geometry, and relative proportions of fiber and matrix) will influence the properties of the final product.

The purpose of this book is to provide a comprehensive methodology to determine the mechanical behaviors, particularly the ultimate load-carrying capacity of fibrous composites from the knowledge of their constituent properties, volume fractions of the constituent materials, geometric arrangement of the reinforcing phase in the matrix, the laminate stacking sequence, etc. The composite forms considered include unidirectional laminae and multidirectional laminates.

1.2 Linear Elasticity

In order to investigate the mechanical behaviors of a material, especially for practical applications, stress and deformation analysis is necessary. The mechanics of elasticity can be considered as the theoretical basis for estimating the elastic stress and deformation of any solid structure or structural material under the action of any general loading (Zhang, 2003). Basic assumptions and concepts of linear elasticity will be briefly summarized here.

Two types of notations are used to designate rectangular coordinates of a point in a material geometry. One is (x, y, z)-notation and another is (x_1, x_2, x_3)-notation. Usually, they refer to two different right-hand coordinate systems, the origins of which may or may not coincide. As long as they refer to the same coordinate system, it is always true that $x_1=x$, $x_2=y$, and $x_3=z$.

When the material under consideration is subjected to some excitation, such as an external load, the initial point P: (x_1, x_2, x_3), will deform to a new point P': $(x_1+u_1, x_2+u_2, x_3+u_3)$, as shown in Fig. 1.1, where u_1, u_2, and u_3 are the displacement components of the point P along x_1-, x_2- and x_3-directions, respectively. One of the fundamental assumptions for linear elasticity is that all of the three displacement components, u_1, u_2, and u_3, are infinitesimal. From these displacement components, we get the infinitesimal strains of the point P as (Timoshenko & Goodier, 1970)

$$\varepsilon_{ij} = \frac{1}{2}\left(\frac{\partial u_i}{\partial x_j} + \frac{\partial u_j}{\partial x_i}\right), \quad i, j = 1, 2, 3 \tag{1.1}$$

Using these strain components, we can define a second-order strain tensor (matrix) $[\varepsilon_{ij}]$, which has a dimension of 3×3. It is seen from Eq. (1.1) that the strain tensor is symmetric. Only six of them are independent. Thus, instead of the strain tensor, we can use a contracted strain vector, $\{\varepsilon_i\}$, to represent the strain state of the point P where

$$\{\varepsilon_i\}^{\mathrm{T}}=\{\varepsilon_1, \varepsilon_2, \varepsilon_3, \varepsilon_4, \varepsilon_5, \varepsilon_6\}=\{\varepsilon_{11}, \varepsilon_{22}, \varepsilon_{33}, 2\varepsilon_{23}, 2\varepsilon_{13}, 2\varepsilon_{12}\} \tag{1.2}$$

The superscript "T" in Eq. (1.2) denotes a transposition. Note that there is a factor "2" before the shear strains, ε_{23}, ε_{13}, and ε_{12}.

As the material has been subjected to the external load, stresses are generated. Let $[\sigma_{ij}]$ denote a 3×3 stress tensor at the point P. By using an infinitesimal volume element containing P and by applying equilibrium conditions, it can be shown (Timoshenko & Goodier, 1970) that the stress tensor is always symmetric. We can thus also use a contracted vector, $\{\sigma_i\}$, to represent the stress state of the point P where

$$\{\sigma_i\}^{\mathrm{T}}= \{\sigma_1, \sigma_2, \sigma_3, \sigma_4, \sigma_5, \sigma_6\}=\{\sigma_{11}, \sigma_{22}, \sigma_{33}, \sigma_{23}, \sigma_{13}, \sigma_{12}\} \tag{1.3}$$

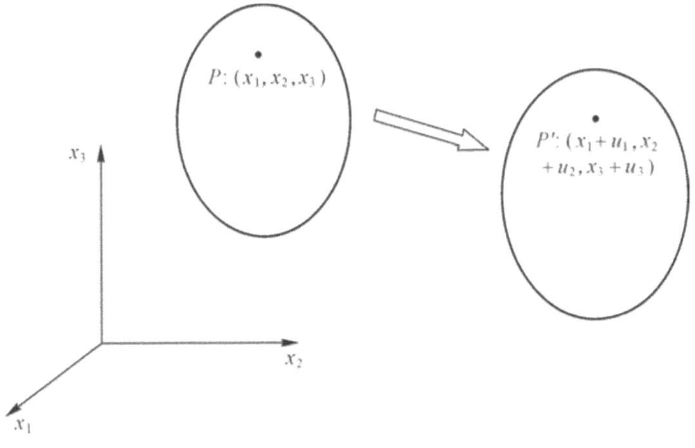

Fig. 1.1 Deformation of a material point

At any point P of the material, the elastic strain $\{\varepsilon_i\}$ is related to the stress $\{\sigma_j\}$ by Hooke's law,

$$\{\varepsilon_i\}=[S_{ij}]\{\sigma_j\} \tag{1.4.1}$$

or

$$\{\sigma_i\}=[C_{ij}]\{\varepsilon_j\} \tag{1.4.2}$$

where the 6×6 matrices $[S_{ij}]$ and $[C_{ij}]$ are named as compliance and stiffness matrices of the material, respectively. Each can be obtained from inverting the other, i.e.,

$$[S_{ij}]=[C_{ij}]^{-1} \tag{1.5.1}$$

$$[C_{ij}]=[S_{ij}]^{-1} \tag{1.5.2}$$

When the deformation of the material is in an elastic range (i.e., all of the displacement components, u_1, u_2, and u_3, vanish if the external load is reduced to zero), there is a strain energy function W such that

$$W=\frac{1}{2}\sigma_{ij}\varepsilon_{ij}=\frac{1}{2}\sum_{i=1}^{3}\sum_{j=1}^{3}\sigma_{ij}\varepsilon_{ij} \tag{1.6}$$

Here and in the following, a summation convention is applied to any repeated subscripts, such as i and j in Eq. (1.6), in their variation range. Therefore,

$$\sigma_{ij}=\frac{\partial W}{\partial \varepsilon_{ij}}, \text{ or } \varepsilon_{ij}=\frac{\partial W}{\partial \sigma_{ij}} \tag{1.7}$$

The function W is always positive for any non-zero stress or strain tensor. This means that the compliance and stiffness matrices, $[S_{ij}]$ and $[C_{ij}]$, are always positive definite. From Eq. (1.7), we can further conclude that the matrices $[S_{ij}]$ and $[C_{ij}]$ are always symmetric because, after substituting Eq. (1.4.1) or Eq. (1.4.2) into Eq. (1.6), the resulting function W is a quadratic equation and the coefficient matrix of a quadratic can always be made to be symmetric. Hence, there are at most 21 independent elastic constants for any material. If, however, the material has some symmetric planes, i.e., the planes with respect to which the material properties are the same, the number of the independent constants can be reduced further (Timoshenko & Goodier, 1970). In engineering practice, there are three kinds of materials that are most commonly encountered. They are isotropic, transversely isotropic and orthotropic materials.

1.2.1 Isotropic Material

If the material is symmetric with respect to every direction, it is said to be isotropic. Most metals, ceramics and polymers are isotropic materials. In general, matrix materials used in composite fabrication are essentially taken as isotropic. For this kind of material, there are only two independent elastic constants. They are usually given in terms of engineering moduli, i.e., Young's modulus, E, and Poisson's ratio, ν. Young's modulus is defined as the slope of a uniaxial stress-strain curve of the material at an initial stage, whereas the Poisson's ratio is defined as the negative of the ratio of the transverse strain over the longitudinal strain when a testing load is applied in the longitudinal direction. The compliance matrix, $[S_{ij}]$, of an isotropic material takes the form

$$[S_{ij}] = \begin{bmatrix} [S_{ij}]_\sigma & 0 \\ 0 & [S_{ij}]_\tau \end{bmatrix} \tag{1.8}$$

where $[S_{ij}]_\sigma$ and $[S_{ij}]_\tau$ are the sub-matrices of the compliance relating normal stresses with elongation strains and shear stresses with shear strains, respectively, and are given by

$$[S_{ij}]_\sigma = \begin{bmatrix} \frac{1}{E} & -\frac{\nu}{E} & -\frac{\nu}{E} \\ & \frac{1}{E} & -\frac{\nu}{E} \\ & \text{symmetry} & \frac{1}{E} \end{bmatrix} \tag{1.9}$$

$$[S_{ij}]_\tau = \begin{bmatrix} \frac{1}{G} & 0 & 0 \\ & \frac{1}{G} & 0 \\ & \text{symmetry} & \frac{1}{G} \end{bmatrix} \tag{1.10}$$

In Eq. (1.10), G is the shear modulus defined as

$$G=0.5E/(1+\nu) \tag{1.11}$$

1.2.2 Transversely Isotropic Material

A material is said to be transversely isotropic if its elastic properties are kept unchanged with respect to an arbitrary rotation around a given axis. For convenience of illustration, this axis is called the symmetric axis (or direction). Such a kind of material is of special importance in the study of fibrous composites, since a unidirectional (abbreviated to "UD") composite, the most important fibrous composite, is generally considered as transversely isotropic. When fibers are uniformly arranged in the matrix in such a way that the axes of the fibers are parallel to each other, the material is said to be an unidirectionally fiber-reinforced composite. A UD composite is also called a UD lamina. Fig. 1.2 shows a high-contrast micrograph of the transverse plane section of such a boron fiber-aluminium matrix composite. The dark dots represent the cross-sections of the boron fibers and the white area designates the continuous aluminium matrix. From the figure, it can be easily concluded that the material properties are best considered as invariant (symmetric) with respect to any rotation about the fiber axis, since both the boron and the aluminium are isotropic. A further conclusion is that the resulting composite is still transversely isotropic even if the fiber material is transversely isotropic but has a symmetric direction along the fiber axis. This is important because a number of commonly used fibers such as graphite, carbon and aramid (Kevlar) are transversely isotropic.

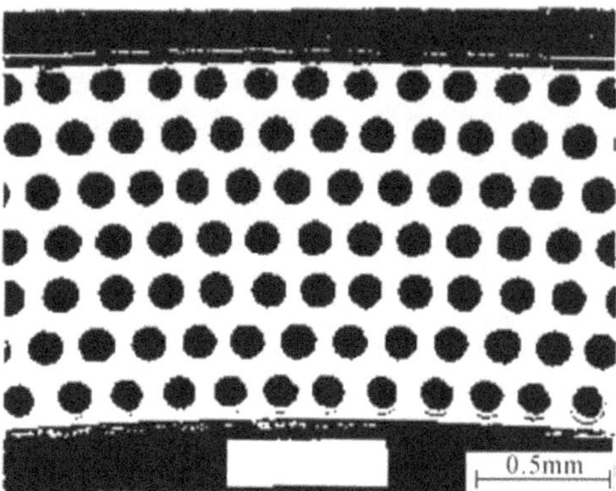

Fig. 1.2 A micrograph of the cross-sectional plane of a UD composite. The black dots are fibers and the white continuum is the matrix

For a transversely isotropic material, let its symmetric axis (the fiber axis in a UD composite) be x_1. The compliance matrix of the material is the same as that given by Eq. (1.8) but with different sub-matrices, which read

$$[S_{ij}]_\sigma = \begin{bmatrix} \dfrac{1}{E_{11}} & -\dfrac{\nu_{12}}{E_{11}} & -\dfrac{\nu_{12}}{E_{11}} \\ & \dfrac{1}{E_{22}} & -\dfrac{\nu_{23}}{E_{22}} \\ \text{symmetry} & & \dfrac{1}{E_{22}} \end{bmatrix} \tag{1.12}$$

$$[S_{ij}]_\tau = \begin{bmatrix} \dfrac{1}{G_{23}} & 0 & 0 \\ & \dfrac{1}{G_{12}} & 0 \\ \text{symmetry} & & \dfrac{1}{G_{12}} \end{bmatrix} \tag{1.13}$$

In Eqs. (1.12) and (1.13), E_{11} and E_{22} are Young's moduli in x_1 and x_2 (or x_3) directions, whereas ν_{12} and ν_{23} are the Poisson's ratios, and G_{12} and G_{23} are shear moduli in the x_1–x_2 (or x_1–x_3) and x_2–x_3 planes, respectively. Note that among the six material constants, E_{11}, E_{22}, ν_{12}, ν_{23}, G_{23}, and G_{12}, only five of them are independent. The constants E_{22}, ν_{23}, and G_{23} are related by

$$G_{23}=E_{22}/(2+2\nu_{23}) \tag{1.14}$$

Therefore, a transversely isotropic material has only five independent elastic constants.

1.2.3 Orthotropic Material

When a material has three mutually orthogonal planes, with respect to which its elastic constants are the same, it is said to be orthotropic. Most composites exhibit essentially orthotropic behaviour. An important feature of an orthotropic material is that an applied shear stress does not result in any elongation strain and *vice versa*. Therefore, the compliance matrix of the orthotropic material still takes the form of Eq. (1.8). The two sub-matrices are amended to

$$[S_{ij}]_\sigma = \begin{bmatrix} \dfrac{1}{E_{11}} & -\dfrac{\nu_{12}}{E_{11}} & -\dfrac{\nu_{13}}{E_{11}} \\ & \dfrac{1}{E_{22}} & -\dfrac{\nu_{23}}{E_{22}} \\ \text{symmetry} & & \dfrac{1}{E_{33}} \end{bmatrix} \tag{1.15}$$

$$[S_{ij}]_\tau = \begin{bmatrix} \dfrac{1}{G_{23}} & 0 & 0 \\ & \dfrac{1}{G_{13}} & 0 \\ \text{symmetry} & & \dfrac{1}{G_{12}} \end{bmatrix} \tag{1.16}$$

where E_{11}, E_{22}, E_{33} are Young's moduli in x_1, x_2, and x_3 directions, respectively; ν_{ij} is Poisson's ratio, which is defined as the negative of the ratio of transverse strain in the x_j-direction over the axial strain in the x_i-direction when a uniaxial testing load is applied in the x_i-direction, i.e., $\nu_{ij}=(-\varepsilon_{jj}/\varepsilon_{ii})$; G_{ij} is shear modulus in the x_i–x_j plane.

Eqs. (1.15) and (1.16) indicate that there are, together, nine independent elastic constants for an orthotropic material. They must be all provided simultaneously. As can be expected, experimental determination of all the nine constants may be difficult or expensive, in general.

Example 1.1 Write out the stiffness matrix of an orthotropic material.

Solution. The stiffness matrix is represented by Eq. (1.5.2), i.e.,

$$[C_{ij}]=\left[S_{ij}\right]^{-1}=\begin{bmatrix}\left[S_{ij}\right]_\sigma & 0\\ 0 & \left[S_{ij}\right]_\tau\end{bmatrix}^{-1}=\begin{bmatrix}\left[S_{ij}\right]_\sigma^{-1} & 0\\ 0 & \left[S_{ij}\right]_\tau^{-1}\end{bmatrix},$$

where $\left[S_{ij}\right]_\sigma$ and $\left[S_{ij}\right]_\tau$ are given by Eqs. (1.15) and (1.16), respectively. Thus, the sub-matrices $\left[S_{ij}\right]_\sigma^{-1}$ and $\left[S_{ij}\right]_\tau^{-1}$ are derived as

$$\left[S_{ij}\right]_\sigma^{-1}=\begin{bmatrix}C_{11} & C_{12} & C_{13}\\ & C_{22} & C_{23}\\ \text{symmetry} & & C_{33}\end{bmatrix} \text{ and } \left[S_{ij}\right]_\tau^{-1}=\begin{bmatrix}G_{23} & 0 & 0\\ & G_{13} & 0\\ \text{symmetry} & & G_{12}\end{bmatrix}.$$

where $C_{11}=E_{11}\dfrac{1-\nu_{23}\nu_{32}}{\varDelta}$, $C_{12}=E_{11}\dfrac{\nu_{21}+\nu_{31}\nu_{23}}{\varDelta}$, $C_{13}=E_{11}\dfrac{\nu_{31}+\nu_{21}\nu_{32}}{\varDelta}$,

$$C_{22}=E_{22}\frac{1-\nu_{13}\nu_{31}}{\varDelta},\quad C_{23}=E_{22}\frac{\nu_{32}+\nu_{12}\nu_{31}}{\varDelta},\quad C_{33}=E_{33}\frac{1-\nu_{12}\nu_{21}}{\varDelta},$$

$$\varDelta=1-\nu_{12}\nu_{21}-\nu_{23}\nu_{32}-\nu_{31}\nu_{13}-2\nu_{21}\nu_{32}\nu_{13},\quad \nu_{ij}/E_{ii}=\nu_{ji}/E_{jj},\ i,j=1,2,3$$

1.3 Basic Concepts

In this section we will introduce the following concepts: representative volume element (RVE), volume averaged stress and strain, and maximum fiber volume fraction.

1.3.1 Representative Volume Element (RVE)

One of the most important concepts in mechanics of composite theory development is that of a representative volume element.

As aforementioned, a fibrous composite is essentially made from two constituent materials, i.e., fibers and matrix. While the matrix is a continuous phase, the fibers are discontinuous from a transverse plane view, as indicated in

Fig. 1.2. Namely, the fibers are isolated but densely distributed in the matrix. As the fiber distribution is not continuous, a composite theory cannot be established point-wisely, in contrast to the theories for isotropic materials. Thus, a fundamental step is to take a representative volume element (abbreviated to "RVE") for the composite. By definition, an RVE is the smallest material element that has two characteristics. Firstly, the whole composite can be constructed by repeating the RVE. Secondly, the composite properties can be completely represented by the properties of the RVE. Roughly speaking, an RVE to the composite is equivalent to a geometric point to an isotropic material. Thus, parallel to an isotropic material theory that is developed point-wisely, a composite theory is developed RVE-wisely. It is noted that an RVE generally has a finite volume whereas a geometric point is infinitesimal in volume.

For a unidirectional fiber reinforced composite (Fig. 1.2), a concentric cylinder can well represent its RVE, as shown in Fig. 1.3, in which the central cylinder stands for the fiber and the outside cylinder for the matrix. For some other kinds of fibrous preform reinforced composites, their representative volume elements are schematically shown in Figs. 5.57 and 5.58.

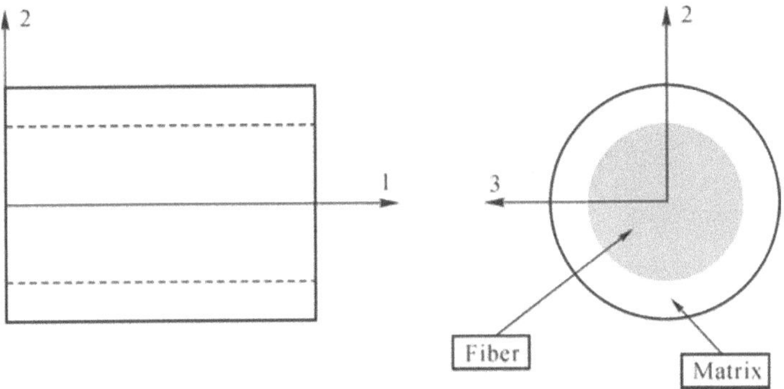

Fig. 1.3 A representative volume element (RVE) of a UD composite

1.3.2 Volume Averaged Stress and Strain

As a composite theory is developed RVE-wisely, any quantity involved must be volume averaged with respect to the RVE. Let us consider volume averaged stress and strain. Due to imperfect fabrication, voids may also occur in a composite. Hence, a two-phase composite geometry possibly contains three kinds of regions, i.e., fiber occupied region, matrix occupied region and void region. Let us denote by V' the volume of the RVE such as that shown in Fig. 1.3. The volumes of the fiber, matrix and voids in the RVE are V'_f, V'_m, and V'_v, respectively. Suppose that the i-th stress and strain in the RVE are σ_i and ε_i , which may be different at a different point. Namely, they are point-wise quantities. The volume-averaged stress $\overline{\sigma_i}$ of the composite is defined as

$$\begin{aligned}\overline{\sigma_i} &= \frac{1}{V'}\int_{V'}\sigma_i \mathrm{d}V = \frac{1}{V'}\left[\int_{V'_f}\sigma_i \mathrm{d}V + \int_{V'_m}\sigma_i \mathrm{d}V + \int_{V'_v}\sigma_i \mathrm{d}V\right] \\ &= \left(\frac{V'_f}{V'}\right)\left(\frac{1}{V'_f}\int_{V'_f}\sigma_i \mathrm{d}V\right) + \left(\frac{V'_m}{V'}\right)\left(\frac{1}{V'_m}\int_{V'_m}\sigma_i \mathrm{d}V\right) \\ &= V_f\overline{\sigma_i^f} + V_m\overline{\sigma_i^m}\end{aligned} \tag{1.17}$$

Here, suppose that no stress is transmitted in the voids. In Eq. (1.17), $V_f = V'_f / V'$ and $V_m = V'_m / V'$ are referred to as volume fractions of the fiber and the matrix and $\overline{\sigma_i^f}$ and $\overline{\sigma_i^m}$ are volume-averaged internal stresses in the fiber and the matrix, respectively. The void volume fraction is calculated from

$$V_v = 1 - V_f - V_m \tag{1.18}$$

Similarly, the volume-averaged strain of the composite is given by

$$\overline{\varepsilon_i} = \frac{1}{V'}\int_{V'}\varepsilon_i \mathrm{d}V = \frac{1}{V'}\left[\int_{V'_f}\varepsilon_i \mathrm{d}V + \int_{V'_m}\varepsilon_i \mathrm{d}V + \int_{V'_v}\varepsilon_i \mathrm{d}V\right] \tag{1.19}$$

Unlike the stress, the strain in voids does not vanish. The void strain is defined in terms of the boundary displacements of the voids. The integral over the volume of the voids can be replaced by an integral over the void boundary based on a divergence theorem, which can be found in any textbook on calculus (Rektorys, 1977). In view of this, Eq. (1.19) can be rewritten as

$$\overline{\varepsilon_i} = V_f\overline{\varepsilon_i^f} + V_m\overline{\varepsilon_i^m} + V_v\overline{\varepsilon_i^v}, \tag{1.20}$$

where $\overline{\varepsilon_i^f}$, $\overline{\varepsilon_i^m}$, and $\overline{\varepsilon_i^v}$ are the volume-averaged strains of the fiber, matrix and voids, respectively. In practice, the void content of a composite is generally small. For example, typical autoclave-cured composites may have void contents in the range of 0.1% – 1%. An engineering composite generally cannot have a void content of more than 5%. Thus, we can neglect the last term on the right-hand side of Eq. (1.20) and use

$$\overline{\varepsilon_i} = V_f\overline{\varepsilon_i^f} + V_m\overline{\varepsilon_i^m} \tag{1.21}$$

in all the subsequent discussions with the understanding that $V_f + V_m = 1$. Here and in the following, the suffixes (either superscripts or subscripts) "f" and "m" refer to the fiber and matrix phases, respectively. A quantity without any suffix designates composite or, sometimes, a special kind of material. Eqs. (1.17) and (1.21) are valid for every $i = 1, 2, \ldots, 6$. As in this book we are solely dealing with volume-averaged quantities, unless otherwise stated, the overbars in Eqs. (1.17) and (1.21) can be omitted. Thus, we obtain the following two fundamental equations:

$$\{\sigma_i\} = V_f\{\sigma_i^f\} + V_m\{\sigma_i^m\}, \tag{1.22}$$

$$\{\varepsilon_i\} = V_f\{\varepsilon_i^f\} + V_m\{\varepsilon_i^m\} \tag{1.23}$$

No other pre-assumptions have been made during the derivation of Eqs. (1.22) and (1.23), except for an implication that the volumes V', V'_f, and V'_m remain constants, together with a negligibly small void content. They must be valid regardless of any kind of load condition as well as for any constituent materials.

Let us further consider constitutive equations relating to the stress and strain vectors in different phases of materials. Suppose that $[S_{ij}^f]$ and $[S_{ij}^m]$ are, respectively, the compliance matrices of the fiber and the matrix materials. These two matrices are invariant with respect to a volume average. Thus,

$$\{\varepsilon_i^f\} = [S_{ij}^f]\{\sigma_j^f\}, \tag{1.24.1}$$

$$\{\varepsilon_i^m\} = [S_{ij}^m]\{\sigma_j^m\}, \tag{1.24.2}$$

and further,

$$\{\varepsilon_i\} = [S_{ij}]\{\sigma_j\}, \tag{1.25}$$

where $[S_{ij}]$ denotes the compliance matrix of the composite.

Substituting Eqs. (1.24.1), (1.24.2) and (1.25) into Eq. (1.23), a relation connecting $[S_{ij}]$ with $[S_{ij}^f]$ and $[S_{ij}^m]$ is given by

$$[S_{ij}]\{\sigma_j\} = V_f[S_{ij}^f]\{\sigma_j^f\} + V_m[S_{ij}^m]\{\sigma_j^m\} \tag{1.26}$$

Eq. (1.26) indicates that as long as the average stresses in the fibers, matrix and composite, i.e., $\{\sigma_j^f\}$, $\{\sigma_j^m\}$ and $\{\sigma_j\}$, together with the compliance matrices of the constituent materials, $[S_{ij}^f]$ and $[S_{ij}^m]$, have been known, the equivalent compliance matrix of the composite $[S_{ij}]$ can be obtained.

1.3.3 Maximum Fiber Volume Fraction

It can be understood that the fiber volume fraction will play a critical role in the composite macroscopic response. The maximum fiber volume fraction depends on fiber arrangements. Let us consider ideal square and triangular arrays shown in

Figs. 1.4(a) and 1.4(b). If we assume that the fiber spacing, s, and the fiber diameter, d, do not change along the fiber length, then the area fractions must be equal to the volume fractions. Indeed, optical determination of area fractions (and hence volume fractions) is possible by using micrograph pictures. The fiber volume fraction for the square array is obtained by dividing the area of fiber enclosed in the square by the total area of the square (Gibson, 1994), i.e.,

$$V_f = \frac{\pi}{4}\left(\frac{d}{s}\right)^2 \tag{1.27}$$

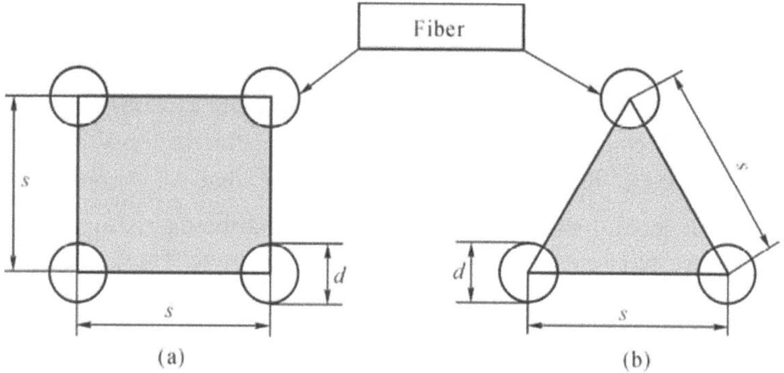

Fig. 1.4 Representative area elements of idealized square and triangular fiber-packing geometry

Note that all fibers are assumed to have a circular cross-section. Clearly, the maximum theoretical fiber volume fraction occurs when s=d. In such a case,

$$V_{f\max} = \pi / 4 = 0.785 \tag{1.28}$$

Hence, a square array arrangement of fibers can give a maximum fiber volume fraction of 0.785. Similarly, the fiber volume fraction of the triangular array arrangement (Fig. 1.4(b)) is given by

$$V_f = \frac{\pi}{2\sqrt{3}}\left(\frac{d}{s}\right)^2 \tag{1.29}$$

and the maximum value results if s=d, giving

$$V_{f\max} = \frac{\pi}{2\sqrt{3}} = 0.907 \tag{1.30}$$

1.4 Micromechanics

Micromechanics is a theory for studying the macroscopic response or the mechanical property (also called "effective property") of a unidirectional fiber reinforced composite based on the properties and the geometrical occupations of its constituent materials. It should be realized that although a micromechanics theory was originally developed for UD composites, it can also be applied to obtain the mechanical properties of other fibrous composites, such as those considered in this book. Details will be described in Chapter 5.

In terms of the micromechanics approach, perhaps the first attempts were carried out by Taylor (1938) and extended by Bishop and Hill (1951) for the response of a polycrystal composed of single crystals. In the last fifty years, micromechanics has been a very active subject in the literature about composites and a number of micromechanical models have been developed. Only several of the simplest ones are summarized here.

The following assumptions are generally made in the development of a micromechanics theory:

(1) The fibers are uniformly distributed throughout the matrix.

(2) The surfaces of the fiber and matrix phases are in direct contact and are bonded perfectly (either chemically or physically) so that there is no slippage at the phase interface before composite failure.

(3) The volume of the voids in the RVE is negligibly small and the constituent volume fractions remain unchanged.

1.4.1 Rule of Mxture Formulae

One of the simplest micromechanical models for predicting the elastic constants of fibrous composites is the rule of mixture approach. Consider a composite lamina with the fiber axis in the x direction, Fig. 1.5. The representative volume element of this composite is chosen to be a rectangular fiber bar embedded in a matrix plate. The fiber is assumed to have a rectangular cross section with the same thickness as the matrix plate. This will simplify the derivation for the plane elastic moduli of the composite.

There are three more assumptions in the rule of mixture approach, i.e.,

(1) When a uniaxial load is applied, only the corresponding internal stress in the constituent materials will be generated, and all the other internal stresses are equal to zero.

(2) The volume averaged longitudinal (i.e., the fiber axis directional) strains in the fiber, matrix and composite are the same when a uniaxial load is applied longitudinally.

(3) The volume averaged transverse and shear stresses in the fiber, matrix and composite are, respectively, equal to each other when any other kind of uniaxial load, except for the longitudinal load, is applied.

Let us apply different uniaxial loads to the composite, separately. Firstly, only a longitudinal stress is applied to the composite. In such a load condition, the above basic assumptions give

$$\varepsilon_{xx} = \varepsilon^f_{xx} = \varepsilon^m_{xx},\ \sigma_{yy} = \sigma^f_{yy} = \sigma^m_{yy} = 0 \text{ and } \sigma_{xy} = \sigma^f_{xy} = \sigma^m_{xy} = 0 \tag{1.31}$$

From Eq. (1.4.2) and using Eqs. (1.22) and (1.31), we obtain (due to the uniaxial stress-state)

$$\sigma_{xx} = E_{xx}\varepsilon_{xx} = V_f\sigma^f_{xx} + V_m\sigma^m_{xx} = V_f E^f_{xx}\varepsilon^f_{xx} + V_m E^m_{xx}\varepsilon^m_{xx} = (V_f E^f_{xx} + V_m E^m_{xx})\varepsilon_{xx}$$

Therefore, the overall longitudinal Young's modulus of the composite is given by

$$E_{xx} = V_f E^f_{xx} + V_m E^m_{xx} \tag{1.32}$$

Similarly, from

$$\varepsilon_{yy} = -v_{xy}\varepsilon_{xx} = V_f\varepsilon^f_{yy} + V_m\varepsilon^m_{yy} = V_f(-v^f_{xy}\varepsilon^f_{xx}) + V_m(-v^m_{xy}\varepsilon^m_{xx}) = -(V_f v^f_{xy} + V_m v^m_{xy})\varepsilon_{xx},$$

we obtain the overall longitudinal Poisson's ratio to be

$$v_{xy} = V_f v^f_{xy} + V_m v^m_{xy} \tag{1.33}$$

Next, apply a transverse stress only. According to the basic assumptions, the stress states generated in the fiber, matrix and composite are

$$\sigma_{xx} = \sigma^f_{xx} = \sigma^m_{xx} = 0,\ \sigma_{yy} = \sigma^f_{yy} = \sigma^m_{yy} \neq 0 \text{ and } \sigma_{xy} = \sigma^f_{xy} = \sigma^m_{xy} = 0 \tag{1.34}$$

The overall strain in the y direction is derived as

$$\varepsilon_{yy} = \frac{\sigma_{yy}}{E_{yy}} = V_f\varepsilon^f_{yy} + V_m\varepsilon^m_{yy} = V_f\left(\frac{\sigma^f_{yy}}{E^f_{yy}}\right) + V_m\left(\frac{\sigma^m_{yy}}{E^m_{yy}}\right) = \left(\frac{V_f}{E^f_{yy}} + \frac{V_m}{E^m}\right)\sigma_{yy}$$

Hence, the resulting transverse Young's modulus is obtained from

$$\frac{1}{E_{yy}} = \frac{V_f}{E^f_{yy}} + \frac{V_m}{E^m} \tag{1.35}$$

Finally, apply a pure shear stress, σ_{xy}, to the composite. The resulting stress states are

$$\sigma_{xx} = \sigma^f_{xx} = \sigma^m_{xx} = 0,\ \sigma_{yy} = \sigma^f_{yy} = \sigma^m_{yy} = 0 \text{ and } \sigma_{xy} = \sigma^f_{xy} = \sigma^m_{xy} \neq 0 \tag{1.36}$$

From the overall shear strain,

$$\varepsilon_{xy} = \frac{\sigma_{xy}}{G_{xy}} = V_f\varepsilon^f_{xy} + V_m\varepsilon^m_{xy} = V_f\left(\frac{\sigma^f_{xy}}{G^f_{xy}}\right) + V_m\left(\frac{\sigma^m_{xy}}{G^m_{xy}}\right) = \left(\frac{V_f}{G^f_{xy}} + \frac{V_m}{G^m}\right)\sigma_{xy}$$

we obtain the longitudinal shear modulus as

$$\frac{1}{G_{xy}} = \frac{V_f}{G_{xy}^f} + \frac{V_m}{G^m} \tag{1.37}$$

In summary, the rule of mixture approach gives the formulae for the engineering elastic constants as follows (using 1, 2, and 3 instead of x, y and z):

$$E_{11} = V_f E_{11}^f + V_m E^m \tag{1.38.1}$$

$$\nu_{12} = V_f \nu_{12}^f + V_m \nu^m \tag{1.38.2}$$

$$E_{22} = \frac{E^m}{1 - V_f (1 - E^m / E_{22}^f)} \tag{1.38.3}$$

$$G_{12} = \frac{G^m}{1 - V_f (1 - G^m / G_{12}^f)} \tag{1.38.4}$$

$$G_{23} = \frac{G^m}{1 - V_f (1 - G^m / G_{23}^f)} \tag{1.38.5}$$

It has been verified through many experiments that the longitudinal Young's modulus and Poisson's ratio formulae, i.e., Eqs. (1.38.1) and (1.38.2), are sufficiently accurate. However, the transverse Young's modulus and shear modulus are much underestimated. See also the evidence shown in Chapter 3. Due to this drawback, many modifications have been proposed to refine the transverse and shear moduli.

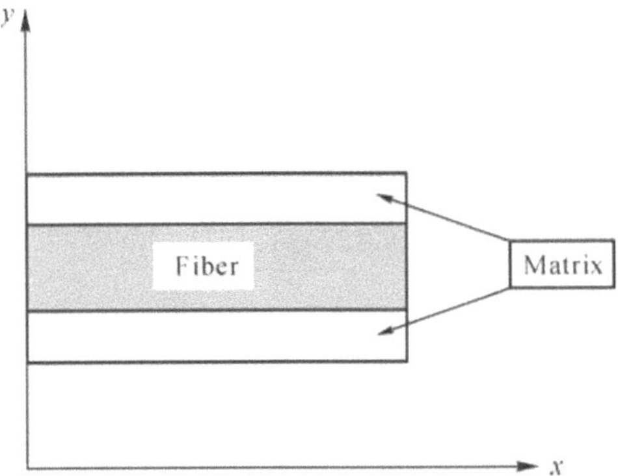

Fig. 1.5 A UD composite model for rule of mixture approach

1.4.2 Chamis Formulae

It is seen that the rule of mixture formulae, Eqs. (1.38.1) – (1.38.5), are derived without referring to any particular fiber-packing geometry. This is the case for E_{11} and ν_{12}. On the other hand, the transverse and shear moduli that are inaccurately estimated by Eqs. (1.38.3) and (1.38.4) might have some dependency on the fiber-packing geometry. Hopkins and Chamis (1988) have developed a refined model for transverse and shear moduli based on a square fiber-packing array and a method of dividing the RVE into sub-regions. The derivation adapted follows Hopkins and Chamis (1988).

A square fiber arrangement is shown in Fig. 1.4(a) and the RVE for such an array is indicated in Fig. 1.6(a). The circular cross section of the fiber is replaced by an equivalent square section, which has the same area as the circular one, and the RVE is divided into several subregions (Fig. 1.6(b)). The square fiber shown in Fig. 1.6(b) must then have the dimension

$$s_f = \sqrt{\pi / 4}d . \tag{1.39}$$

On the other hand, Eq. (1.27) gives

$$s = \sqrt{\frac{\pi}{4V_f}}d \tag{1.40}$$

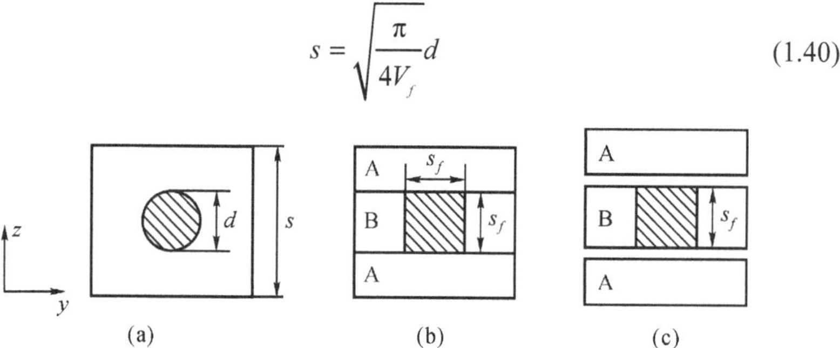

Fig. 1.6 A representative area used in Hopkins and Chamis model derivation

The RVE is divided into subregions A and B, Fig. 1.6(c). Let the subregion B be subjected to the stress state of Eq. (1.34). The effective transverse modulus for this subregion, E_{Byy}, is found from (Eq. (1.35))

$$\frac{1}{E_{Byy}} = \frac{(s_f / s)}{E_{yy}^f} + \frac{(s_m / s)}{E^m} \tag{1.41}$$

where $s_m = s - s_f$ is the matrix dimension and E_{yy}^f represents the transverse modulus of the fiber (the fiber can be transversely isotropic but the matrix is isotropic). From Eqs. (1.39) and (1.40) it is seen that

$$\frac{s_f}{s} = \sqrt{V_f} \quad \text{and} \quad \frac{s_m}{s} = 1 - \sqrt{V_f} \tag{1.42}$$

Substituting Eq. (1.42) into Eq. (1.41) we get

$$E_{Byy} = \frac{E^m}{1 - \sqrt{V_f}(1 - E^m / E_{yy}^f)} \tag{1.43}$$

As the subregions A and B (using E_{Byy} to denote an equivalent "single phase" material) along the y direction have the same feature as that shown in Fig. 1.5, the stress-state of Eq. (1.31) should be applicable. Therefore, the effective Young's modulus in this direction should take the following form

$$E_{yy} = E_{Byy}\frac{s_f}{s} + E^m \frac{s_m}{s} \tag{1.44}$$

Combining Eqs. (1.42) and (1.43), Eq. (1.44) is rewritten as

$$E_{yy} = E^m \left((1 - \sqrt{V_f}) + \frac{\sqrt{V_f}}{1 - \sqrt{V_f}(1 - E^m / E_{yy}^f)} \right) \tag{1.45}$$

A similar result may be obtained for G_{xy}. The detailed derivation by Hopkins and Chamis (1988) also included the effect of a fiber/matrix interphase material, which was assumed to be an annular volume surrounding the fiber. The complete set of equations for effective moduli of the three-phase model is given in Hopkins and Chamis's publication (1988).

In separate publications, Chamis (1984, 1989) presented the so-called "simplified micromechanical equations", which are based on the same method of subregions, except that only the terms for subregion B (Fig. 1.6) are retained. Thus, the simplified micromechanical equation for E_{yy} would be the same as that for E_{Byy} in Eq. (1.43). The whole set of Chamis's (1989) equations are given below (using 1, 2, and 3 instead of x, y, and z):

$$E_{11} = V_f E_{11}^f + V_m E^m \tag{1.46.1}$$

$$\nu_{12} = V_f \nu_{12}^f + V_m \nu^m \tag{1.46.2}$$

$$E_{22} = E_{33} = \frac{E^m}{1 - \sqrt{V_f}(1 - E^m / E_{22}^f)} \tag{1.46.3}$$

$$G_{12} = G_{13} = \frac{G^m}{1 - \sqrt{V_f}(1 - G^m / G_{12}^f)} \tag{1.46.4}$$

$$G_{23} = \frac{G^m}{1-\sqrt{V_f}(1-G^m / G_{23}^f)} \tag{1.46.5}$$

It is noted that the simplified Chamis formulae can give slightly more accurate results than the corresponding Hopkins and Chamis's modifications (also Chapter 3). Compared with Eqs. (1.38.1) – (1.38.5), it can be seen that when V_f in Eqs. (1.38.3) – (1.38.5) is replaced by $\sqrt{V_f}$ the rule of mixture formulae becomes exactly the same as Chamis's formulae.

1.4.3 Hill-Hashin-Christensen-Lo Formulae

Hill (1964, 1965a, 1965b) and Hashin (1964, 1965, 1979) independently obtained the following formulae for four of the five effective moduli of a UD composite.

$$E_{11}=V_f E_{11}^f + V_m E^m + \frac{4(\nu_{12}^f - \nu^m)^2 V_f (1-V_f)}{\frac{V_f}{k^m} + \frac{1-V_f}{k^f} + \frac{1}{G^m}} \tag{1.47.1}$$

$$\nu_{12} = V_f \nu_{12}^f + V_m \nu^m + \frac{(\nu_{12}^f - \nu^m) V_f (1-V_f)}{\frac{V_f}{k^m} + \frac{1-V_f}{k^f} + \frac{1}{G_{12}}} \left(\frac{1}{k^m} - \frac{1}{k^f}\right) \tag{1.47.2}$$

$$G_{12} = G^m \frac{(G_{12}^f + G^m) + V_f (G_{12}^f - G^m)}{(G_{12}^f + G^m) - V_f (G_{12}^f - G^m)} \tag{1.47.3}$$

$$E_{22} = \frac{2}{0.5 / K_L + 0.5 / G_{23} + 2\nu_{12}^2 / E_{11}} \tag{1.47.4}$$

where $k^f = \frac{E_{11}^f}{3(1-2\nu_{12}^f)}$, $k^m = \frac{E^m}{3(1-2\nu^m)}$

and $K_L = k^m + \frac{G^m}{3} + \frac{V_f}{\frac{1}{k^f - k^m + (G_{12}^f - G^m)/3} + \frac{1-V_f}{k^m + 4G^m/3}}$

The Hill and Hashin's micromechanical model was made complete by Christensen and Lo (1979; see also (Christensen, 1991)), by presenting the last formula for the transverse shear modulus of the composite. Unfortunately,

Christensen and Lo's formula is only applicable to composites made from two isotropic materials (Swanson, 1997; Berthelot, 1999). Christensen & Lo's formula for the transverse shear modulus is given by (Berthelot, 1999):

$$G_{23} = G^m \left(1 + \frac{V_f}{G^m / (G^f - G^m) + (k^m + 7G^m / 3)(1 - V_f) / (2k^m + 8G^m / 3)} \right) \quad (1.47.5)$$

1.5 Eshelby's Problem

By definition, micromechanics is a theory for composites based on the properties and geometrical structures of their constituent materials. Thus, in order to develop micromechanical models, it is of importance to understand the stress and strain fields in the constituent materials. Some theories, e.g., rule of mixture and Chamis's models, make use of more or less simplified assumptions for the stress and strain distributions. However, in general one does not know how the distributions are obtained. On the other hand, efforts have been made to derive the stress and strain fields based on the theory of elasticity. Amongst them, Eshelby's work (Eshelby, 1957, 1959) was one of the most outstanding approaches and has served as a basis for the generation of many other micromechanical models, such as the self-consistent scheme (Hill, 1964, 1965a, 1965b; Chou, 1980), the generalized self-consistent scheme (Christensen & Lo, 1979), and the Mori-Tanaka method (Mori & Tanaka, 1973; Mura, 1982). Eshelby's problem and its basic characteristics are briefly summarized in this section.

1.5.1 Eshelby's Approach

In order to describe Eshelby's problem more easily, let us use the concept of an eigenstrain, which is referred to as a non-elastic strain, such as thermal expansion, phase transformation, initial strain or misfit strain. Eigenstrain was introduced by Mura (Mura, 1982) and is usually denoted by ε_{ij}^*.

Eshelby's problem can be expressed as follows. An inclusion defined as a sub-domain Ω is embedded in an infinite domain D, as shown in Fig 1.7. An eigenstrain ε_{ij}^* is assigned to Ω, which is assumed to be zero in D–Ω. Elastic moduli of the Ω and D–Ω are assumed to be the same. By definition, Eshelby's problem is to determine the stress and strain fields caused by the eigenstrain. Due to this reason, Eshelby's problem is sometimes called an eigenstrain problem.

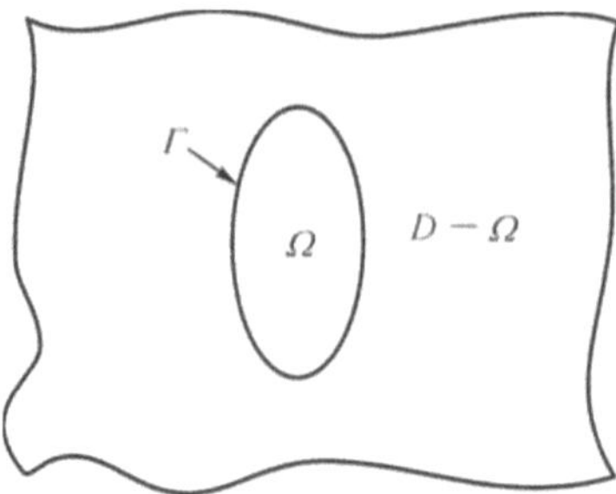

Fig. 1.7 Inclusion field

Due to the constraint by the part surrounding Ω, stresses σ_{ij} are induced both in the domain Ω and in D–Ω. For the same reason, the actual strain ε_{ij} is no longer equal to the eigenstrain ε_{ij}^*, and should be expressed as the sum of the eigenstrain and the elastic strain caused by the stresses. In order to deal with this problem, it can be divided into three sub-problems as follows.

(1) Divide the whole region into two separated domains, i.e., Ω and D–Ω. Suppose that the domain Ω can deform freely and an eigenstrain ε_{ij}^* occurs in it. In this case there is no stress in both the domain Ω and D–Ω.

(2) Apply a load p_i^{out} on the boundary of the domain Ω to push it back to the original shape. The applied load is given by

$$p_i^{\text{out}} = -\sigma_{ij}^* n_j \tag{1.48}$$

As mentioned before, a summation is applied to the repeated subscripts j. In Eq. (1.48), n_j is the j-th component of the outward normal to the boundary Γ, and σ_{ij}^* are called the eigenstresses corresponding to the eigenstrains ε_{ij}^*, which are given by

$$\sigma_{ij}^* = C_{ijkl}\varepsilon_{kl}^* \tag{1.49}$$

When the domain Ω comes back to the original shape, there is no strain in this domain, but the residual stresses $-\sigma_{ij}^*$.

(3) Put the inclusion Ω back into the domain D, and let the domain Ω and D–Ω deform together. Then, apply the load $p_i = -p_i^{\text{out}}$ on the boundary Γ. The displacement field for this problem can be derived by using Green's function approach (Eshelby, 1957)

$$u_i(\mathbf{x}) = \int_\Gamma U_{ij}(\mathbf{x}-\mathbf{x}')\, p_j \,\mathrm{d}s \tag{1.50}$$

where Γ is the boundary of both the domain Ω and the domain D–Ω with $\mathbf{x}'$ on Γ and $U_{ij}(\mathbf{x}-\mathbf{x}')$ is called a Green's function, which is the solution given by Kelvin for a unit force applied on a material with an unbounded boundary. The point $\mathbf{x}$ in Eq. (1.50) can be within or outside Ω. The function $U_{ij}(\mathbf{x}-\mathbf{x}')$ represents the displacement component in the x_i-direction at point $\mathbf{x}$ when a unit force in the x_j-direction is applied at a point $\mathbf{x}'$. For an isotropic material, Green's function can be explicitly expressed as

$$U_{ij}(\mathbf{x}-\mathbf{x}') = \frac{1}{4\pi\mu}\frac{\delta_{ij}}{|\mathbf{x}-\mathbf{x}'|} - \frac{1}{16\pi\mu(1-\nu)}\frac{\partial^2}{\partial x_i \partial x_j}|\mathbf{x}-\mathbf{x}'| \tag{1.51}$$

where μ and ν are the shear modulus and the Poisson's ratio of the material, respectively. It should be noted that Green's functions are explicitly obtainable only for isotropic and transversely isotropic materials. For other materials such as an arbitrary anisotropic material, the explicit usage of Green's function is limited. Some other expressions, such as Fourier integral expressions (Mura, 1982), can be used in this regard. However, the expressions for anisotropic materials are much more complicated than those for isotropic and transversely isotropic materials. Substituting Eq. (1.50) into geometric equations

$$\varepsilon_{ij} = \frac{1}{2}(u_{i,j} + u_{j,i}), \tag{1.52}$$

the actual strains ε_{ij} can be obtained. Then, by Hooke's law, the stresses in the domain Ω and $D-\Omega$ can be derived as

$$\sigma_{ij} = C_{ijkl}\varepsilon_{kl} + (-\sigma_{ij}^*) = C_{ijkl}\varepsilon_{kl} - C_{ijkl}\varepsilon_{kl}^* \quad \text{in } \Omega \tag{1.53.1}$$

$$\sigma_{ij} = C_{ijkl}\varepsilon_{kl} \quad \text{in } D-\Omega \tag{1.53.2}$$

1.5.2 Eshelby's Tensor

Suppose that a material under consideration is isotropic and the eigenstrain ε_{ij}^* in the domain Ω is homogeneous. Substituting Eqs. (1.48) and (1.49) into Eq. (1.50) gives

$$u_i = -C_{jkmn}\varepsilon_{mn}^* \int_\Gamma U_{ij}(\mathbf{x}-\mathbf{x}')n_k \,\mathrm{d}s \tag{1.54}$$

Let us first consider the case where the point $\mathbf{x}$ is within the inclusion Ω. Substituting Eq. (1.51) into Eq. (1.54) and after some manipulation by using the divergence theorem (Rektorys, 1977) for an integral leads to (Eshelby, 1957)

$$u_i(\mathbf{x}) = \frac{\varepsilon_{jk}^*}{8\pi(1-\nu)} \int_\Omega \frac{g_{ijk}(\mathbf{l})}{r^2} \mathrm{d}\mathbf{x}' \tag{1.55}$$

where $g_{ijk}(\mathbf{l}) = (1-2\nu)\left(\delta_{ij}l_k + \delta_{ik}l_j - \delta_{jk}l_i\right) + 3l_i l_j l_k$ with $l_i = \frac{1}{r}\left(x_i - x'_i\right)$ and $r = \left|\mathbf{x} - \mathbf{x}'\right|$.

Suppose that the inclusion Ω is an ellipse, which is defined as

$$\left(\frac{x_1}{a_1}\right)^2 + \left(\frac{x_2}{a_2}\right)^2 + \left(\frac{x_3}{a_3}\right)^2 \le 1 \tag{1.56}$$

For the elliptic inclusion, the integral Eq. (1.55) can be carried out following Eshelby's approach (Eshelby, 1957; Mura, 1982). When a point $\mathbf{x} = \mathbf{x}(x_1, x_2, x_3)$ is located inside the inclusion Ω, as shown in Fig. 1.8, the volume element $\mathrm{d}\mathbf{x}'$ in Eq. (1.55) can be rewritten as

$$\mathrm{d}\mathbf{x}' = \mathrm{d}x'_1 \mathrm{d}x'_2 \mathrm{d}x'_3 = \mathrm{d}r\mathrm{d}\Gamma = \mathrm{d}r r^2 \mathrm{d}\omega \tag{1.57}$$

where r is the distance between point $\mathbf{x}$ and $\mathbf{x}'$, $\mathrm{d}\Gamma$ is an area element of the inclusion and $\mathrm{d}\omega$ is a non-dimensional area element of a unit sphere (Fig. 1.8), which is introduced to evaluate the integral Eq. (1.55). Substituting Eq. (1.57) into (1.55) and integrating the resulting equation with respective to r gives

$$u_i(\mathbf{x}) = \frac{-\varepsilon^*_{jk}}{8\pi(1-\nu)} \int_{\Sigma} r' g_{ijk}(\mathbf{l})\mathrm{d}\omega \tag{1.58}$$

where Σ is the boundary to the elliptical inclusion Ω, r' is the distance between the point $\mathbf{x}$ and the point on the boundary Σ, which is a positive root to the following equation

$$\frac{\left(x_1 + r'l_1\right)^2}{a_1^2} + \frac{\left(x_2 + r'l_2\right)^2}{a_2^2} + \frac{\left(x_3 + r'l_3\right)^2}{a_3^2} = 1 \tag{1.59}$$

From Eq. (1.59), r' is obtained as

$$r' = \frac{-b + \sqrt{b^2 - 4ac}}{2a} \tag{1.60}$$

where

$$a = \frac{l_1^2}{a_1^2} + \frac{l_2^2}{a_2^2} + \frac{l_3^2}{a_3^2} \tag{1.60.1}$$

$$b = 2\left(\frac{l_1 x_1}{a_1^2} + \frac{l_2 x_2}{a_2^2} + \frac{l_3 x_3}{a_3^2}\right) \tag{1.60.2}$$

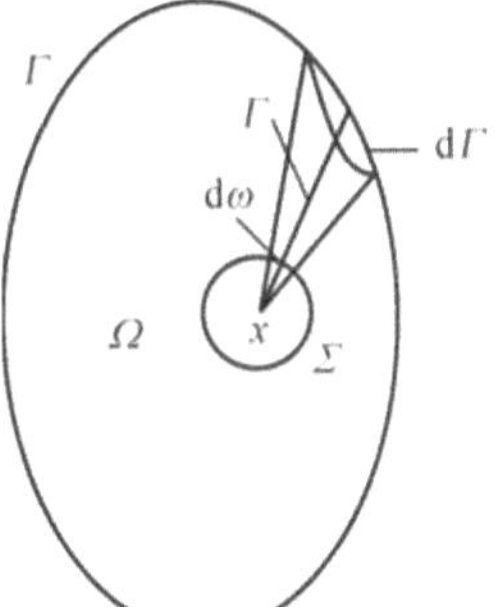

Fig. 1.8 Elliptical inclusion and integral sphere

$$c = \frac{x_1^2}{a_1^2} + \frac{x_2^2}{a_2^2} + \frac{x_3^2}{a_3^2} - 1 \tag{1.60.3}$$

For convenience of expression, λ_1, λ_2 and λ_3 are introduced as

$$\lambda_1 = l_1 / a_1^2 , \quad \lambda_2 = l_2 / a_2^2 , \quad \lambda_3 = l_3 / a_3^2 \tag{1.61}$$

Substituting Eqs. (1.60) and (1.61) into Eq. (1.58) (noticing that when Eqs. (1.60) are substituted into Eq. (1.58), the term $\sqrt{b^2 - 4ac}$ in Eq. (1.60) should be omitted since it is even with respect to **l**, whereas g_{ijk} is odd (Eshelby, 1957; Mura, 1982)) gives

$$u_i(\mathbf{x}) = \frac{x_m \varepsilon_{jk}^*}{8\pi(1-\nu)} \int_\Sigma \frac{\lambda_m g_{ijk}}{a} \mathrm{d}\omega \tag{1.62}$$

Substituting Eq. (1.62) into Eq. (1.52), the corresponding strain is expressed as

$$\varepsilon_{ij}(x) = \frac{\varepsilon_{kl}^*}{16\pi(1-\nu)} \int_\Sigma \frac{\lambda_i g_{jkl} + \lambda_j g_{ikl}}{a} \mathrm{d}\omega \tag{1.63}$$

which was first obtained by Eshelby (1957). From Eq. (1.63), it can be seen that the integral is independent of x. In other words, the strain inside the inclusion is homogeneous. Eq. (1.63) is rewritten in the following form,

$$\varepsilon_{ij} = L_{ijkl} \varepsilon_{kl}^* \tag{1.64}$$

where $L_{ijkl} = \dfrac{1}{16\pi(1-\nu)} \displaystyle\int_\Sigma \frac{\lambda_i g_{jkl} + \lambda_j g_{ikl}}{a} \mathrm{d}\omega$ is called an Eshelby's tensor. By means of the work of Routh (Eshelby, 1957; Mura, 1982), the surface integral in Eshelby's tensor can be reduced to a simpler integral from which explicit expressions are obtained as

$$L_{1111} = \frac{3}{8\pi(1-\nu)} a_1^2 I_{11} + \frac{1-2\nu}{8\pi(1-\nu)} I_1 \quad (1.65.1)$$

$$L_{1122} = \frac{3}{8\pi(1-\nu)} a_2^2 I_{12} - \frac{1-2\nu}{8\pi(1-\nu)} I_1 \quad (1.65.2)$$

$$L_{1133} = \frac{3}{8\pi(1-\nu)} a_3^2 I_{13} - \frac{1-2\nu}{8\pi(1-\nu)} I_1 \quad (1.65.3)$$

$$L_{1212} = \frac{a_1^2 + a_2^2}{16\pi(1-\nu)} I_{12} + \frac{1-2\nu}{16\pi(1-\nu)} (I_1 + I_2) \quad (1.65.4)$$

where
$$I_1 = 2\pi a_1 a_2 a_3 \int_0^\infty \frac{\mathrm{d}q}{(a_1^2 + q)q'} \quad (1.66.1)$$

$$I_{11} = 2\pi a_1 a_2 a_3 \int_0^\infty \frac{\mathrm{d}q}{(a_1^2 + q)^2 q'} \quad (1.66.2)$$

$$I_{12} = 2\pi a_1 a_2 a_3 \int_0^\infty \frac{\mathrm{d}q}{(a_1^2 + q)(a_2^2 + q)q'} \quad (1.66.3)$$

with $q' = \left\{(a_1^2 + q)(a_2^2 + q)(a_3^2 + q)\right\}^{1/2}$. The other coefficients I_i and I_{ij} are determined by a cyclic permutation on subscripts (1, 2, 3). For instance,

$$I_2 = 2\pi a_1 a_2 a_3 \int_0^\infty \frac{\mathrm{d}q}{(a_2^2 + q)q'} \quad (1.66.4)$$

$$I_{22} = 2\pi a_1 a_2 a_3 \int_0^\infty \frac{\mathrm{d}q}{(a_2^2 + q)^2 q'} \quad (1.66.5)$$

$$I_{23} = 2\pi a_1 a_2 a_3 \int_0^\infty \frac{\mathrm{d}q}{(a_2^2 + q)(a_3^2 + q)q'} \quad (1.66.6)$$

Similarly, other elements L_{ijkl} of Eshelby's tensor can be obtained by cyclic permutations on subscripts (1, 2, 3) from Eq. (1.65), as long as these elements are non-zero. Examples are

$$L_{2222} = \frac{3}{8\pi(1-\nu)} a_2^2 I_{22} + \frac{1-2\nu}{8\pi(1-\nu)} I_2 \quad (1.67.1)$$

$$L_{2233} = \frac{3}{8\pi(1-\nu)} a_3^2 I_{23} - \frac{1-2\nu}{8\pi(1-\nu)} I_2 \tag{1.67.2}$$

$$L_{1313} = \frac{a_1^2 + a_3^2}{16\pi(1-\nu)} I_{13} + \frac{1-2\nu}{16\pi(1-\nu)} (I_1 + I_3) \tag{1.67.3}$$

$$L_{2323} = \frac{a_2^2 + a_3^2}{16\pi(1-\nu)} I_{23} + \frac{1-2\nu}{16\pi(1-\nu)} (I_2 + I_3) \tag{1.67.4}$$

It is noted that any element which cannot be obtained by a cyclic permutation is zero, e.g., $L_{1112} = L_{1223} = L_{1232} = 0$. Moreover, if the inclusion becomes an infinitely long cylinder with $a_2 = a_3$ and $a_1 = \infty$, Eshelby's tensor can be further simplified to

$$[L] = \begin{bmatrix} L_{1111} & L_{1122} & L_{1133} & 0 & 0 & 0 \\ L_{2211} & L_{2222} & L_{2233} & 0 & 0 & 0 \\ L_{3311} & L_{3322} & L_{3333} & 0 & 0 & 0 \\ 0 & 0 & 0 & 2L_{2323} & 0 & 0 \\ 0 & 0 & 0 & 0 & 2L_{1313} & 0 \\ 0 & 0 & 0 & 0 & 0 & 2L_{1212} \end{bmatrix} \tag{1.68}$$

where $L_{2211} = L_{3311} = \dfrac{\nu}{2(1-\nu)}$, $L_{2222} = L_{3333} = \dfrac{1}{2(1-\nu)}\left[\dfrac{3}{4} + \dfrac{(1-2\nu)}{2}\right]$,

$L_{2233} = L_{3322} = \dfrac{1}{2(1-\nu)}\left[\dfrac{1}{4} - \dfrac{(1-2\nu)}{2}\right]$, $L_{2323} = \dfrac{1}{2(1-\nu)}\left[\dfrac{1}{4} - \dfrac{(1-2\nu)}{2}\right]$,

$$L_{1212} = L_{1313} = 1/4,\ L_{1111} = L_{1122} = L_{1133} = 0 \tag{1.69}$$

In Eq. (1.69), ν is Poisson's ratio of the material.

For a point outside the inclusion Ω, the solution is much more complicated and explicit expressions for Eshelby's tensor elements are difficult to obtain. For more information refer to e.g., Eshelby (1959), Timoshenko (1970) and Mura and Cheng (1977).

1.5.3 Equivalent Inclusion

In the previous approach, the elastic moduli of the subdomain Ω were the same as those of the remaining domain (also called the matrix domain) $D-\Omega$. If the subdomain Ω in a material D has different elastic moduli from those of the matrix domain $D{-}\Omega$, Ω is called an inhomogeneity (Mura, 1982).

A material containing inhomogeneities is free from any stress unless a load is applied. However, if the material is subjected to a load, the stress field will be disturbed by the existence of the inhomogeneity. Eshelby (1957, 1959) first pointed out that the stress disturbance in such a problem could be simulated by an eigenstrain problem when the eigenstrain was properly chosen. This equivalency is called the equivalent inclusion (Mura, 1982).

Suppose that an inclusion domain Ω with elastic moduli $C_{ijkl}^{(1)}$ is embedded in an infinite material D with elastic moduli $C_{ijkl}^{(0)}$. A load applied at infinity is denoted by σ_{ij}^{0} and the corresponding strain is denoted by ε_{ij}^{0}. The disturbances of the stress and strain fields caused by the existence of inhomogeneity Ω are represented by σ'_{ij} and ε'_{ij}. Thus, the actual stresses and strains are $\sigma_{ij}^{0}+\sigma'_{ij}$ and $\varepsilon_{ij}^{0}+\varepsilon'_{ij}$, respectively. Interrelations between them are written as

$$\sigma_{ij}^{0}+\sigma'_{ij}=C_{ijkl}^{(1)}\left(\varepsilon_{kl}^{0}+\varepsilon'_{kl}\right) \quad \text{in } \Omega \tag{1.70.1}$$

$$\sigma_{ij}^{0}+\sigma'_{ij}=C_{ijkl}^{(0)}\left(\varepsilon_{kl}^{0}+\varepsilon'_{kl}\right) \quad \text{in } D{-}\Omega \tag{1.70.2}$$

Consider an infinite homogeneous material with the elastic moduli $C_{ijkl}^{(0)}$ everywhere (i.e., both the domains Ω and $D{-}\Omega$ have the same elastic moduli). A homogeneous eigenstrain ε_{ij}^{*} is assigned to the subdomain Ω. The stress and strain fields induced can be expressed as

$$\sigma_{ij}^{0}+\sigma'_{ij}=C_{ijkl}^{(0)}\left(\varepsilon_{kl}^{0}+\varepsilon'_{kl}-\varepsilon_{kl}^{*}\right) \quad \text{in } \Omega \tag{1.71.1}$$

$$\sigma_{ij}^{0}+\sigma'_{ij}=C_{ijkl}^{(0)}\left(\varepsilon_{kl}^{0}+\varepsilon'_{kl}\right) \quad \text{in } D{-}\Omega \tag{1.71.2}$$

where σ'_{ij} and ε'_{ij} are the disturbances of the stresses and strains caused by the eigenstrain ε_{ij}^{*}. From the preceding section we have known that the disturbances of the strains in the domain Ω can be related to the eigenstrain ε_{ij}^{*} by Eshelby's tensor through

$$\varepsilon'_{ij}=L_{ijkl}\varepsilon_{kl}^{*} \quad \text{in } \Omega \tag{1.72}$$

Eshelby has pointed out that when the ε_{ij}^{*} is properly chosen, the stress disturbances given by Eqs. (1.70.1) and (1.71.1) are equivalent to each other, i.e.,

$$C_{ijkl}^{(1)}\left(\varepsilon_{kl}^{0}+\varepsilon'_{kl}\right)=C_{ijkl}^{(0)}\left(\varepsilon_{kl}^{0}+\varepsilon'_{kl}-\varepsilon_{kl}^{*}\right) \qquad \text{in} \quad \Omega \tag{1.73}$$

Substituting Eq. (1.72) into Eq. (1.73), the proper eigenstrains ε_{ij}^{*} can be obtained by resolving the resulting equations. Thus far, the problem with an inhomogeneity has been successfully converted to an eigenstrain problem of isotropic materials as described in the previous Subsections 1.5.1 and 1.5.2. The stress distributions in the domains Ω and $D-\Omega$ can be derived accordingly.

It should be pointed out that the equivalent inclusion method is efficient only when an inhomogeneity is embedded in an infinite domain. It usually cannot be used to characterize a composite where a great many reinforcing fibers are arranged in a matrix material with a finite domain. Thus, many different approaches have been proposed to modify the method. Examples include the self-consistent scheme (Hill, 1964, 1965a, 1965b; Chou et al., 1980), the generalized self-consistent scheme (Christensen & Lo, 1979) and the Mori-Tanaka approach (Mori & Tanaka, 1973; Mura, 1982). The Mori and Tanaka's approach will be illustrated in Chapter 3 of this book.

1.6 Coordinate Transformation

In any practical application, a transformation between different coordinate systems is essentially inevitable. This is especially true for composites, which are anisotropic in nature.

As aforementioned, the material principal coordinate system (also called "local coordinate system"), (x_1, x_2, x_3), of a UD composite is always established in such a way that x_1 is along the fiber axial (i.e., longitudinal) direction. Such a system may not coincide with the global coordinate system. The mechanical quantities given in the local coordinate system may have to be transformed into those in the global one or *vice versa*. This can be accomplished based upon the tensor transformation rules governing the stresses and strains between any two rectangular coordinate systems (Timoshenko & Goodier, 1970; Reddy, 1988).

Let (x, y, z) represent the global coordinate system. Suppose that the directional cosines between the local coordinates O_{x_1}, O_{x_2}, O_{x_3} and the global ones O_x, O_y, O_z are denoted by (l_i, m_i, n_i) where

$$l_i=\cos(x_i, x),\ m_i=\cos(x_i, y), \quad n_i=\cos(x_i, z), \quad i=1, 2, 3 \tag{1.74}$$

With these coefficients, the two sets of coordinates are correlated by

$$\begin{Bmatrix} x_1 \\ x_2 \\ x_3 \end{Bmatrix} = \begin{bmatrix} l_1 & m_1 & n_1 \\ l_2 & m_2 & n_2 \\ l_3 & m_3 & n_3 \end{bmatrix} \begin{Bmatrix} x \\ y \\ z \end{Bmatrix} = [e_{ij}] \begin{Bmatrix} x \\ y \\ z \end{Bmatrix} \tag{1.75}$$

Let $[\sigma_{ij}^G]=\begin{bmatrix}\sigma_{xx} & \sigma_{xy} & \sigma_{xz}\\ \sigma_{yx} & \sigma_{yy} & \sigma_{yz}\\ \sigma_{zx} & \sigma_{zy} & \sigma_{zz}\end{bmatrix}$ represent the stress tensor in the global coordinate system. According to the tensor transformation formula (Reddy, 1988; Cristescu et al., 2004) between the local and the global stresses, i.e., $\sigma_{kl}^G = e_{ik}e_{jl}\sigma_{ij}$, we can easily derive the following equations

$$\begin{aligned}\sigma_{11}^G = \sigma_{xx} &= e_{11}e_{11}\sigma_{11}+ e_{11}e_{21}\sigma_{12}+e_{11}e_{31}\sigma_{13}+e_{21}e_{11}\sigma_{21}+\\ &e_{21}e_{21}\sigma_{22}+e_{21}e_{31}\sigma_{23}+e_{31}e_{11}\sigma_{31}+e_{31}e_{21}\sigma_{32}+e_{31}e_{31}\sigma_{33}\\ &\cdots,\end{aligned} \tag{1.76.1}$$

or in a matrix form,

$$\{\sigma_i^G\} = [T_{ij}]_c\{\sigma_j\} \tag{1.76.2}$$

where $\{\sigma_i^G\} = \{\sigma_1^G,\sigma_2^G,\sigma_3^G,\sigma_4^G,\sigma_5^G,\sigma_6^G\}^{\mathrm{T}} = \{\sigma_{xx},\sigma_{yy},\sigma_{zz},\sigma_{yz},\sigma_{xz},\sigma_{xy}\}^{\mathrm{T}}$ is the contracted stress vector in the global system and $\{\sigma_j\}= \{\sigma_{11},\sigma_{22},\sigma_{33},\sigma_{23},\sigma_{13},\sigma_{12}\}^{\mathrm{T}}$ is that in the local one and

$$[T_{ij}]_c = \begin{bmatrix} l_1^2 & l_2^2 & l_3^2 & 2l_2l_3 & 2l_3l_1 & 2l_1l_2\\ m_1^2 & m_2^2 & m_3^2 & 2m_2m_3 & 2m_3m_1 & 2m_1m_2\\ n_1^2 & n_2^2 & n_3^2 & 2n_2n_3 & 2n_3n_1 & 2n_1n_2\\ m_1n_1 & m_2n_2 & m_3n_3 & m_2n_3+m_3n_2 & n_3m_1+n_1m_3 & m_1n_2+m_2n_1\\ n_1l_1 & n_2l_2 & n_3l_3 & l_2n_3+l_3n_2 & n_3l_1+n_1l_3 & l_1n_2+l_2n_1\\ l_1m_1 & l_2m_2 & l_3m_3 & l_2m_3+l_3m_2 & l_1m_3+l_3m_1 & l_1m_2+l_2m_1 \end{bmatrix} \tag{1.77}$$

Similarly, from the strain transformation formula, $\varepsilon_{kl}^G = e_{ik}e_{jl}\varepsilon_{ij}$, we obtain

$$\{\varepsilon_i^G\} = [T_{ij}]_s\{\varepsilon_j\} \tag{1.78}$$

in which $\{\varepsilon_i^G\} = \{\varepsilon_1^G,\varepsilon_2^G,\varepsilon_3^G,\varepsilon_4^G,\varepsilon_5^G,\varepsilon_6^G\}^{\mathrm{T}} = \{\varepsilon_{xy},\varepsilon_{yy},\varepsilon_{zz},2\varepsilon_{yz},2\varepsilon_{xz},2\varepsilon_{xy}\}^{\mathrm{T}}$ is the global strain vector, $\{\varepsilon_j\}= \{\varepsilon_{11},\varepsilon_{22},\varepsilon_{33},2\varepsilon_{23},2\varepsilon_{13},2\varepsilon_{12}\}^{\mathrm{T}}$ is the local one and

$$[T_{ij}]_s = \begin{bmatrix} l_1^2 & l_2^2 & l_3^2 & l_2 l_3 & l_3 l_1 & l_1 l_2 \\ m_1^2 & m_2^2 & m_3^2 & m_2 m_3 & m_3 m_1 & m_1 m_2 \\ n_1^2 & n_2^2 & n_3^2 & n_2 n_3 & n_3 n_1 & n_1 n_2 \\ 2m_1 n_1 & 2m_2 n_2 & 2m_3 n_3 & m_2 n_3 + m_3 n_2 & n_3 m_1 + n_1 m_3 & m_1 n_2 + m_2 n_1 \\ 2n_1 l_1 & 2n_2 l_2 & 2n_3 l_3 & l_2 n_3 + l_3 n_2 & n_3 l_1 + n_1 l_3 & l_1 n_2 + l_2 n_1 \\ 2l_1 m_1 & 2l_2 m_2 & 2l_3 m_3 & l_2 m_3 + l_3 m_2 & l_1 m_3 + l_3 m_1 & l_1 m_2 + l_2 m_1 \end{bmatrix} \quad (1.79)$$

By using the condition that a strain energy at a point is invariant regardless of any coordinate system used, i.e., $\{\sigma_i^G\}^{\mathrm{T}}\{\varepsilon_i^G\} = \{\sigma_i\}^{\mathrm{T}}\{\varepsilon_i\}$, it is easy to show that the two transformation matrices given by Eqs. (1.77) and (1.79) satisfy

$$[T_{ij}]_s^{\mathrm{T}} = [T_{ij}]_c^{-1} \quad (1.80.1)$$

$$[T_{ij}]_c^{\mathrm{T}} = [T_{ij}]_s^{-1} \quad (1.80.2)$$

Suppose that the global compliance and stiffness matrices are denoted by $[S_{ij}^G]$ and $[C_{ij}^G]$, respectively. From the relation

$$\{\varepsilon_i^G\} = [S_{ij}^G]\ \{\sigma_j^G\}$$

and Eqs. (1.76), (1.78) and (1.80), we obtain

$$\{\varepsilon_i^G\} = [T_{ij}]_s\ [S_{ij}]\ [T_{ij}]_s^{\mathrm{T}}\ \{\sigma_j^G\}$$

Hence, the transformation formula for the compliance matrix from the local coordinate system to the global one is given by

$$[S_{ij}^G] = [T_{ij}]_s\ [S_{ij}]\ [T_{ij}]_s^{\mathrm{T}} \quad (1.81)$$

Similarly, we obtain the stiffness transformation formula as

$$[C_{ij}^G] = [T_{ij}]_c\ [C_{ij}]\ [T_{ij}]_c^{\mathrm{T}} \quad (1.82)$$

Example 1.2 Suppose that the local x_3-axis is parallel to the global z-coordinate. Furthermore, suppose that the local x_1-axis has an inclined angle θ with respect to the global x-coordinate, as indicated in Fig. 1.9. Write out the coordinate transformation matrices, $[T_{ij}]_s$ and $[T_{ij}]_c$, for these two systems.

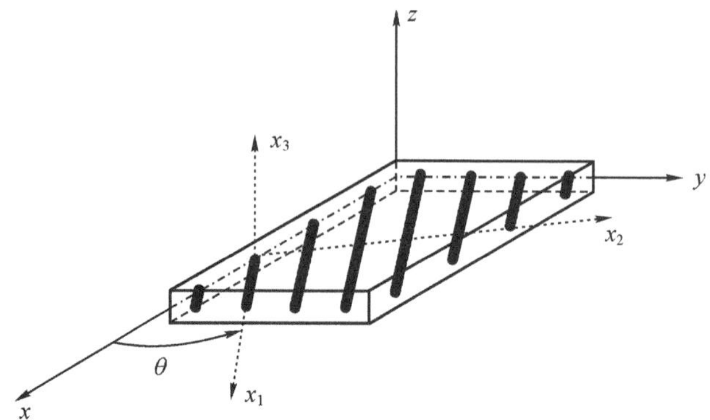

Fig. 1.9 Coordinate transformation between a local, (x_1, x_2, x_3) and the global (x, y, z) systems

Solution According to Eq. (1.74) we have

$$l_1=\cos(\theta),\ m_1=\sin(\theta),\ n_1=0,\ l_2=-\sin(\theta),\ m_2=\cos(\theta),\ n_2=0,\ l_3=0,\ m_3=0,\ n_3=1. \quad (1.83)$$

Therefore, $$\left[T_{ij}\right]_c=\begin{bmatrix}\cos^2(\theta) & \sin^2(\theta) & 0 & 0 & 0 & -\sin(2\theta)\\ \sin^2(\theta) & \cos^2(\theta) & 0 & 0 & 0 & \sin(2\theta)\\ 0 & 0 & 1 & 0 & 0 & 0\\ 0 & 0 & 0 & \cos(\theta) & \sin(\theta) & 0\\ 0 & 0 & 0 & -\sin(\theta) & \cos(\theta) & 0\\ \dfrac{\sin(2\theta)}{2} & -\dfrac{\sin(2\theta)}{2} & 0 & 0 & 0 & \cos(2\theta)\end{bmatrix} \quad (1.84)$$

and $$\left[T_{ij}\right]_s=\begin{bmatrix}\cos^2(\theta) & \sin^2(\theta) & 0 & 0 & 0 & -\sin(2\theta)\\ \sin^2(\theta) & \cos^2(\theta) & 0 & 0 & 0 & \sin(2\theta)\\ 0 & 0 & 1 & 0 & 0 & 0\\ 0 & 0 & 0 & \cos(\theta) & \sin(\theta) & 0\\ 0 & 0 & 0 & -\sin(\theta) & \cos(\theta) & 0\\ \sin(2\theta) & -\sin(2\theta) & 0 & 0 & 0 & \cos(2\theta)\end{bmatrix} \quad (1.85)$$

Example 1.3 Write out coordinate transformation formulae for a plane stress problem.

Solution For a plane stress problem, we only have at most three non-zero stress components which are in the same plane, i.e., the local stress vector $\{\sigma_i\}=\{\sigma_{11}, \sigma_{22}, \sigma_{12}\}^{\mathrm{T}}$ and the global one $\{\sigma_i^G\}=\{\sigma_{xx}, \sigma_{yy}, \sigma_{xy}\}^{\mathrm{T}}$. It is important to realize that the two coordinate systems, i.e., (x_1, x_2) and (x, y), must obey the same right-hand screwing rule and the third coordinates of them, x_3 and z, must be oriented in the same direction, as indicated in Fig. 1.10(a) or Fig. 1.10(b). The planar coordinate transformation formulae are then adapted from Eqs. (1.77) and (1.79) as follows.

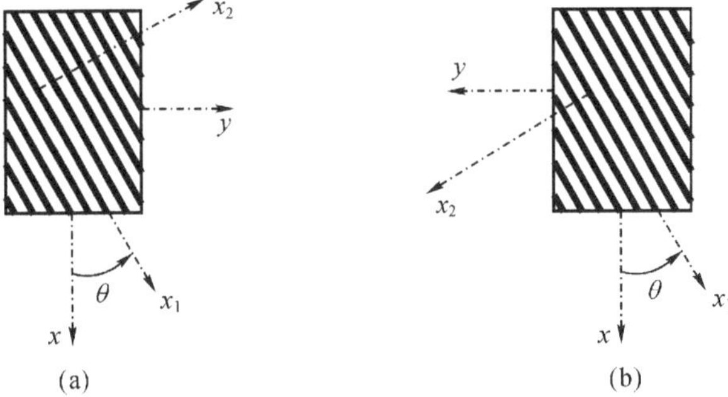

Fig. 1.10 Coordinate transformation between a local, (x_1, x_2) and a global (x, y) system

$$\left[T_{ij}\right]_c = \begin{bmatrix} l_1^2 & l_2^2 & 2l_1l_2 \\ m_1^2 & m_2^2 & 2m_1m_2 \\ l_1m_1 & l_2m_2 & l_1m_2 + l_2m_1 \end{bmatrix} \tag{1.86}$$

$$\left[T_{ij}\right]_s = \begin{bmatrix} l_1^2 & l_2^2 & l_1l_2 \\ m_1^2 & m_2^2 & m_1m_2 \\ 2l_1m_1 & 2l_2m_2 & l_1m_2 + l_2m_1 \end{bmatrix} \tag{1.87}$$

l_i=cos(x, x_i) and m_i=cos(y, x_i) with i=1 and 2. Thus, for the coordinate systems of Fig. 1.10(a), we have l_1=cos(θ), l_2=–sin(θ), m_1=sin(θ) and m_2=cos(θ), whereas for the systems of Fig. 1.10(b) it is true that l_1=cos(θ), l_2=sin(θ), m_1=–sin(θ) and m_2=cos(θ).

References

Berthelot, J.M. (1999) Composite Materials, Mechanical Behavior and Structural Analysis. Berlin, New York: Springer.

Bishop, J.F.W. & Hill, R. (1951) A theoretical derivation of plastic properties of a polycrystalline face-center metal. Philos. Mag. 42, 1298- 1307.

Chamis, C.C. (1984) Simplified composite micromechanics equations for hygral, thermal and mechanical properties. SAMPE Quarterly 15, 14-23.

Chamis, C.C. & Hopkins, D.A. (1988) Thermoviscoplastic nonlinear constitutive relationships for structural analysis of high temperature metal matrix composites, in Testing Technology of Metal Matrix Composites, ASTM STP 964, P.R. DiGiovanni. & N.R. Adsit (eds.). American Society for Testing and Materials, Philadelphia, 177-196.

Chamis, C.C. (1989) Mechanics of composite materials: Past, present, and future.

J. Comp. Technol. Res. 11, 3-14.
Chou T.W., Nomura S. & Taya M. (1980) A self-consistent approach to the elastic stiffness of short-fiber composites. Journal of Composite Materials 14, 178-188.
Christensen, R.M. & Lo, K.H. (1979) Solutions for effective shear properties in three phase shere and cylinder models. J. Mech. Phys. Solids 27, 315-330.
Christensen, R.M. (1991) Mechanics of Composite Materials. Malabar: Krieger Pub. Co.
Cristescu, N.D., Craciun, E.M. & Soos, E. (2004) Mechanics. of Elastic Composites. Boca Raton: Chapman & Hall/CRC Press.
Eshelby, J.D. (1957) The determination of the elastic field of an ellipsoidal inclusion and related problems. Proceeding of Royal Society A240, 367-396.
Eshelby, J.D. (1959) The elastic field outside an ellipsoidal inclusion. Proceeding of Royal Society A252, 561-569.
Gibson, R.F. (1994) Principles of Composite Material Mechanics. New York, NY: McGraw-Hill, Inc.
Gordon, J.E. (1976) The New Science of Strong Materials (2nd ed.). Princeton, NJ: Princeton University Press.
Griffith, A.A. (1920) The phenomena of rupture and flow in solids. Philosophical Transactions of the Royal Society 221A, 163-198.
Hashin, Z. & Rosen, B. W. (1964) The elastic moduli of fiber-reinforced materials. J. Appl. Mech. 31, 223-232.
Hashin, Z. (1965) On elastic behaviour of fiber reinforced materials of arbitrary transverse phase geometry. J. Mech. Phys. Solids 13, 119-134.
Hashin, Z. (1979) Analysis of properties of fiber composites with anisotropic constituents. J. Applied Mechanics 46, 453-550.
Hill, R. (1964) Theory of mechanical properties of fiber-strengthened materials: I. Elastic behaviour. J. Mech. Phys. Solids 12, 199-212.
Hill, R. (1965a) A self-consistent mechanics of composite materials. J Mech. & Phys. Solids 13, 213-222.
Hill, R. (1965b) Theory of mechanical properties of fiber-strengthened materials, III. Self-consistent model. J. Mech. Phys. Solids 13, 189-198.
Hopkins, D.A. & Chamis, C.C. (1988) A unique set of micromechanics equations for high temperature metal matrix composites, in Testing Technology of Metal Matrix Composites. P.R. & Adsit, N.R. (eds.). ASTM STP 964, DiGiovanni, American Society for Testing and Materials, Philadelphia, 159-176.
Lerch, B.A. & Saltsman, J.F. (1993) Tensile Deformation of SiC/Ti-15-3 Laminates, in Composite Materials: Fatigue and Fracture (Vol. 4), ASTM STP 1156, W. Stinchcomb & N.E. Ashbaugh (eds.). American Society for Testing and Materials, Philadelphia, 161-175.
Mori, T. & Tanaka K. (1973) Average stress in matrix and average energy of materials with misfitting inclusions. Act. Metall 21, 571-574.
Mura, T. & Cheng, P.C. (1977) The elastic field outside an ellipsoidal inclusion. J. Applied Mechanics 44, 591-594.
Mura, T. (1982) Micromechanics of Defects in Solids. Dordrecht: Martinus

Nijhoff Publisher.
Reddy, J.N. (1988) Theory and Analysis of Laminated Composite Plates and Shells. New York, NY: John Wiley & Sons.
Rektorys, K. (1977) Variational Methods in Mathematics, Science and Engineering. Boston, MA: Dordrecht, D. Reidel Pub. Co.
Swanson, S.R. (1997) Introduction to Design and Analysis with Advanced Composite Materials. NJ: Prentice-Hall International, Inc.
Taylor, G.I. (1938) Plastic strains in Metals. J. Inst. Metals 62, 307-324.
Timoshenko, S.P. & Goodier, J.N. (1970) Theory of Elasticity (3rd ed.). New York, NY: McGraw Hill.
Zhang, W.H. (2003) Fundamentals Elasticity Mechanics and Finite Element Method. Hangzhou: Zhejiang University Press (in Chinese).

2

Plastic Theories of Isotropic Media

2.1 Introduction

Most materials in bulk form display inelastic deformation before failure. The same is true for fibrous composites. The inelastic deformation of a fibrous composite mainly comes from two sources: the material inelastic deformation and the damage evolution. As far as the material inelastic deformation is concerned, it is believed to be attributed to the effect of the matrix inelastic deformation. This is because the fiber material used in the composite is generally very thin and stiff, and hence can well exhibit linearly elastic properties until rupture in most cases. On the other hand, the matrix material used is a continuous phase, much resembling its bulk form, and hence can generate inelastic deformation. As mentioned in Chapter 1, most matrix materials including metals, ceramics and polymers are isotropic. It is thus necessary to review or establish constitutive theories for the inelastic stress-strain response of isotropic media. In this chapter, two typical ones are described. One is Prandtl-Reuss elasto-plastic theory and another is called Bodner-Parton unified plasticity theory.

2.2 Prandtl-Reuss Elasto-Plastic Theory

In a linear-elastic deformation range, Hooke's law Eq. (1.4.1) links the elongation strains with the normal stresses in the following set of equations

$$\varepsilon_{11}^{e} = [\sigma_{11} - \nu(\sigma_{22} + \sigma_{33})] / E \tag{2.1.1}$$

$$\varepsilon_{22}^{e} = [\sigma_{22} - \nu(\sigma_{11} + \sigma_{33})] / E \tag{2.1.2}$$

$$\varepsilon_{33}^{e} = [\sigma_{33} - \nu(\sigma_{11} + \sigma_{22})] / E \tag{2.1.3}$$

where the superscript e indicates that the material under consideration is in an elastic deformation. Note that the compliance matrix, Eq. (1.8), for isotropic materials has been employed. By eliminating *E*, Eq. (2.1) can be rewritten as

$$\frac{\varepsilon_{11}^{e}}{\sigma_{11}-\nu(\sigma_{22}+\sigma_{33})}=\frac{\varepsilon_{22}^{e}}{\sigma_{22}-\nu(\sigma_{11}+\sigma_{33})}=\frac{\varepsilon_{33}^{e}}{\sigma_{33}-\nu(\sigma_{11}+\sigma_{22})}=\frac{1}{E} \tag{2.2}$$

Once a material yields, the relationship between the strain and stress in the material alters in the sense that the unique correspondence between the two parameters E and ν, as shown in Eq. (2.2), may no longer apply. A general observation of an isotropic material after yielding indicates that, within the plastic range of deformation, an incremental change in strain is associated with a finite stress but not with an incremental change in stress that might have produced the incremental strain. Another unique phenomenon observed in plasticity is that the volumetric strain in the plastic regime vanishes. This latter behavior is usually characterized as a plastic incompressibility condition, which means that $\varepsilon_{11}^{p}+\varepsilon_{22}^{p}+\varepsilon_{33}^{p}=0$, where "*p*" indicates the plastic deformation when the plastic strains involved are small.

If the magnitude of the plastic deformation is so large that the elastic component of a strain is numerically insignificant, the relationship between strain and stress can be expressed in a manner similar to that of Hooke's law but with a different coefficient. Considering the above-mentioned first observation, the so-called Levy-Mises (Hill, 1950; Blazynski, 1983) relationships are given by

$$\mathrm{d}\varepsilon_{11}^{p}=\lambda[\sigma_{11}-\nu(\sigma_{22}+\sigma_{33})] \tag{2.3.1}$$

$$\mathrm{d}\varepsilon_{22}^{p}=\lambda[\sigma_{22}-\nu(\sigma_{11}+\sigma_{33})] \tag{2.3.2}$$

$$\mathrm{d}\varepsilon_{33}^{p}=\lambda[\sigma_{33}-\nu(\sigma_{11}+\sigma_{22})] \tag{2.3.3}$$

where λ is a positive parameter that depends on the yield, equilibrium and boundary conditions of the material. By using the plastic incompressibility condition, $\mathrm{d}\varepsilon_{11}^{p}+\mathrm{d}\varepsilon_{22}^{p}+\mathrm{d}\varepsilon_{33}^{p}=0$, one obtains $\nu=0.5$. Eliminating λ from Eq. (2.3), an expression comparable to Eq. (2.2) is given by

$$\frac{\mathrm{d}\varepsilon_{11}^{p}}{\sigma_{11}-\nu(\sigma_{22}+\sigma_{33})}=\frac{\mathrm{d}\varepsilon_{22}^{p}}{\sigma_{22}-\nu(\sigma_{11}+\sigma_{33})}=\frac{\mathrm{d}\varepsilon_{33}^{p}}{\sigma_{33}-\nu(\sigma_{11}+\sigma_{22})}=\lambda \tag{2.4}$$

To simplify the above equations, let us define the deviatoric stresses

$$\sigma'_{ij}=\sigma_{ij}-\frac{1}{3}\sigma_{kk}\delta_{ij}=\sigma_{ij}-\frac{1}{3}(\sigma_{11}+\sigma_{22}+\sigma_{33})\delta_{ij},\quad i,j=1,2,3 \tag{2.5}$$

where δ_{ij} is the Kronecker delta having the characteristic that

$$\delta_{ij}=\begin{cases}0, & \text{if } i\neq j\\ 1, & \text{if } i=j\end{cases} \tag{2.6}$$

By using Eq. (2.5), Eq. (2.4) becomes (and remembering $\nu = 0.5$)

$$\frac{\mathrm{d}\varepsilon_{11}^{p}}{\sigma'_{11}} = \frac{\mathrm{d}\varepsilon_{22}^{p}}{\sigma'_{22}} = \frac{\mathrm{d}\varepsilon_{33}^{p}}{\sigma'_{33}} = \frac{2}{3}\lambda \tag{2.7}$$

which is the Levy-Mises plasticity theory.

A more general plastic flow theory, called the Prandtl-Reuss flow rule, can be stated as (Blazynski, 1983)

$$\mathrm{d}\varepsilon_{ij}^{p} = \lambda \sigma'_{ij} \tag{2.8}$$

The parameter λ in Eq. (2.8) must be defined in order for the rule to be applicable. One method of defining the parameter λ is to multiply Eq. (2.8) by itself (Adams, 1974). This gives (with the summention being applied to the repeated subscripts)

$$\lambda = \left[\mathrm{d}\varepsilon_{ij}^{(p)} \mathrm{d}\varepsilon_{ij}^{(p)}\right]^{1/2} / (\sigma'_{ij}\sigma'_{ij})^{1/2} \tag{2.9}$$

In a situation in which both the elastic and plastic strains are of the same order of magnitude, a more general stress-strain relationship is required. The equation of this type is known as the Prandtl-Reuss elasto-plastic constitutive relationship and contains elements of both the elastic and the plastic strains. The incremental overall strain can thus be expressed as

$$\mathrm{d}\varepsilon_{ij} = \mathrm{d}\varepsilon_{ij}^{(e)} + \mathrm{d}\varepsilon_{ij}^{(p)} \tag{2.10}$$

where the elastic component is related to the stress increment through (Eq. (1.4.1) and (1.8)).

$$\mathrm{d}\varepsilon_{ij}^{(e)} = \frac{1-2\nu}{3E}\mathrm{d}\sigma_{kk}\delta_{ij} + \frac{1+\nu}{E}\mathrm{d}\sigma'_{ij} \tag{2.11}$$

which is the same as Eq. (1.4.1) providing that the total strains and stresses in Eq. (1.4.1) are changed to the incremental ones. Defining the octahedral plastic shear strain increment as

$$\begin{aligned}\mathrm{d}\varepsilon_{0}^{(p)} &= \left[\frac{1}{3}\mathrm{d}\varepsilon_{ij}^{(p)}\mathrm{d}\varepsilon_{ij}^{(p)}\right]^{1/2} \\ &= \frac{1}{\sqrt{3}}\left[(\mathrm{d}\varepsilon_{11}^{(p)})^2 + (\mathrm{d}\varepsilon_{22}^{(p)})^2 + (\mathrm{d}\varepsilon_{33}^{(p)})^2 + 2\{(\mathrm{d}\varepsilon_{12}^{(p)})^2 + (\mathrm{d}\varepsilon_{13}^{(p)})^2 + (\mathrm{d}\varepsilon_{23}^{(p)})^2\}\right]^{1/2}\end{aligned} \tag{2.12}$$

and the octahedral shear stress as

$$\tau_0 = \left[\frac{1}{3}\sigma'_{ij}\sigma'_{ij}\right]^{1/2} = \frac{1}{\sqrt{3}}\left[(\sigma'_{11})^2 + (\sigma'_{22})^2 + (\sigma'_{33})^2 + 2\{(\sigma'_{12})^2 + (\sigma'_{13})^2 + (\sigma'_{23})^2\}\right]^{1/2} \tag{2.13}$$

Eq. (2.9) becomes

$$\lambda = \mathrm{d}\varepsilon_0^{(p)} / \tau_0 \tag{2.14}$$

Substituting the last equation into Eq. (2.8), one has

$$\mathrm{d}\varepsilon_{ij}^{(p)} = \left(\frac{\mathrm{d}\varepsilon_0^{(p)}}{\tau_0} \right) \sigma'_{ij} \tag{2.15}$$

In order to establish a relationship between $\mathrm{d}\varepsilon_0^{(p)}$ and τ_0, let us apply Eq. (2.15) to a uni-axial tensile test in which only the stress component in the loading direction (taken as x_1 direction) does not vanish, i.e.,

$$\sigma_{22}=\sigma_{33}=\sigma_{12}=\sigma_{13}=\sigma_{23}=0, \text{ and } \sigma_{11}\neq 0 \tag{2.16}$$

Supposing the testing load has caused the material to deform plastically, the incremental plastic strains are expressed as

$$\mathrm{d}\varepsilon_{12}^{(p)} = \mathrm{d}\varepsilon_{23}^{(p)} = \mathrm{d}\varepsilon_{13}^{(p)} = 0, \quad \mathrm{d}\varepsilon_{22}^{(p)} = \mathrm{d}\varepsilon_{33}^{(p)} = -\frac{1}{2}\mathrm{d}\varepsilon_{11}^{(p)} \quad \text{and} \quad \mathrm{d}\varepsilon_{11}^{(p)} \neq 0 \tag{2.17}$$

where the plastic incompressibility condition has been utilized. Substituting Eq. (2.17) into Eq. (2.12) gives

$$\mathrm{d}\varepsilon_{11}^{(p)} = \sqrt{2}\mathrm{d}\varepsilon_0^{(p)} \tag{2.18}$$

while substituting Eq. (2.16) into Eq. (2.5) and then into Eq. (2.13) yields

$$\tau_0 = \frac{\sqrt{2}}{3}\sigma_{11} \tag{2.19}$$

Suppose that the tensile stress-strain curve of the material is composed of piece-wise linear segments. A typical bilinear curve is shown in Fig. 2.1.

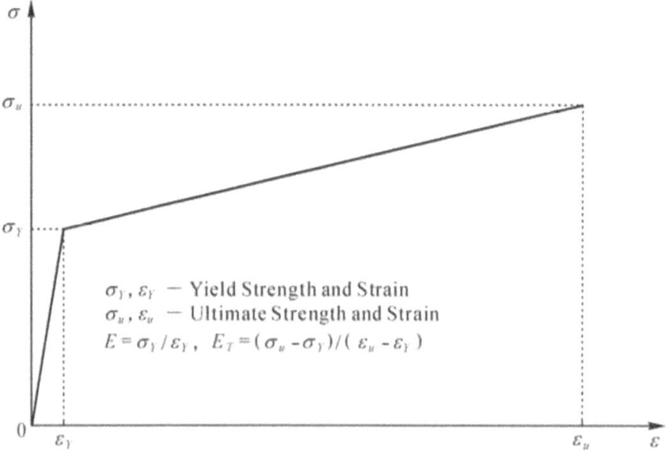

Fig. 2.1 A typical bilinear elastic-plastic stress-strain curve

Let us use E_T to represent a hardening modulus of the material, which is the tangent to the stress-strain curve at a plastic region (Fig. 2.1). Due to the well-known property in the plastic unloading process (which states that an unloading stress-strain curve is parallel to the linearly elastic segment of the material), the relationship between the octahedral plastic shear strain increment and the tensile stress increment can be denoted by (noting that $E_{\mathrm{T}} = \frac{\mathrm{d}\sigma}{\mathrm{d}\varepsilon}$, $E = \frac{\mathrm{d}\sigma}{\mathrm{d}\varepsilon^{(e)}}$ and $\mathrm{d}\varepsilon = \mathrm{d}\varepsilon^{(e)} + \mathrm{d}\varepsilon^{(p)}$)

$$\mathrm{d}\varepsilon_0^{(p)} = \frac{1}{\sqrt{2}}\left(\frac{1}{E_{\mathrm{T}}} - \frac{1}{E}\right)\mathrm{d}\sigma$$

or

$$\mathrm{d}\tau_0 = \frac{2M_{\mathrm{T}}}{3}\mathrm{d}\varepsilon_0^{(p)} \tag{2.20}$$

where

$$M_{\mathrm{T}} = \frac{EE_{\mathrm{T}}}{E - E_{\mathrm{T}}} \tag{2.21}$$

On the other hand, differentiating Eq. (2.13) gives

$$\mathrm{d}\tau_0 = \frac{\sigma'_{ij}}{3\tau_0}\mathrm{d}\sigma'_{ij}$$

Substituting Eq. (2.20) into the last equation, one gets

$$\mathrm{d}\varepsilon_0^{(p)} = \frac{\sigma'_{ij}}{2M_T\tau_0}\mathrm{d}\sigma'_{ij} \tag{2.22}$$

From Eqs. (2.22) and (2.15), one obtains

$$\mathrm{d}\varepsilon_{ij}^{(p)} = \frac{\sigma'_{kl}\mathrm{d}\sigma'_{kl}}{2M_T\tau_0^2}\sigma'_{ij} \tag{2.23}$$

By making use of the assumption that no plastic work can be done by the hydrostatic component of an applied stress field (as a result of the plastic incompressibility condition since, otherwise, a plastic volumetric strain would have been generated), i.e.,

$$\sigma'_{ij}\mathrm{d}\sigma'_{ij} = \sigma'_{ij}\left(\mathrm{d}\sigma_{ij} - \frac{1}{3}\mathrm{d}\sigma_{kk}\delta_{ij}\right) = \sigma'_{ij}\mathrm{d}\sigma_{ij}$$

Eq. (2.23) is rewritten as

$$\mathrm{d}\varepsilon_{ij}^{(p)} = \frac{\sigma'_{kl}\sigma'_{ij}}{2M_T\tau_0^2}\mathrm{d}\sigma_{kl} \tag{2.24}$$

Substituting Eqs. (2.11) and (2.24) into Eq. (2.10), we finally arrive at

$$\mathrm{d}\varepsilon_{ij} = \frac{1-2\nu}{3E}\mathrm{d}\sigma_{kk}\delta_{ij} + \frac{1+\nu}{E}\mathrm{d}\sigma'_{ij} + \frac{\sigma'_{kl}\sigma'_{ij}}{2M_T\tau_0^2}\mathrm{d}\sigma_{kl} \tag{2.25}$$

Eq. (2.25) is the general constitutive relation governing isotropic elastic-plastic material behavior that relates total strain increments with stress increments in terms of the octahedral shear stress τ_0, which is defined by Eq. (2.13). Rewriting Eq. (2.25) in a vector form, the incremental elastic-plastic constitutive equations of an isotropic material are denoted by

$$\{\mathrm{d}\varepsilon_i\}=[S_{ij}]\{\mathrm{d}\sigma_j\} \tag{2.26}$$

in which the compliance matrix has the form

$$[S_{ij}] = \begin{cases} [S_{ij}]^e, & \text{when } \tau_0 \le \frac{\sqrt{2}}{3}\sigma_Y \\ [S_{ij}]^e + [S_{ij}]^p, & \text{when } \tau_0 > \frac{\sqrt{2}}{3}\sigma_Y \end{cases} \tag{2.27}$$

where σ_Y is the uniaxial yield strength of the material (Fig. 2.1), $[S_{ij}]^e$ is the elastic component of the compliance matrix given by Eqs. (1.8), (1.9) and (1.10), and $[S_{ij}]^p$ is the plastic component defined as

$$[S_{ij}]^p = \frac{1}{2M_T\tau_0^2}\begin{bmatrix} \sigma'_{11}\sigma'_{11} & \sigma'_{22}\sigma'_{11} & \sigma'_{33}\sigma'_{11} & 2\sigma'_{23}\sigma'_{11} & 2\sigma'_{13}\sigma'_{11} & 2\sigma'_{12}\sigma'_{11} \\ & \sigma'_{22}\sigma'_{22} & \sigma'_{33}\sigma'_{22} & 2\sigma'_{23}\sigma'_{22} & 2\sigma'_{13}\sigma'_{22} & 2\sigma'_{12}\sigma'_{22} \\ & & \sigma'_{33}\sigma'_{33} & 2\sigma'_{23}\sigma'_{33} & 2\sigma'_{13}\sigma'_{33} & 2\sigma'_{12}\sigma'_{33} \\ & & & 4\sigma'_{23}\sigma'_{23} & 4\sigma'_{13}\sigma'_{23} & 4\sigma'_{12}\sigma'_{23} \\ & & & & 4\sigma'_{13}\sigma'_{13} & 4\sigma'_{12}\sigma'_{13} \\ & \text{symmetry} & & & & 4\sigma'_{12}\sigma'_{12} \end{bmatrix} \tag{2.28}$$

Remark 2.1

When the plastic curve (either a tensile or a compressive curve) of a material consists of more than one linear segments as indicated in Fig. 2.2 for a bulk (monolithic) epoxy material under uniaxial tension, its hardening modulus is different in a different region (corresponding to a different range of octahedral shear stress). Supposing that whole stress-strain curve consists of n segments, including the initial linear elasticity segment, the current hardening modulus is defined as

$$E_T = (E_T)_i \text{ when } (\sigma_Y)_i \leq \frac{3\tau_0}{\sqrt{2}} \leq (\sigma_Y)_{i+1}, \quad i = 0,1, \ldots, n-1 \qquad (2.29.1)$$

$$(E_T)_0 = E = \text{Young's modulus}, \ (\sigma_Y)_0 = 0 \text{ and } (\sigma_Y)_1 = \sigma_Y = \text{yield strength} \qquad (2.29.2)$$

For the material shown in Fig. 2.2, we have $(\sigma_Y)_1$=15.8 MPa, $(\sigma_Y)_2$=33.3 MPa, $(\sigma_Y)_3$=42.5 MPa, $(\sigma_Y)_4$=44.1 MPa, $(E_T)_0$=2.38 GPa, $(E_T)_1$=1.67 GPa, $(E_T)_2$=1.10 GPa and $(E_T)_3$=0.22 GPa.

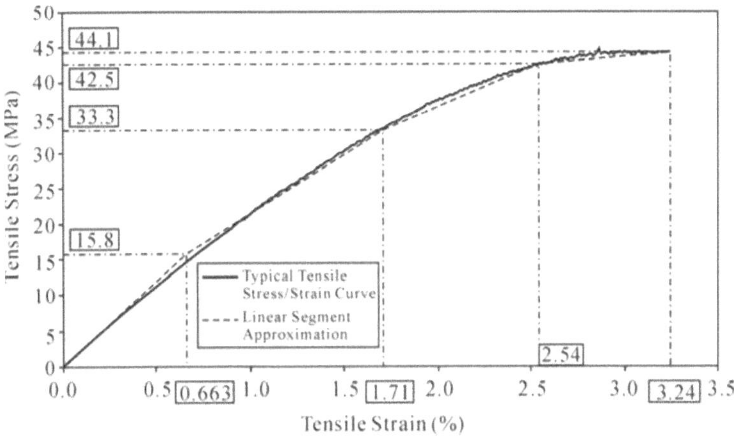

Fig. 2.2 Tensile stress-strain curve of an epoxy polymer consisting of four linear segments

Remark 2.2

In the elastic-plastic constitutive Eq. (2.26), the plastic component of the compliance matrix, given by Eq. (2.28), can only occur under a loading condition. As long as there is an unloading, the compliance matrix is simply given by its elastic component.

Remark 2.3

Many materials, especially polymers, exhibit a different stress-strain response at compression from that at tension. Accordingly, different material parameters should be used in the Prandtl-Reuss equations. As such, an essential step is to differentiate a compressive load condition from a tensile one. In practice the material, especially the constituent matrix material in a fibrous composite, will be generally subjected to a multiaxial stress state. It is highly possible that some stress component is positive (in tension) and the others are negative (in compression). A general criterion, based on which the material of any stress state can be easily understood to be under an essential tension or to be under an essential compression, is necessary. Using the three principal stresses of the material, this criterion is simply expressed as (Huang, 2001; Huang & Ramakrishna, 2002) in the following:

(1) If $\sigma^1+\sigma^2+\sigma^3<0$, the material is under an essential compression;

(2) if $\sigma^1+\sigma^2+\sigma^3\geq 0$, the material is under an essential tension.

In the above, σ^1, σ^2, and σ^3 stand for the three principal stresses in the material.

Remark 2.4

In applying the Prandtl-Reuss plastic flow theory, an important consideration is on a measure used for stress and strain. If a nominal measure, which is generally adopted in conducting material tests in engineering, is used, the plastic incompressibility condition is expressed as $(1+\varepsilon_{11}^p)(1+\varepsilon_{22}^p)(1+\varepsilon_{33}^p)=1$, where 1, 2 and 3 are the three principal stretch directions. In such a case, the expression $\varepsilon_{11}^p+\varepsilon_{22}^p+\varepsilon_{33}^p=0$ does not represent the plastic incompressibility condition and a conversion of nominal strain to logarithmic strain and nominal stress to "true" stress is required before applying the Prandtl-Reuss theory. However, supposing that the plastic strain involved is small, the situation will be much simplified. We can simply use the nominal stress and strain in this theory, which only gives rise to an acceptable error. For example, if the ultimate nominal plastic strain of a material, ε_u^p, is less than 10%, the error between usage of ε_u^p and $\ln(1+\varepsilon_u^p)$ is less than 5% (4.68%), which might be acceptable in engineering.

Remark 2.5

Supposing that one has nominal stress-strain data for a uniaxial test and that the material is isotropic, a simple conversion of these nominal data to true stress and logarithmic plastic strain is given by (ABAQUS, 1995)

$$\sigma_{\text{true}}=\sigma_{\text{nom}}(1+\varepsilon_{\text{nom}})$$

$$\varepsilon_{\ln}^p=\ln(1+\varepsilon_{\text{nom}})-\frac{\sigma_{\text{true}}}{E}$$

where ε_{nom} is the total nominal strain, σ_{nom} is the nominal stress and E is the Young's modulus.

2.3 2D Prandtl-Reuss formulae

For a plane stress problem we have $\sigma_{13}=\sigma_{23}=\sigma_{33}=0$, whereas σ_{11}, σ_{22} and σ_{12} are not necessarily zero. Substituting these stress components into the Prandtl-Reuss elasto-plastic theory, the planar incremental strain-stress constitutive equations are expressed as

$$\begin{Bmatrix} \mathrm{d}\varepsilon_{11} \\ \mathrm{d}\varepsilon_{22} \\ 2\mathrm{d}\varepsilon_{12} \end{Bmatrix} = \begin{bmatrix} S_{11} & S_{12} & S_{13} \\ & S_{22} & S_{23} \\ \text{symmetry} & & S_{33} \end{bmatrix} \begin{Bmatrix} \mathrm{d}\sigma_{11} \\ \mathrm{d}\sigma_{22} \\ \mathrm{d}\sigma_{12} \end{Bmatrix} = [S_{ij}]_{3\times3} \begin{Bmatrix} \mathrm{d}\sigma_{11} \\ \mathrm{d}\sigma_{22} \\ \mathrm{d}\sigma_{12} \end{Bmatrix} \tag{2.30}$$

where
$$[S_{ij}]_{3\times3} = \begin{cases} [S_{ij}]^e_{3\times3}, \text{when } \sigma_e \le \sigma_y \\ [S_{ij}]^e_{3\times3} + [S_{ij}]^p_{3\times3}, \text{when } \sigma_e > \sigma_Y \end{cases} \tag{2.31}$$

$$[S_{ij}]^e_{3\times3} = \begin{bmatrix} \dfrac{1}{E} & -\dfrac{\nu}{E} & 0 \\ & \dfrac{1}{E} & 0 \\ & \text{symmetry} & \dfrac{1}{G} \end{bmatrix} \tag{2.32}$$

$$\sigma_e = \sqrt{(\sigma_{11})^2 + (\sigma_{22})^2 - (\sigma_{11})(\sigma_{22}) + 3(\sigma_{12})^2}\,, \tag{2.33}$$

$$[S_{ij}]^p_{3\times3} = \frac{9}{4M_T(\sigma_e)^2} \begin{bmatrix} \sigma'_{11}\sigma'_{11} & \sigma'_{22}\sigma'_{11} & 2\sigma'_{12}\sigma'_{11} \\ & \sigma'_{22}\sigma'_{22} & 2\sigma'_{12}\sigma'_{22} \\ & \text{symmetry} & 4\sigma'_{12}\sigma'_{12} \end{bmatrix} \tag{2.34}$$

$$\sigma'_{ij} = \sigma_{ij} - \frac{1}{3}(\sigma_{11}+\sigma_{22})\delta_{ij}, \quad i, j = 1, 2 \tag{2.35}$$

It is noted that the plastic yielding condition has been expressed in terms of von Mises effective stress, σ_e, in Eq. (2.33), rather than in terms of the octahedral shear stress τ_0. This is because in a planar problem the von Mises effective stress can be easily calculated, see Eq. (2.33). Furthermore, it is noticed that $\tau_0=\sigma_e\sqrt{2/3}$.

Example 2.1 Suppose that the uniaxial tensile stress-strain curve of an isotropic material is given by Fig. 2.2, with a Poisson's ratio of ν=0.3. The material is subjected to a planar stress state. At a particular load level, the stress components are found to be $\{\sigma_{11}, \sigma_{22}, \sigma_{12}\}=\{29.5, -18.0, 0\}$. Find out the instantaneous compliance matrix of the material at this load level.

Solution

(1) Calculate the principal stresses based on the current stress state $\{\sigma_{11}, \sigma_{22}, \sigma_{12}\} = \{29.5, -18.0, 0\}$. As no shear stress is involved, the principal stresses can be easily obtained from the normal stresses, i.e., $\sigma^1=29.5$, $\sigma^2=0$ and $\sigma^3=-18.0$.

(2) As $\sigma^1+\sigma^2+\sigma^3= (29.5)+(0)+(-18.0)=11.5>0$, the material is under an essential tension corresponding to the given stress state. The given stress-strain curve is applicable.

(3) von Mises effective stress is $\sigma_e= \sqrt{(29.5)^2+(-18)^2-(29.5)(-18)} = \sqrt{1419.25} = 37.7$ (MPa). This means that the material has already yielded and the effective stress is on the third linear segment of the stress-strain curve.

(4) According to the von Mises effective stress, the hardening modulus of the material is found to be $E_T=(E_T)_3=1.10$ GPa, whereas the elastic modulus is $E=2.38$ GPa.

(5) The instantaneous compliance matrix at the given load level consists of an elastic component and a plastic one. Non-zero elements of the elastic compliance $[S_{ij}]^e$ are found to be

$S_{11}^e=S_{22}^e=1/E=0.420\ (\text{GPa})^{-1}$, $S_{12}^e=S_{21}^e=-\nu/E=-0.126\ (\text{GPa})^{-1}$, and $S_{33}^e=(2+2\nu)/E=1.092\ (\text{GPa})^{-1}$.

(6) The plastic compliance component, $[S_{ij}]^p$, is calculated as follows.

$$\sigma'_{11}=\sigma_{11}-\frac{1}{3}(\sigma_{11}+\sigma_{22})=29.5-11.5/3=25.67\ (\text{MPa}),$$

$$\sigma'_{22}=\sigma_{22}-\frac{1}{3}(\sigma_{11}+\sigma_{22})=-18-11.5/3=-21.83\ (\text{MPa}),$$

$$\sigma'_{12}=\sigma_{12}=0,$$

$$M_T=\frac{EE_T}{E-E_T}=(2.38)(1.1)/(2.38-1.1)=2.05\ (\text{GPa}),$$

$$9/[4M_T(\sigma_e)^2]=9/[4\times2.05\times10^3\times37.7^2]=7.72\times10^{-7}\ (\text{MPa})^{-3},$$

$$S_{11}^p=9\sigma'_{11}\sigma'_{11}/[4M_T(\sigma_e)^2]=5.087\times10^{-4}\ (\text{MPa})^{-1}=0.509\ (\text{GPa})^{-1},$$

$$S_{22}^p=9\sigma'_{22}\sigma'_{22}/[4M_T(\sigma_e)^2]=3.679\times10^{-4}\ (\text{MPa})^{-1}=0.368\ (\text{GPa})^{-1},$$

$$S_{12}^p=S_{21}^p=9\sigma'_{22}\sigma'_{11}/[4M_T(\sigma_e)^2]=-0.433\ (\text{GPa})^{-1},$$

and $S_{13}^p=S_{31}^p=S_{23}^p=S_{32}^p=S_{33}^p=0$.

(7) The instantaneous compliance matrix is given by

$$[S_{ij}] = [S_{ij}]^e + [S_{ij}]^p = \begin{bmatrix} 0.929 & -0.126 & 0 \\ -0.126 & 0.788 & 0 \\ 0 & 0 & 1.092 \end{bmatrix} \text{(GPa)}^{-1}$$

2.4 Bodner-Partom Unified Plasticity Theory

Although the preceding incremental form of the Prandtl-Reuss plastic flow theory is quite general and easily applicable, it can only represent the material elasto-plastic behavior at some specific conditions, such as a fixed temperature, a prescribed strain rate (i.e., loading speed) and so on. It is well-known that the material plasticity is very sensitive to environmental conditions. Significant factors that can affect the material plastic response include working temperature, strain rate, loading history, creep and stress relaxation, etc. The incremental form of the Prandtl-Reuss theory has not explicitly incorporated the influence of such different factors on the material response. Therefore, it is not a "unified" form: the material plastic parameters, i.e., hardening modulus and yield strength, must be separately provided with a different environmental condition.

Many attempts have been made to address the aforementioned problem and to try to establish a "unified" theory for, at least, some kinds of materials. One such theory is the Bodner-Partom unified plasticity theory (Bodner & Partom, 1975). The theory has found much success in describing the mechanical behavior of titanium alloys which are widely used in making titanium-matrix-based ceramic fiber reinforced composites for advanced aerospace applications (Mall and Nicholas, 1997). The latest version of the Bodner-Partom unified plasticity theory is summarized in the following (Stouffer & Bodner, 1979; Chan & Lindholm, 1990, Aboudi, 1991; Robertson & Mall, 1997; Nicholas & Kroupa, 1998).

The total strain rate, $\dot{\varepsilon}_{ij}$, of a material can be decomposed into the sum of two parts, i.e., an elastic and an inelastic components (Eq. (2.10))

$$\dot{\varepsilon}_{ij} = \dot{\varepsilon}_{ij}^{(e)} + \dot{\varepsilon}_{ij}^{I} \tag{2.36}$$

where the elastic component is obtained by the time derivative of Hooke's law (with reference to Eq. (2.11)). For the inelastic strain component, it is still assumed that the isotropic form of the Prandtl-Reuss flow law be applicable, but in a different type, i.e.,

$$\dot{\varepsilon}_{ij}^{I} = \Lambda \sigma'_{ij} \tag{2.37}$$

Here, the coefficient Λ is a flow rule function. It is evident that a different flow theory assumes a different function form for the coefficient Λ. The Bodner-Partom flow law gives the inelastic strain rate as (Bodner & Partom, 1975; Robertson &

Mall, 1997; Nicholas & Kroupa, 1998)

$$\dot{\varepsilon}_{ij}^{I}=D_0\exp\left[-\frac{1}{2}\left(\frac{(Z^I+Z^D)^2}{3J_2}\right)^n\right]\frac{\sigma'_{ij}}{\sqrt{J_2}} \tag{2.38}$$

where

$$J_2=\frac{1}{2}\sigma'_{ij}\sigma'_{ij}=\frac{3}{2}\tau_0^2 \tag{2.39.1}$$

$$\dot{Z}^I=m_1\dot{W}_p(Z_1-Z^I)-A_1Z_1\left(\frac{Z^I-Z_2}{Z_1}\right)^{r_1}+\dot{T}\left[\left(\frac{Z^I-Z_2}{Z_1-Z_2}\right)\frac{\partial Z_1}{\partial T}+\left(\frac{Z_1-Z^I}{Z_1-Z_2}\right)\frac{\partial Z_2}{\partial T}\right] \tag{2.39.2}$$

$$\dot{W}_p=\sigma_{ij}\dot{\varepsilon}_{ij}^{I},\quad Z^D=\beta_{ij}u_{ij},\quad u_{ij}=\frac{\sigma_{ij}}{\sqrt{\sigma_{kl}\sigma_{kl}}},\quad Z^I(0)=Z_0 \tag{2.39.3}$$

$$\dot{\beta}_{ij}=m_2\dot{W}_p(Z_3u_{ij}-\beta_{ij})-A_2Z_1\frac{\beta_{ij}}{\sqrt{\beta_{kl}\beta_{kl}}}\left(\frac{\sqrt{\beta_{kl}\beta_{kl}}}{Z_1}\right)^{r_2}+\dot{T}\frac{\beta_{ij}}{Z_3}\frac{\partial Z_3}{\partial T},\quad \beta_{ij}(0)=0 \tag{2.39.4}$$

In the above, T denotes temperature and the overhead dot of a quantity designates a differentiation with respect to time t. D_0, n, m_1, Z_1, A_1, r_1, Z_2, Z_0, m_2, Z_3, A_2, and r_2 are material-dependent parameters, which control the time and temperature-dependent effects and hardening characteristics.

As can be expected, the Bodner-Partom flow law is much more complicated than the incremental form of the Prandtl-Reuss theory. There are many more material parameters involved in the Bodner-Partom model, which must be determined through experiments on the material under various possible conditions (different temperatures, strain rates and different creep and stress relaxation actions). However, once these parameters have been determined, the material response under any load/environmental condition can be reasonably described using the Bodner-Partom unified model.

As an example, the material parameters of the Bodner-Partom model for a β-Titanium alloy (Nicholas et al., 1996), TIMETAL 21S, have been obtained by Neu (1993). The parameters are listed in Table 2.1 (Kroupa et al., 1996; Robertson & Mall, 1997). As derivatives of the parameters Z_1, Z_2, and Z_3 with respect to temperature T are also required in Eqs. (2.39.2) and (2.39.4), variations of them with respect to T have been polynomially interpolated from the data shown in Table 2.1. Thus, the temperature-derivatives can be expressed as (Huang, 2002)

$$\frac{\partial z_1}{\partial T}=0 \tag{2.40.1}$$

$$\frac{\partial z_2}{\partial T}=a_1+a_2(T-T_p)+\ldots+a_6(T-T_p)^5 \tag{2.40.2}$$

$$\frac{\partial z_3}{\partial T}=b_1+b_2(T-T_p)+\ldots+b_7(T-T_p)^6 \tag{2.40.3}$$

where a_1=−0.81916E−0, a_2=0.58491E−2, a_3=−0.88078E−4, a_4=−0.45170E−6, a_5=0.58263E−9, a_6=0.26124E−11, b_1=0.35176E+2, b_2=0.12348E−0, b_3=−0.13405E−2, b_4=−0.30590E−5, b_5=0.20623E−7, b_6=0.12014E−10, b_7=−0.78467E−13 and T_p =512.56 °C. All the units of a_i and b_i are in MPa/(°C)^i.

Table 2.1 Bodner-Partom material parameters for TIMETAL 21S (ν^m=0.34)

T (°C)	E (GPa)	α^* (×10^{-6} °C^{-1})	n	$Z_0=Z_2$ (MPa)	Z_3 (MPa)	m_2 (MPa^{-1})	$A_1=A_2$ (s^{-1})
23	112	6.31	4.8	1550	100	0.35	0
260	108	7.26	3.5	1300	300	0.35	0
315	106	7.48	3.05	1250	390	1.50	0
365	104	7.68	2.65	1205	500	2.55	0.0003
415	101	7.88	2.24	1160	660	3.60	0.0013
465	99.1	8.08	1.84	1115	960	4.64	0.0050
482	98.1	8.15	1.7	1100	1100	5.00	0.00764
500	97.0	8.22	1.5	1089	1300	5.76	0.0116
525	95.5	8.32	1.28	1074	1670	6.82	0.0203
550	93.9	8.43	1.1	1059	2100	7.88	0.0342
575	92.2	8.53	0.97	1045	2600	8.94	0.0559
600	90.4	8.63	0.82	1030	3700	10.0	0.0887
650	86.6	8.83	0.74	1000	3800	10.0	0.208
760	77.2	9.27	0.58	600	4000	15.0	1.01
815	72.0	9.49	0.55	300	4100	30.0	1.97
m_1=0 MPa^{-1};		r_1=3;		r_2=3;	Z_1=1600 MPa;		D_0=10,000 s^{-1}

* = thermal expansion coefficient

2.5 Conversion of Bodner-Partom Model into Prandtl-Reuss Equations

It is seen that the Bodner-Partom model depends on load history and an integration of the model is expressed in total stress and total strain forms, which may be inconvenient for some structural analyses. Furthermore, a general micromechanics model presented in the next chapter is essentially based on an incremental description for the material strain-stress relationship. For this reason, a conversion of the Bodner-Partom model into the Prandtl-Reuss model is necessary. Namely, the constitutive equations in the total stress-total strain form, integrated from the Bodner-Partom flow law, are required to be converted to the equations in an incremental stress-strain form, represented by the Prandtl-Reuss theory.

To do this, the material uniaxial stress-strain curve at every level (i.e., a pre-assumed loading condition of temperature, strain rate, loading history, etc.) is plotted based on the Bodner-Partom model. From this curve, the yield strength and the hardening modulus are obtained. Then, the incremental Prandtl-Reuss constitutive relationship, Eq. (2.26), can be defined. Therefore, we only need to integrate the one-dimensional Bodner-Partom equation.

Under a uniaxial load condition, the general Bodner-Partom Eq. (2.38) reads

$$\dot{\varepsilon}^I = D_0 \exp\left[-\frac{1}{2}\left(\frac{Z^I+\beta}{\sigma}\right)^{2n}\right]\frac{2}{\sqrt{3}}\operatorname{sgn}(\sigma),\quad \operatorname{sgn}(\sigma)=\begin{cases}1, & \text{if } \sigma>0\\ -1, & \text{if } \sigma<0\end{cases} \tag{2.41}$$

where

$$\dot{Z}^I = m_1\sigma\dot{\varepsilon}^I(Z_1-Z^I)-A_1Z_1\left(\frac{Z^I-Z_2}{Z_1}\right)^{r_1}+\dot{T}\left[\left(\frac{Z^I-Z_2}{Z_1-Z_2}\right)\frac{\partial Z_1}{\partial T}+\left(\frac{Z_1-Z^I}{Z_1-Z_2}\right)\frac{\partial Z_2}{\partial T}\right] \tag{2.42.1}$$

$$Z^I(0)=Z_0 \tag{2.42.2}$$

$$\dot{\beta} = m_2\sigma\dot{\varepsilon}^I(Z_3-\beta)-A_2Z_1\left(\frac{\beta}{Z_1}\right)^{r_2}+\dot{T}\frac{\beta}{Z_3}\frac{\partial Z_3}{\partial T},\quad \beta(0)=0 \tag{2.42.3}$$

In the above, σ and ε^I are, respectively, the stress and inelastic strain of the material under a uniaxial load. Closed form integration of Eq. (2.41) together with Eqs. (2.42.1) and (2.42.3) is generally not easy. However, we can easily integrate it using a numerical technique. The simplest and also an unconditionally stable integration scheme is the Euler trapezoidal method (Huang, 1991). Suppose that

$$\dot{Z} = f(Z,t) \quad \text{with } Z(0)=Z_0 \tag{2.43}$$

has been specified. Then, the Euler trapezoidal method gives the following discretized equations

$$Z_k = Z_{k-1}+0.5(t_k-t_{k-1})[f(Z_k, t_k)+f(Z_{k-1}, t_{k-1})],\ k=1, 2, \ldots, \tag{2.44}$$

where $t_0=0<t_1<t_2<\ldots$ are discretized time moments and Z_k is a discretized value taken at the time moment t_k. Using Eq. (2.44), a set of discretized evolution equations for Eqs. (2.42.1) and (2.42.3) are obtained as (Huang, 2002)

$$
\begin{aligned}
Z_k^I &= Z_{k-1}^I + \frac{\Delta t m_1 \dot{\varepsilon}^I}{2}[\sigma_{k-1}(Z_{1,k-1} - Z_{k-1}^I) + \sigma_k(Z_{1,k} - Z_k^I)] \\
&- \frac{\Delta t A_1}{2}\left[Z_{1,k}\left(\frac{Z_k^I - Z_{2,k}}{Z_{1,k}}\right)^{r_1} + Z_{1,k-1}\left(\frac{Z_{k-1}^I - Z_{2,k-1}}{Z_{1,k-1}}\right)^{r_1}\right] \\
&+ \frac{\Delta t}{2}\left\{ \dot{T}_{k-1}\left[\left(\frac{Z^I - Z_2}{Z_1 - Z_2}\right)_{k-1}\left(\frac{\partial Z_1}{\partial T}\right)_{k-1} + \left(\frac{Z_1 - Z^I}{Z_1 - Z_2}\right)_{k-1}\left(\frac{\partial Z_2}{\partial T}\right)_{k-1}\right] \right. \\
&\left. + \dot{T}_k\left[\left(\frac{Z^I - Z_2}{Z_1 - Z_2}\right)_k\left(\frac{\partial Z_1}{\partial T}\right)_k + \left(\frac{Z_1 - Z^I}{Z_1 - Z_2}\right)_k\left(\frac{\partial Z_2}{\partial T}\right)_k\right]\right\}
\end{aligned}
\tag{2.45.1}
$$

$$
\begin{aligned}
\beta_k &= \beta_{k-1} + \frac{\Delta t m_2 \dot{\varepsilon}^I}{2}[\sigma_{k-1}(Z_{3,k-1} - \beta_{k-1}) + \sigma_k(Z_{3,k} - \beta_k)] \\
&- \frac{\Delta t A_2}{2}\left[Z_{1,k-1}\left(\frac{\beta_{k-1}}{Z_{1,k-1}}\right)^{r_2} + Z_{1,k}\left(\frac{\beta_k}{Z_{1,k}}\right)^{r_2}\right] \\
&+ \frac{\Delta t}{2}\left[\dot{T}_{k-1}\frac{\beta_{k-1}}{Z_{3,k-1}}\left(\frac{\partial Z_3}{\partial T}\right)_{k-1} + \dot{T}_k\frac{\beta_k}{Z_{3,k}}\left(\frac{\partial Z_3}{\partial T}\right)_k\right]
\end{aligned}
\tag{2.45.2}
$$

In the above, $\Delta t=(t_k-t_{k-1})$ is a time increment. It is noted that when a quantity, such as Z_1, depends explicitly on temperature, the subscript k indicates that this quantity ($Z_{1,k}$) takes the value at the current temperature T_k, which is the temperature at the current time. Eqs. (2.45.1) and (2.45.2) can be solved by iteration using a Newton type method.

As the plastic strain rate, $\dot{\varepsilon}^I$, is constant, it follows from Eq. (2.41) that

$$(Z^I + \beta) = \alpha\sigma \quad \text{or} \quad (Z_k^I + \beta_k) = \alpha\sigma_k \tag{2.46}$$

where $\alpha = \left[-2\ln\{\sqrt{3}\dot{\varepsilon}^I / [2D_0 \operatorname{sgn}(\sigma)]\}\right]^{\frac{1}{2n}}$. From Eqs. (2.45.1) and (2.45.2), we can solve Z_k^I and β_k and further obtain σ_k from Eq. (2.46). The total strain, ε_k, is then calculated from

$$\varepsilon_k = \varepsilon_k^I + \sigma_k/E = \varepsilon_{k-1}^I + \Delta t\,\dot{\varepsilon}^I + \sigma_k/E. \tag{2.47}$$

In this way, the uniaxial stress-strain curve can be obtained incrementally. Sometimes a total strain rate $\dot{\varepsilon}$ rather than a plastic strain rate is given. In such a case we have

$$\dot{\varepsilon}^I \approx \dot{\varepsilon} - \Delta\sigma / (\Delta t E) \approx \dot{\varepsilon} \tag{2.48}$$

This is because $\Delta\sigma$ would be very small after the saturation value of stress is attained.

As an example, consider the uniaxial tensile stress-strain curves of the TIMETAL 21S material, as described in Table 2.1 at 566 °C with strain rates of $\dot{\varepsilon}$=8.3×10^{-4} s^{-1}, $\dot{\varepsilon}$=8.3×10^{-5} s^{-1} and $\dot{\varepsilon}$=8.3×10^{-6} s^{-1}. The calculated results using the above procedure are plotted in Fig. 2.3. Experimental data taken from Kroupa et al. (1996) are also shown in the figure for comparison. Good correlation has been found between the model and the experiments.

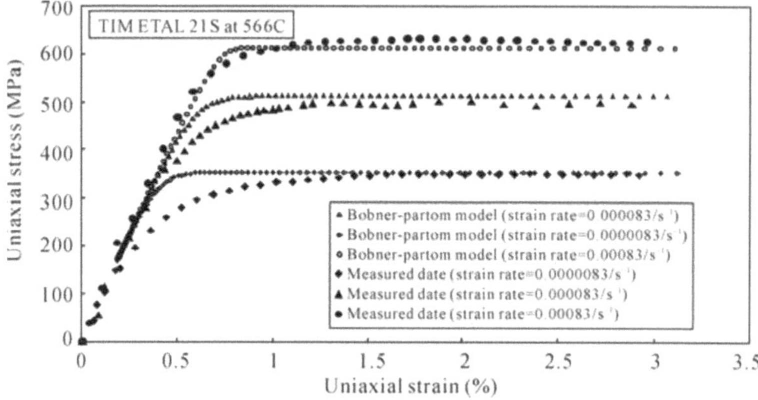

Fig. 2.3 Measured (Kroupa et al., 1996) and predicted uniaxial stress/strain curves for TIMETAL 21S at 566 °C with different strain rates (from Huang, 2002)

References

ABAQUS (1995) Theory manual, Version 5.5. Pawtucket: Hibbitt, Karlsson & Sorensen, Inc.

Aboudi, J.A. (1991) Mechanics of Composite Materials: A Unified Micromechanical Approach. Amsterdam: Elsevier.

Adams, D.F. (1974) Elastoplastic behavior of composites, in Mechanics of Composite Materials, G.P. Sendeckyj, (ed.) New York, NY: Academic Press.

Blazynski, T.Z. (1983) Applied Elasto-Plasticity of Solids. London: Macmillan Press.

Bodner, S.R. & Partom, Y. (1975) Constitutive Equations for Elastic Viscoplastic Strain Hardening Materials. ASME J. Appl. Mech. 42, 385-389.

Chan, K.S. & Lindholm, U.S. (1990) Inelastic deformation under nonisothermal loading. ASME J. Eng. Mater. & Tech. 112, 15-25.

Hill, R. (1950) The Mathematical Theory of Plasticity. Qxford: Clarendon Press.

Huang, Z.M. (1991) A set of accurate and unconditionally stable step-by-step integration algorithms for structural dynamics analysis. Transactions of the 11th International Conference on Structural Dynamics in Reactor Technology (B, 63-70), Tokyo, Japan.

Huang, Z.M. (2001) Simulation of the mechanical properties of fibrous composites by the bridging micromechanics model. Composites Part A 32, 143-172.

Huang, Z.M. (2002) Cyclic response of metal matrix composite laminates subjected to thermomechanical fatigue loads. International Journal of Fatigue 24, 463-475.

Huang, Z.M. & Ramakrishna, S. (2002) Towards Automatic Designing of 2D Biaxial Woven and Braided Fabric Reinforced Composites. J. Comp. Mater. 36, 1541-1579.

Kroupa, J.L., Neu, R.W., Nicholas, T., Coker, D., Robertson, D.D. & Mall, S. (1996) A comparison of analysis tools for predicting the inelastic cyclic response of cross-ply titanium matrix composites, in Life Prediction Methodology for Titanium Matrix Composites. ASTM STP 1253, W.S. Johnson, J.M. Larsen & B.N. Cox (eds.) American Society for Testing and Materials, Philadelphia, 297-327.

Mall, S. & Nicholas, T. (eds.) (1997) Titanium Matrix Composites— Mechanical Behavior. Lancaster, Basel: Technomic Publishing Co., Inc.

Neu, R. W. (1993) Nonisothermal material parameters for the Bodner-Partom model, in *Material Parameter Estimation for Modern Constitutive Equations*. MD-Vol. 43/AMD-Vol. 168, L. A. Bertram, S. B. Brown & A. D. Freed (eds.) ASME, New York, 211-226.

Nicholas, T., Russ, S.M., Neu, R.W., & Schehl, N. (1996) Life prediction of a [0/90] metal matrix composite under isothermal and thermomechanical fatigue, in Life Prediction Methodology for Titanium Matrix Composites, ASTM STP 1253, W.S. Johnson, J.M. Larsen & B.N. Cox (eds.) American Society for Testing and Materials, Philadelphia, 596-617.

Nicholas, T. & Kroupa, J.L. (1998) Micromechanics analysis and life prediction of titanium matrix composites. J. Comp. Tech. & Res., JCTRER 20, 79-88.

Robertson, D. & Mall, S. (1997) Micromechanical analysis and modeling, in Titanium Matrix Composites—Mechanical Behavior (eds.). S. Mall & T. Nicholas. Technomic Publishing Co., Inc., Lancaster, Basel, 397-464

Stouffer, D.C. & Bodner, S.R. (1979) A constitutive model for the deformation induced anisotropic plastic flow of metals. Int. J. Eng. Sci. 17, 757-764.

3

Bridging Micromechanics Model

3.1 Introduction

The conventional approach to the mechanical properties of a material is based on a pragmatic philosophy that demands an experimental determination of material response characteristics. This approach certainly offers the best advantage for isotropic materials. However, the mechanical properties of anisotropic composite materials are of a highly directional dependence. The introduction of this dependence together with the variation of the properties according to composition, orientation and packing geometry of the components greatly magnifies the labor and expense involved in the experimental determination of the overall material response characteristics for composite materials. These complications provide a strong motivation for the development of a constitutive relationship in terms of (1) composition, (2) the properties of the components and (3) the internal micro-structure as reflected by the relative orientation, size, shape and packing geometry of each of the components of a composite material. Such a constitutive relationship can be incorporated into, e.g., a FEM technique (Huang, 2007), to predict and design the structural performance. In developing the constitutive relationship, the macroscopic view underlying the experimental investigation must be complemented with a microscopic view that takes into account recognition of the multiphase nature of composite materials.

As will be seen in Chapter 5, any fibrous composite can be essentially simulated, based on its subdivision into a number of unidirectional (UD) composites. Therefore, the constitutive relationship of a UD composite becomes fundamental. Section 1.4 briefly outlined the approach of several micromechanics models to the elastic properties of UD composites. When the component materials are confined to linear and elastic deformation, a great number of micromechanical models, in addition to those given in Section 1.4, exist in the literature, with which the constitutive relationship of unidirectional fiber-reinforced composites can be well described. A thorough review of these models is not a concern of the present book and they can be found in e.g., Hashin (1964, 1983), Chamis & Sendeckyj

(1968), McCullough (1990), Halpin (1992), Bogdanovich & Pastore (1996) and Bogdanovich & Sierakowski (1999).

On the whole, many of those models can predict the elastic stiffness of the composite with a successful accuracy. However, without significant modifications, most of them cannot predict the composite inelastic properties such as yield strength, failure strength, ultimate strain, the inelastic stress-strain curve, etc.

In general, the stress sharing capacities of the constituents (fibers and matrix) in the composite should be determined only by the material parameters of the constituents and by the geometry of the fibers embedded in the matrix. The stresses shared by the matrix should therefore have a fixed and explicit relationship with those of the fibers, as long as both the constituents are in the same deformation (elastic or inelastic) region. When the constituent materials deform continuously from one region (such as a linear elastic region) to another (such as a plastic region), the coefficients correlating the two stress shares should only be influenced by the varied material properties, providing that the deformation of the materials is not very large. These considerations have drawn the author's attention to look for an explicit coefficient matrix, called a bridging matrix, to correlate the internal stress vector of the matrix with that of the fibers in the composite. Once this bridging matrix is determined, the internal stresses in the constituents are easily derived and so are all the effective properties of the composite, from the knowledge of the constituent properties (Huang, 2000a, 2000b, 2000c, 2000d, 2001a, 2001b, 2001c). The micromechanics model thus developed is called a Bridging Model. A detailed description of the development of this model is given in the following sections. Its applications to the determination of various mechanical properties of fibrous composites, including unidirectional laminae, multidirectional laminates and woven and braided fabric reinforced composites, will be shown or highlighted in the subsequent chapters.

3.2 Model Development

Let x_1, x_2, and x_3 be a rectangular co-ordinate system of a three-dimensional composite geometry. The fibers are unidirectionally arranged in the composite with their axes parallel to the x_1-direction. By using contracted notations, the volume averaged stresses and strains in the RVE satisfy the following relationships (Subsection 1.3.2)

$$\{\sigma_i\} = V_f\{\sigma_i^f\} + V_m\{\sigma_i^m\} \tag{3.1}$$

and

$$\{\varepsilon_i\} = V_f\{\varepsilon_i^f\} + V_m\{\varepsilon_i^m\} \tag{3.2}$$

where $\{\sigma_i\}=\{\sigma_{11}, \sigma_{22}, \sigma_{33}, \sigma_{23}, \sigma_{13}, \sigma_{12}\}^{\mathrm{T}}$ and $\{\varepsilon_j\}=\{\varepsilon_{11}, \varepsilon_{22}, \varepsilon_{33}, 2\varepsilon_{23}, 2\varepsilon_{13}, 2\varepsilon_{12}\}^{\mathrm{T}}$. We use $[S_{ij}^f]$, $[S_{ij}^m]$ and $[S_{ij}]$ to denote the compliance matrices of the fiber, matrix and the composite respectively. Let us temporally assume that both the constituent materials are in linear elastic deformation range. This assumption can be later released. Furthermore, the fibers used can be isotropic or transversely isotropic, whereas the matrix is isotropic. In such a case, the resulting composite is transversely isotropic and the material compliance matrices are defined as

$$\left[S_{ij}^{f}\right]=\begin{bmatrix}\left[S_{ij}^{f}\right]_{\sigma} & 0\\ 0 & \left[S_{ij}^{f}\right]_{\tau}\end{bmatrix} \tag{3.3.1}$$

$$\left[S_{ij}^{m}\right]=\begin{bmatrix}\left[S_{ij}^{m}\right]_{\sigma} & 0\\ 0 & \left[S_{ij}^{m}\right]_{\tau}\end{bmatrix} \tag{3.3.2}$$

$$\left[S_{ij}\right]=\begin{bmatrix}\left[S_{ij}\right]_{\sigma} & 0\\ 0 & \left[S_{ij}\right]_{\tau}\end{bmatrix} \tag{3.3.3}$$

In Eqs. (3.3.1) – (3.3.3), $[S_{ij}^{f}]_{\sigma}$, $[S_{ij}^{m}]_{\sigma}$ and $[S_{ij}]_{\sigma}$ are the sub-matrices of the corresponding compliances relating normal stresses to elongation strains while $[S_{ij}^{f}]_{\tau}$, $[S_{ij}^{m}]_{\tau}$ and $[S_{ij}]_{\tau}$ are the sub-matrices relating shear stresses to shear strains (refer to Eq. (1.8)). Explicit expressions of them are

$$[S_{ij}^{f}]_{\sigma}=\begin{bmatrix}\dfrac{1}{E_{11}^{f}} & -\dfrac{\nu_{12}^{f}}{E_{11}^{f}} & -\dfrac{\nu_{12}^{f}}{E_{11}^{f}}\\ & \dfrac{1}{E_{22}^{f}} & -\dfrac{\nu_{23}^{f}}{E_{22}^{f}}\\ & \text{symmetry} & \dfrac{1}{E_{22}^{f}}\end{bmatrix} \tag{3.4.1}$$

$$[S_{ij}^{f}]_{\tau}=\begin{bmatrix}\dfrac{1}{G_{23}^{f}} & 0 & 0\\ & \dfrac{1}{G_{12}^{f}} & 0\\ & \text{symmetry} & \dfrac{1}{G_{12}^{f}}\end{bmatrix} \tag{3.4.2}$$

$$[S_{ij}^m]_\sigma = \begin{bmatrix} \dfrac{1}{E^m} & -\dfrac{\nu^m}{E^m} & -\dfrac{\nu^m}{E^m} \\ & \dfrac{1}{E^m} & -\dfrac{\nu^m}{E^m} \\ \text{symmetry} & & \dfrac{1}{E^m} \end{bmatrix} \tag{3.4.3}$$

$$[S_{ij}^m]_\tau = \begin{bmatrix} \dfrac{1}{G^m} & 0 & 0 \\ & \dfrac{1}{G^m} & 0 \\ \text{symmetry} & & \dfrac{1}{G^m} \end{bmatrix} \tag{3.4.4}$$

$$[S_{ij}]_\sigma = \begin{bmatrix} \dfrac{1}{E_{11}} & -\dfrac{\nu_{12}}{E_{11}} & -\dfrac{\nu_{13}}{E_{11}} \\ & \dfrac{1}{E_{22}} & -\dfrac{\nu_{23}}{E_{22}} \\ \text{symmetry} & & \dfrac{1}{E_{33}} \end{bmatrix} \tag{3.4.5}$$

$$[S_{ij}]_\tau = \begin{bmatrix} \dfrac{1}{G_{23}} & 0 & 0 \\ & \dfrac{1}{G_{13}} & 0 \\ \text{symmetry} & & \dfrac{1}{G_{12}} \end{bmatrix} \tag{3.4.6}$$

In Eqs. (3.4.1) and (3.4.2), E_{11}^f, ν_{12}^f, and G_{12}^f are the longitudinal Young's modulus, Poisson's ratio and shear modulus of the fiber, respectively; E_{22}^f, ν_{23}^f and G_{23}^f are its transverse Young's modulus, out-of-plane Poisson's ratio and shear modulus, with $G^f_{23} = \dfrac{E^f_{22}}{2(1+\nu^f_{23})}$. In Eqs. (3.4.3) and (3.4.4) E^m, ν^m and G^m are the Young's modulus, Poisson's ratio and shear modulus of the matrix with $G^m = \dfrac{E^m}{2(1+\nu^m)}$. In Eqs. (3.4.5) and (3.4.6) $E_{33}=E_{22}$, $\nu_{13}=\nu_{12}$ and $G_{13}=G_{12}$ with

$$G_{23} = \frac{E_{22}}{2(1+\nu_{23})} \tag{3.5}$$

as a result of transverse isotropy. Using $[S^f_{ij}]$, $[S^m_{ij}]$ and $[S_{ij}]$, the volume averaged stresses and strains are connected by (Eqs. (1.24) and (1.25))

$$\{\varepsilon^f_i\} = [S^f_{ij}]\{\sigma^f_j\} \tag{3.6.1}$$

$$\{\varepsilon^m_i\} = [S^m_{ij}]\{\sigma^m_j\} \tag{3.6.2}$$

and

$$\{\varepsilon_i\}=[S_{ij}]\{\sigma_j\} \tag{3.6.3}$$

Our main goal is to find out $[S_{ij}]$ from the given $[S^f_{ij}]$ and $[S^m_{ij}]$. Suppose that there exists a bridging matrix $[A_{ij}]$ such that

$$\{\sigma^m_i\} = [A_{ij}]\{\sigma^f_j\} \tag{3.7}$$

The bridging matrix represents the load sharing capacity of the matrix material with respect to the fiber material. Substituting Eq. (3.7) into Eq. (3.1) and inverting the resulting equation gives

$$\{\sigma^f_i\} =(V_f[I]+V_m[A_{ij}])^{-1}\{\sigma_j\} \tag{3.8.1}$$

where $[I]$ is a unit matrix. Substituting Eq. (3.8.1) into Eq. (3.7) yields

$$\{\sigma^m_i\} =[A_{ij}](V_f[I]+V_m[A_{ij}])^{-1}\{\sigma_j\} \tag{3.8.2}$$

Furthermore, substituting Eqs. (3.6.1) and (3.6.2) into Eq. (3.2) and making use of Eq. (3.7), we obtain

$$\{\varepsilon_i\}=(V_f[S^f_{ij}]+V_m[S^m_{ij}][A_{ij}])\{\sigma^f_j\}$$

Substituting Eq. (3.8.1) into the last equations and comparing the resulting equations with Eq. (3.6.3), the compliance matrix, $[S_{ij}]$, of the composite is seen to be

$$[S_{ij}]=(V_f[S^f_{ij}]+V_m[S^m_{ij}][A_{ij}])(V_f[I]+V_m[A_{ij}])^{-1} \tag{3.9}$$

whereas the stiffness matrix, $[C_{ij}]$, of the composite is simply given by

$$[C_{ij}]=[S_{ij}]^{-1} \tag{3.10}$$

Therefore, once the bridging matrix, $[A_{ij}]$, is determined, the overall compliance and stiffness matrices of the composite can be calculated from Eqs. (3.9) and (3.10). In addition, as Eqs. (3.8.1) and (3.8.2) explicitly correlate the internal stresses shared by the fiber and the matrix respectively with those applied overall on the composite, an optimal design for composite load carrying capacity can be achieved based upon the knowledge of the constituent properties. For instance, one can choose properly the constituent materials and/or the fiber volume fraction to make the load to strength ratios in both the constituents be comparable.

It is seen from the above derivation that the most important task of the model is to properly define the bridging matrix $[A_{ij}]$. The only prerequisite for the definition is that the overall compliance matrix, $[S_{ij}]$, given by Eq. (3.9) must be symmetric. However, as can be expected, the material properties of the constituents, the relative position of the fiber arrangement, the fiber volume fraction in the composite, as well as the fiber cross sectional shape, may also influence the bridging matrix.

3.3 Characterization of Bridging Matrix

As aforementioned, when a transversely isotropic fiber material is unidirectionally arranged in an isotropic matrix, the resulting composite is transversely isotropic, which has a compliance matrix in the form of Eq. (3.3.3). It is evident that the bridging matrix $[A_{ij}]$ in such a case can be sub-divided into

$$\left[A_{ij}\right]=\begin{bmatrix} [A_{ij}]_I & [0] \\ [0] & [A_{ij}]_{II} \end{bmatrix} \tag{3.11}$$

where $[A_{ij}]_I$ and $[A_{ij}]_{II}$ are 3×3 sub-matrices such that

$$[S_{ij}]_\sigma=(V_f[S_{ij}^f]_\sigma+V_m[S_{ij}^m]_\sigma[A_{ij}]_I)(V_f[I]+V_m[A_{ij}]_I)^{-1} \tag{3.12}$$

$$[S_{ij}]_\tau=(V_f[S_{ij}^f]_\tau+V_m[S_{ij}^m]_\tau[A_{ij}]_{II})(V_f[I]+V_m[A_{ij}]_{II})^{-1} \tag{3.13}$$

Clearly, $[A_{ij}]_{II}$ should be diagonal since $[S_{ij}^f]_\tau$, $[S_{ij}^m]_\tau$ and $[S_{ij}]_\tau$ are all diagonal. As the composite is transversely isotropic, there are only five independent elastic constants in its compliance matrix $[S_{ij}]$. Consequently, only five elements of $[A_{ij}]$ can be independent. One of them is $A_{55}=A_{66}$ in the sub-matrix $[A_{ij}]_{II}$. The other four are arranged among $[A_{ij}]_I$, whose locations can be assigned by considering those of the independent elastic constants in $[S_{ij}]_\sigma$. Hence, we may take A_{11}, $A_{22}=A_{33}$, $A_{31}=A_{21}$ and A_{32} to be independent. The remaining elements, A_{12}, A_{13}, and A_{23} are dependent, which should be determined by requesting that the resulting compliance matrix, $[S_{ij}]_\sigma$, given by Eq. (3.12), be symmetric, i.e.,

$$S_{21}=S_{12},\ S_{31}=S_{13} \text{ and } S_{32}=S_{23} \tag{3.14}$$

The last bridging matrix element, A_{44}, is to be obtained from the following equation

$$S_{44}=1/G_{23}=2(1+\nu_{23})/E_{22}=2(S_{22}-S_{23}) \tag{3.15}$$

Therefore, the element A_{44} is not independent.

Let us then consider determination of the independent bridging matrix elements. Before going into any detail, let us first characterize qualitatively their behavior. It has been pointed out that the stress sharing capacity of a constituent, and thus the bridging matrix, is influenced by the material parameters of the constituents and by the geometry of the fiber embedded in the matrix. As such, all of the factors affecting the independent elements, A_{11}, $A_{22}=A_{33}$, $A_{31}=A_{21}$, A_{32} and $A_{55}=A_{66}$, can be divided into two classes. The first class consists of the material properties of the constituent fiber and matrix materials, whereas the second class is related with fiber packing geometries which include the fiber volume fraction, the fiber arrangement pattern in the matrix, the fiber cross-sectional shape and the interface bounding situation between the fiber and the matrix. The independent elements can be well regarded as multi-variable functions of the two class parameters. From elementary calculus, we have known that any continuous multi-variable function can be expanded into a series upon a selected number of the variables. Let us do such expansions for the independent bridging matrix elements with respect to the constituent material properties. Recognizing that the bridging matrix $[A_{ij}]$ must be identical when the properties of the two materials become the same, the general forms of the independent elements can be always expressed as a power series of the elastic constants of the constituents, i.e.,

$$A_{11}=1+\lambda_{11}(1-E^m/E^f_{11})+\cdots \tag{3.16.1}$$

$$A_{21}=A_{31}=\lambda_{21}(1-\nu^m/\nu^f_{12})+\cdots \tag{3.16.2}$$

$$A_{22}=A_{33}=1+\lambda_{31}(1-E^m/E^f_{22})+\cdots \tag{3.16.3}$$

$$A_{32}=\lambda_{41}(1-\nu^m/\nu^f_{23})+\cdots \tag{3.16.4}$$

$$A_{55}=A_{66}=1+\lambda_{51}(1-G^m/G^f_{12})+\cdots \tag{3.16.5}$$

as long as both the fiber and the matrix in the composite are under linearly elastic deformations. In Eqs. (3.16.1) – (3.16.5), λ_{ij} are the expansion coefficients which may depend on the fiber packing geometries but are independent of the constituent material properties. In this way, the independent elements are explicitly correlated with the two classes of the influencing parameters, the material properties and the fiber packing geometries. One of the best benefits for us in expressing the independent elements in Eqs. (3.16.1) – (3.16.5) is that once the expansion coefficients λ_{ij} are determined within an elastic deformation range they will remain unchanged in an inelastic one. This is because a fiber packing geometry such as the fiber volume fraction or the fiber cross-sectional shapes can be regarded as unchanged when the composite deforms from an elastic region to a plastic one. Only the corresponding elastic moduli of the constituents may need to vary during this deformation.

The coefficients λ_{ij} in Eqs. (3.16.1) – (3.16.5) can be determined through experiments or other methods such as numerical simulation. In fact, any micromechanical model, e.g., the rule of mixture or the Chamis model introduced in Chapter 1, can be essentially regarded as achieved in a particular way in the determination of a special bridging matrix. This is because by substituting the resulting compliance matrix of the UD composite defined by the model into the left hand side of Eq. (3.9), one can always get a bridging matrix by solving the equations. In the following, we will make use of Mori-Tanaka's approach to determine the independent bridging matrix elements.

3.4 Mori-Tanaka Approach

Mori and Tanaka used a micromechanical approach to determine the elastic properties of a UD composite (Mori & Tanaka, 1973) using a formulation of Eshelby's problem, which has become one of the most successful micromechanics models (Weng, 1984; Benveniste, 1987). However, a concept called "average stresses" was introduced in the Mori-Tanaka approach to represent internal stresses in the constituent materials whereas Eshelby's problem focused on stress and strain fields in the inclusion and matrix materials. With this approach, the two fundamental relationships, Eqs. (3.1) and (3.2), correlating the averaged stresses and strains in an unbounded composite with those in the infinitely long fiber cylinder and the unbounded matrix material, i.e.,

$$\{\sigma_i\} = V_f\{\sigma_i^f\} + V_m\{\sigma_i^m\}$$

and

$$\{\varepsilon_i\} = V_f\{\varepsilon_i^f\} + V_m\{\varepsilon_i^m\}$$

are applicable.

Now, apply a load σ_{ij}^0 at the infinite boundary of the composite. Supposing that an isotropic material whose elastic properties are the same as those of the matrix material is subjected to the same load, the corresponding strain is expressed as

$$\varepsilon_{ij}^{0} = S_{ijkl}^{m} \sigma_{kl}^{0} \tag{3.17.1}$$

or

$$\sigma_{ij}^{0} = C_{ijkl}^{m} \varepsilon_{kl}^{0} \tag{3.17.2}$$

Due to the existence of the inclusion (i.e., the fiber), the average stresses and strains in the matrix are disturbed by $\tilde{\sigma}_{ij}$ and $\tilde{\varepsilon}_{ij}$ respectively, i.e.,

$$\sigma_{ij}^{m} = \sigma_{ij}^{0} + \tilde{\sigma}_{ij} = C_{ijkl}^{m} \left(\varepsilon_{kl}^{0} + \tilde{\varepsilon}_{kl} \right) \tag{3.18}$$

where $\varepsilon_{ij}^{0} + \tilde{\varepsilon}_{ij} = \varepsilon_{ij}^{m}$ are the average strains in the matrix.

On the other hand, the average stresses and strains in the fiber are different from those in the matrix, and the differences can be expressed as σ'_{ij} and ε'_{ij}, i.e.,

$$\sigma_{ij}^{f} = \sigma_{ij}^{0} + \tilde{\sigma}_{ij} + \sigma'_{ij} = C_{ijkl}^{f} \left(\varepsilon_{kl}^{0} + \tilde{\varepsilon}_{kl} + \varepsilon'_{kl} \right) \tag{3.19}$$

where $\varepsilon_{ij}^{0} + \tilde{\varepsilon}_{ij} + \varepsilon'_{ij} = \varepsilon_{ij}^{f}$ are the average strains in the fiber.

Similarly as the equivalent inclusion problem illustrated in Subsection 1.5.3, the current problem can be equivalent to a problem in which a homogeneous material is subjected to an eigenstrain. This equivalence is written as

$$\sigma_{ij}^{f} = C_{ijkl}^{f} (\varepsilon_{kl}^{0} + \tilde{\varepsilon}_{kl} + \varepsilon'_{kl}) = C_{ijkl}^{m} (\varepsilon_{kl}^{0} + \tilde{\varepsilon}_{kl} + \varepsilon'_{kl} - \varepsilon_{kl}^{*}) \tag{3.20.1}$$

$$\varepsilon'_{ij} = L_{ijkl} \varepsilon_{kl}^{*} \tag{3.20.2}$$

where L_{ijkl} is the Eshelby's tensor (Section 1.5). By solving this equation, the eigenstrain is derived as

$$\varepsilon_{ij}^{*} = \left(C_{ijkl}^{m} \right)^{-1} \left(C_{klpq}^{m} - C_{klpq}^{f} \right) \varepsilon_{pq}^{f} \tag{3.21}$$

and then ε'_{ij} can also be obtained as

$$\varepsilon'_{ij} = L_{ijkl} \left(C_{klpq}^{m} \right)^{-1} \left(C_{pqrs}^{m} - C_{pqrs}^{f} \right) \varepsilon_{rs}^{f} \tag{3.22}$$

By comparing the expressions for ε_{ij}^{f} and ε_{ij}^{m}, a relation between them is derived as

$$\varepsilon_{ij}^{f} = \varepsilon_{ij}^{m} + \varepsilon'_{ij}$$

Substituting Eq. (3.22) into the last equation, a relationship can be found,

$$\varepsilon_{ij}^{f} = T_{ijkl} \varepsilon_{kl}^{m} \tag{3.23.1}$$

where $$T_{ijkl} = \left[I_{ijkl} + L_{ijpq} \left(C_{pqrs}^{m} \right)^{-1} \left(C_{rskl}^{f} - C_{rskl}^{m} \right) \right]^{-1} \tag{3.23.2}$$

Making use of the constitutive equations for the fiber and matrix materials, Eqs. (3.6.1) and (3.6.2), a relationship correlating σ_{ij}^{f} and σ_{ij}^{m} can be deduced from Eqs. (3.23.1) and (3.23.2) as

$$\sigma_{ij}^{f} = W_{ijkl} \sigma_{kl}^{m} \tag{3.24.1}$$

where $$W_{ijkl} = C_{ijpq}^{f} T_{pqrs} S_{rskl}^{m} \tag{3.24.2}$$

Substituting Eq. (3.23.1) into Eq. (3.2), the average strains in the constituent materials are expressed with respect to those in the composite through

$$\varepsilon_{ij}^{m} = \left[V_f I_{ijkl} + V_m T_{ijkl} \right]^{-1} \varepsilon_{kl} \tag{3.25.1}$$

$$\varepsilon_{ij}^{f} = T_{ijpq} \left[V_f I_{pqkl} + V_m T_{pqkl} \right]^{-1} \varepsilon_{kl} \tag{3.25.2}$$

Similarly, by using Eqs. (3.24.1) and (3.1), one obtains

$$\sigma_{ij}^{m} = \left[V_f I_{ijkl} + V_m W_{ijkl} \right]^{-1} \sigma_{kl} \tag{3.26.1}$$

$$\sigma_{ij}^{f} = W_{ijpq} \left[V_f I_{pqkl} + V_m W_{pqkl} \right]^{-1} \sigma_{kl} \tag{3.26.2}$$

Furthermore, by substituting Eqs. (3.6.1) and (3.6.2) into Eq. (3.1), making use of Eqs. (3.26.1) and (3.26.2) and comparing the resulting equations with Eq. (3.6.3), the following stiffness and compliance tensors for the composite can be obtained

$$C_{ijkl} = C_{ijkl}^{m} + V_f (C_{ijpq}^{f} - C_{ijpq}^{m}) T_{pqrs} \left[V_m I_{rskl} + V_f T_{rskl} \right]^{-1} \tag{3.27.1}$$

$$S_{ijkl} = S_{ijkl}^{m} + V_f (S_{ijpq}^{f} - S_{ijpq}^{m}) W_{pqrs} \left[V_m I_{rskl} + V_f W_{rskl} \right]^{-1} \tag{3.27.2}$$

3.5 Determination of Bridging Matrix

Using a contracted notation, Eq. (3.24.1) can be rewritten as

$$\left\{ \sigma_i^{f} \right\} = \left[W'_{ij} \right] \left\{ \sigma_j^{m} \right\} \tag{3.28}$$

Comparing Eq. (3.28) with Eq. (3.7), it is seen that the bridging matrix $[A_{ij}]$ is given by

$$[A_{ij}]=[W'_{ij}]^{-1}=[C^m_{ik}]\left([I_{kq}]+[L_{kl}][S^m_{lp}]([C^f_{pq}]-[C^m_{pq}])\right)[S^f_{qj}] \tag{3.29}$$

where $[L_{kl}]$ is a second-order Eshelby's tensor, a contracted form from the fourth-order Eshelby's tensor L_{ijkl}. When both the fiber and the matrix are isotropic, the Eshelby's tensor is given by Eq. (1.68) providing that the Poisson's ratio, ν, in Eq. (1.69) is replaced by that of the matrix, i.e., by ν^m. Substituting all of the known matrices into the right hand side of Eq. (3.29) and after a thorough manipulation and simplification, the non-zero elements of Eq. (3.29) are found to be (Zhang & Huang, 2008)

$$A_{11}=\frac{E^m}{E^f}\left(1+\frac{\nu^m(\nu^m-\nu^f)}{(1+\nu^m)(1-\nu^m)}\right) \tag{3.30.1}$$

$$A_{12}=\frac{E^m}{E^f}\left(\frac{\nu^m-\nu^f}{2(1+\nu^m)(1-\nu^m)}-\frac{\nu^f}{2(1-\nu^m)}\right)+\frac{\nu^m}{2(1-\nu^m)}=A_{13} \tag{3.30.2}$$

$$A_{21}=\frac{E^m}{E^f}\frac{\nu^m-\nu^f}{2(1+\nu^m)(1-\nu^m)}=A_{31} \tag{3.30.3}$$

$$A_{22}=\frac{E^m}{E^f}\left(\frac{\nu^m-\nu^f}{2(1+\nu^m)(1-\nu^m)}+\frac{1+\nu^f}{1+\nu^m}(1-L_{2222})\right)+L_{2222} \tag{3.30.4}$$

$$A_{23}=\frac{E^m}{E^f}\left(\frac{\nu^m-\nu^f}{2(1-\nu^m)(1+\nu^m)}-\frac{1+\nu^f}{1+\nu^m}L_{2233}\right)+L_{2233} \tag{3.30.5}$$

$$A_{32}=\frac{E^m}{E^f}\left(\frac{\nu^m-\nu^f}{2(1-\nu^m)(1+\nu^m)}-\frac{1+\nu^f}{1+\nu^m}L_{3322}\right)+L_{3322} \tag{3.30.6}$$

$$A_{33}=\frac{E^m}{E^f}\left(\frac{\nu^m-\nu^f}{2(1+\nu^m)(1-\nu^m)}+\frac{1+\nu^f}{1+\nu^m}(1-L_{3333})\right)+L_{3333} \tag{3.30.7}$$

$$A_{44}=\frac{G^m}{G^f}+2L_{2323}\frac{G^f-G^m}{G^f} \tag{3.30.8}$$

$$A_{55}=\frac{G^m}{G^f}+2L_{1313}\frac{G^f-G^m}{G^f} \tag{3.30.9}$$

$$A_{66}=\frac{G^m}{G^f}+2L_{1212}\frac{G^f-G^m}{G^f} \tag{3.30.10}$$

where L_{2222}, L_{2233}, L_{2323}, L_{1313} and L_{1212} are elements of the Eshebly's tensor given by

$$L_{2222} = L_{3333} = \frac{1}{2(1-\nu^m)}\left[\frac{3}{4}+\frac{(1-2\nu^m)}{2}\right] \tag{3.31.1}$$

$$L_{2233} = L_{3322} = L_{2323} = \frac{1}{2(1-\nu^m)}\left[\frac{1}{4}-\frac{(1-2\nu^m)}{2}\right] \tag{3.31.2}$$

$$L_{1212}=L_{1313}=1/4. \tag{3.31.3}$$

It must be pointed out that after substituting the bridging matrix defined by Eqs. (3.29) – (3.31) into Eq. (3.9) the resulting compliance matrix of a UD composite may not be symmetric (Wang and Weng, 1992). This is not acceptable in reality and the bridging matrix elements must be modified. The modification is made by requesting that the resulting compliance matrix of a UD composite be always symmetric and that, except for only five independent elements, all of the other elements be dependent. As pointed out in Section 3.3, the elements A_{11}, A_{21}, $A_{22}=A_{33}$, A_{32} and $A_{55}=A_{66}$ are chosen as independent, whereas the other dependent elements are determined by solving Eqs. (3.14) and (3.15).

For simplicity, let us first consider the determination of a two dimensional bridging matrix. The general form of a planar bridging matrix is given by

$$[A_{ij}] = \begin{bmatrix} a_{11} & a_{12} & 0 \\ a_{21} & a_{22} & 0 \\ 0 & 0 & a_{33} \end{bmatrix} \tag{3.32}$$

In a plane problem, a UD composite has only four independent elastic constants and hence there are four independent elements in Eq. (3.32) which are chosen to be a_{33}, a_{11}, a_{22} and a_{21}. By requesting that the resulting compliance matrix, Eq. (3.9), be symmetric, the dependent element, a_{12}, is found to be

$$a_{12} = \frac{(S_{12}^f - S_{12}^m)(a_{22}-a_{11})+(S_{22}^m - S_{22}^f)a_{21}}{S_{11}^m - S_{11}^f} \tag{3.33}$$

It is apparent that the four independent elements a_{11}, a_{21}, a_{22} and a_{33} should have a close connection with the elements A_{11}, A_{21}, A_{22} and A_{66} defined by Eqs. (3.30.1), (3.30.3), (3.30.4) and (3.30.10), respectively. However, the latter elements have been derived based on the Eshelby's tensor given by Eq. (1.68), which was obtained on an assumption that an infinitely long circular fiber cylinder was embedded in an unbounded matrix material. In reality, the matrix and the fiber in a representative volume element, e.g., Fig. 1.3, of a UD composite have comparable dimensions. This implies that the estimated properties of the UD composite using Eqs. (3.30.1), (3.30.3), (3.30.4) and (3.30.10) would be by nature in error. As such, it deserves for us to make necessary modifications on the corresponding Eshelby's tensor elements.

Taking the above comments into consideration and by neglecting unimportant parameters, a set of independent elements are chosen to be (Huang, 2000a, 2001b, 2001c)

$$a_{11} = \frac{E^m}{E^f} \tag{3.34.1}$$

$$a_{21} = 0 \tag{3.34.2}$$

$$a_{22} = \beta + (1-\beta)\frac{E^m}{E^f} \quad (0<\beta<1) \tag{3.34.3}$$

$$a_{33} = \alpha + (1-\alpha)\frac{G^m}{G^f} \quad (0<\alpha<1) \tag{3.34.4}$$

Clearly, when the parameter $2L_{1212}$ in Eq. (3.30.10) is replaced by α, the expressions for a_{33} and A_{66} are the same. In fact, in many cases, the parameter α in Eq. (3.34.4) can take a value of 0.5, which is equal to $2L_{1212}$ (Eq. (3.31.3)). Comparing Eqs. (3.34.1), (3.34.2) and (3.34.3) with Eqs. (3.30.1), (3.30.3) and (3.30.4) respectively, it is seen that only higher order smaller quantities are omitted. For instance, when ν^f=0.22 and ν^m=0.35, which correspond to Poisson's ratios of the fiber and matrix materials in a typical composite, the resulting A_{11}, A_{21} and A_{22} are given by

$$A_{11} = \frac{E^m}{E^f}\left(1+\frac{\nu^m(\nu^m-\nu^f)}{(1+\nu^m)(1-\nu^m)}\right) = \frac{E^m}{E^f}(1+0.052) \approx \frac{E^m}{E^f}$$

$$A_{21} = \frac{E^m}{E^f}\frac{\nu^m-\nu^f}{2(1+\nu^m)(1-\nu^m)} = 0.074\frac{E^m}{E^f} \approx 0$$

$$A_{22} = \frac{E^m}{E^f}\left(\frac{\nu^m-\nu^f}{2(1+\nu^m)(1-\nu^m)}+\frac{1+\nu^f}{1+\nu^m}(1-L_{2222})\right)+L_{2222}$$
$$= \beta + \frac{E^m}{E^f}(0.074+0.9(1-\beta)) \approx \beta+(1-\beta)\frac{E^m}{E^f}$$

as long as the Eshelby's parameter L_{2222} is replaced by β. In Eqs. (3.34.3) and (3.34.4), the parameters β and α, called bridging parameters, are introduced in order that the predicted elastic properties of a UD composite correlate better with measured ones. As will be shown subsequently, the bridging matrix element a_{22} will affect the predicted transverse modulus, E_{22}, whereas a_{33} will influence the predicted in-plane shear modulus, G_{12}, of the UD composite. In reality, the transverse and in-plane shear moduli, E_{22} and G_{12}, of a UD composite are sensitive to a processing condition and a fiber packing geometry. Experiments for them exhibit relatively large deviations (Huang, 2004). Hence, the bridging parameters, β and α, provide a chance to capture a specific fiber packing geometry for a UD composite. If, however, there are no experimental data available, they can take a value within the following range (Huang, 2001c, 2004)

$$\beta=0.3\sim0.6 \tag{3.35.1}$$

$$\alpha=0.3\sim0.6 \tag{3.35.2}$$

The relatively large difference between the bridging parameter β and the Eshelby's one L_{2222} (which is 0.6923 when ν^m=0.35) can be attributed to the unbounded domain used in obtaining the Eshelby's tensor. More investigation on the effect of choosing different β and α on predicted properties of UD composites will be given in the subsequent sections.

Further modifications can be made for the independent bridging matrix elements when the fiber material becomes transversely isotropic. Recognizing that the elements a_{11}, a_{22} and a_{33} will be directly related to the longitudinal, transverse and in-plane shear moduli of a UD composite, i.e., E_{11}, E_{22} and G_{12} respectively, the modifications are given by (Huang, 2000a, 2000b, 2001c, 2004)

$$a_{11}=\frac{E^m}{E^f_{11}} \tag{3.36.1}$$

$$a_{21}=0 \tag{3.36.2}$$

$$a_{22}=\beta+(1-\beta)\frac{E^m}{E^f_{22}} \quad (0<\beta<1) \tag{3.36.3}$$

$$a_{33}=\alpha+(1-\alpha)\frac{G^m}{G^f_{12}} \quad (0<\alpha<1) \tag{3.36.4}$$

Having determined a 2D bridging matrix, it can now be extended to a 3D case. Undoubtedly, the four independent bridging matrix elements defined by Eqs. (3.36.1) – (3.36.4) will remain unchanged. The last independent element, A_{32}, is simply set to zero (Huang, 2001c, 2004) as, from Eq. (3.30.6) and by using ν^f=0.22 and ν^m=0.35, we have

$$A_{32}=\frac{E^m}{E^f}\left(\frac{\nu^m-\nu^f}{2(1-\nu^m)(1+\nu^m)}-\frac{1+\nu^f}{1+\nu^m}L_{3322}\right)+L_{3322}=0.077+0.0046\frac{E^m}{E^f}\approx 0$$

The dependent elements are determined using Eqs. (3.14) and (3.15). This gives

$$[A_{ij}] = \begin{bmatrix} A_{11} & A_{12} & A_{13} & 0 & 0 & 0 \\ 0 & A_{22} & 0 & 0 & 0 & 0 \\ 0 & 0 & A_{33} & 0 & 0 & 0 \\ 0 & 0 & 0 & A_{44} & 0 & 0 \\ 0 & 0 & 0 & 0 & A_{55} & 0 \\ 0 & 0 & 0 & 0 & 0 & A_{66} \end{bmatrix} \tag{3.37}$$

where

$$A_{11} = \frac{E^m}{E_{11}^f} \tag{3.38.1}$$

$$A_{22} = A_{33} = A_{44} = \beta + (1-\beta)\frac{E^m}{E_{22}^f} \quad (0<\beta<1) \tag{3.38.2}$$

$$A_{55} = A_{66} = \alpha + (1-\alpha)\frac{G^m}{G_{12}^f} \quad (0<\alpha<1) \tag{3.38.3}$$

$$A_{12} = A_{13} = \frac{S_{12}^f - S_{12}^m}{S_{11}^f - S_{11}^m}(A_{11} - A_{22}) \tag{3.38.4}$$

Finally, rewriting the independent bridging elements into the following forms

$$A_{11} = 1 - (1 - \frac{E^m}{E_{11}^f}) \tag{3.39.1}$$

$$A_{21} = 0 \tag{3.39.2}$$

$$A_{22} = A_{33} = 1 + (\beta - 1)(1 - \frac{E^m}{E_{22}^f}) \quad (0<\beta<1) \tag{3.39.3}$$

$$A_{32} = 0 \tag{3.39.4}$$

$$A_{55} = A_{66} = 1 + (\alpha - 1)(1 - \frac{G^m}{G_{12}^f}) \quad (0<\alpha<1) \tag{3.39.5}$$

and comparing Eqs. (3.39.1) – (3.39.5) with Eqs. (3.16.1) – (3.16.5), we can see that the expansion coefficients λ_{ij} assume the following values

$$\lambda_{11}=-1,\ \lambda_{31}=\beta-1,\ \lambda_{51}=\alpha-1, \text{ and all of the others } \lambda_{ij}=0 \tag{3.40}$$

Remark 3.1

Other explicit formulae instead of Eqs. (3.37) and (3.38) or Eqs. (3.32), (3.33) and (3.36) may also be obtainable. However, as long as $A_{22}=A_{33}$, the solutions to the symmetric Eq. (3.14) will be unique and are given by:

$$A_{23}=A_{32} \tag{3.41.1}$$

$$A_{31}=A_{21} \tag{3.41.2}$$

$$A_{13}=A_{12}=\frac{(S_{12}^{f}-S_{12}^{m})(A_{11}-A_{22})+(S_{22}^{f}-S_{22}^{m}+S_{23}^{f}-S_{23}^{m})A_{21}+(S_{12}^{f}+S_{12}^{m})A_{23}}{(S_{11}^{f}-S_{11}^{m})} \tag{3.41.3}$$

On the other hand, if A_{12} and A_{13} are chosen, rather than using A_{21} and A_{31} for the independent elements, A_{21}, A_{31}, A_{32} and A_{23} can be derived. In other words, for a unidirectionally fiber-reinforced composite, either taking A_{12} and A_{13} or using A_{21} and A_{31} as independent elements in the bridging matrix $[A_{ij}]$ will give the same composite compliance matrix, providing that the other independent elements used are kept unchanged.

3.6 Effective Elastic Moduli

Based on the bridging matrix given by Eqs. (3.37) and (3.38), the five engineering moduli of the UD composite can be obtained from Eq. (3.9). However, they are more easily derived by applying some special uniaxial loads (i.e., longitudinal, transverse and in-plane shear loads) to the composites, respectively.

Let us consider the stress balance in the transverse direction, i.e.,

$$\sigma_{22}=V_f\sigma_{22}^{f}+V_m\sigma_{22}^{m} \tag{3.42}$$

Furthermore, from the relation Eqs. (3.7) and (3.37), the transverse stresses in the fiber and matrix are related by

$$\sigma_{22}^{m}=A_{22}\sigma_{22}^{f} \tag{3.43}$$

Substituting Eq. (3.43) into Eq. (3.42) and solving σ_{22}^{f} with respect to σ_{22}, we obtain

$$\sigma_{22}^{f}=\frac{\sigma_{22}}{V_f+V_mA_{22}} \tag{3.44.1}$$

Combining Eqs. (3.43) and (3.44.1) gives

$$\sigma_{22}^{m}=\frac{A_{22}\sigma_{22}}{V_f+V_mA_{22}} \tag{3.44.2}$$

Similarly, we have

$$\sigma_{33}^{f}=\frac{\sigma_{33}}{V_f+V_mA_{22}} \tag{3.44.3}$$

$$\sigma_{33}^{m}=\frac{A_{22}\sigma_{33}}{V_f+V_mA_{22}} \tag{3.44.4}$$

$$\sigma_{23}^{f}=\frac{\sigma_{23}}{V_f+V_mA_{22}} \tag{3.44.5}$$

$$\sigma_{23}^{m}=\frac{A_{22}\sigma_{23}}{V_f+V_mA_{22}} \tag{3.44.6}$$

$$\sigma_{12}^{f}=\frac{\sigma_{12}}{V_f+V_mA_{66}} \tag{3.44.7}$$

$$\sigma_{12}^{m}=\frac{A_{66}\sigma_{12}}{V_f+V_mA_{66}} \tag{3.44.8}$$

$$\sigma_{13}^{f}=\frac{\sigma_{13}}{V_f+V_mA_{66}} \tag{3.44.9}$$

$$\sigma_{13}^{m}=\frac{A_{66}\sigma_{13}}{V_f+V_mA_{66}} \tag{3.44.10}$$

$$\sigma_{11}^{f}=\frac{\sigma_{11}}{V_f+V_mA_{11}}-\frac{V_mA_{12}(\sigma_{22}+\sigma_{33})}{(V_f+V_mA_{11})(V_f+V_mA_{22})} \tag{3.44.11}$$

and

$$\sigma_{11}^{m}=\frac{A_{11}\sigma_{11}}{V_f+V_mA_{11}}+\frac{V_fA_{12}(\sigma_{22}+\sigma_{33})}{(V_f+V_mA_{11})(V_f+V_mA_{22})} \tag{3.44.12}$$

To obtain the overall longitudinal Young's modulus and Poisson's ratio of the composite, we apply to it a stress state where $\sigma_{11}\neq 0$ and all the others $\sigma_{ij}=0$. Substituting these stresses into Eqs. (3.44.11) and (3.44.12) yields

$$\sigma_{11}^{f} = \frac{\sigma_{11}}{V_f + V_m A_{11}} \quad \text{and} \quad \sigma_{11}^{m} = \frac{A_{11}\sigma_{11}}{V_f + V_m A_{11}} \tag{3.45}$$

From $\varepsilon_{11} = \frac{\sigma_{11}}{E_{11}} = V_f \frac{\sigma_{11}^{f}}{E_{11}^{f}} + V_m \frac{\sigma_{11}^{m}}{E^{m}} = \left(\frac{V_f / E_{11}^{f} + V_m A_{11} / E^{m}}{V_f + V_m A_{11}} \right) \sigma_{11}$ it follows that

$$E_{11} = \left(\frac{V_f / E_{11}^{f} + V_m A_{11} / E^{m}}{V_f + V_m A_{11}} \right)^{-1} \tag{3.46}$$

Substituting Eq. (3.38.1) into the last equation gives

$$E_{11} = V_f E_{11}^{f} + V_m E^{m} \tag{3.47}$$

Furthermore,

$$\varepsilon_{22} = -\nu_{12}\varepsilon_{11} = V_f \varepsilon_{22}^{f} + V_m \varepsilon_{22}^{m} = V_f(-\nu_{12}^{f}\varepsilon_{11}^{f}) + V_m(-\nu^{m}\varepsilon_{11}^{m}) \tag{3.48}$$

since no other stress component except for the longitudinal one exists in the fiber, matrix and the composite. Now,

$$\varepsilon_{11}^{f} = \frac{\sigma_{11}^{f}}{E_{11}^{f}} = \frac{\sigma_{11}}{E_{11}^{f} V_f + E_{11}^{f} V_m A_{11}} = \frac{\sigma_{11}}{E_{11}^{f} V_f + E^{m} V_m} = \frac{\sigma_{11}}{E_{11}} = \varepsilon_{11} \tag{3.49.1}$$

and

$$\varepsilon_{11}^{m} = \frac{\sigma_{11}^{m}}{E^{m}} = \frac{\sigma_{11}}{E^{m} V_f / A_{11} + E^{m} V_m} = \frac{\sigma_{11}}{E_{11}^{f} V_f + E^{m} V_m} = \frac{\sigma_{11}}{E_{11}} = \varepsilon_{11} \tag{3.49.2}$$

Substituting the last two equations into Eq. (3.48) yields

$$\nu_{12} = V_f \nu_{12}^{f} + V_m \nu^{m} \tag{3.50}$$

Next, let us apply to the composite a loading condition that $\sigma_{12}\neq 0$ (or $\sigma_{13}\neq 0$) and all the other $\sigma_{ij} = 0$. It is easy to derive, using $\sigma_{12} = V_f \sigma_{12}^{f} + V_m \sigma_{12}^{m}$, $\varepsilon_{12} = V_f \varepsilon_{12}^{f} + V_m \varepsilon_{12}^{m} = \sigma_{12} / G_{12}$, and Eqs. (3.44.7) and (3.44.8), that

$$G_{12} = \frac{V_f + V_m A_{66}}{V_f / G_{12}^{f} + V_m A_{66} / G^{m}} \tag{3.51}$$

Similarly, by applying to the composite the condition that $\sigma_{23}\neq 0$ and all the other $\sigma_{ij}=0$, we have

$$G_{23} = \frac{V_f + V_m A_{22}}{V_f / G_{23}^f + V_m A_{22} / G^m} \tag{3.52}$$

Finally, we apply a transverse stress to the composite, i.e., $\sigma_{22} \neq 0$ (or $\sigma_{33} \neq 0$) and all the other $\sigma_{ij} = 0$. In such a case, in addition to the internal transverse stresses in the fiber and the matrix materials given by Eqs. (3.44.1) and (3.44.2) respectively, the internal longitudinal stresses are also found to be

$$\sigma_{11}^f = -\frac{V_m A_{12} \sigma_{22}}{(V_f + V_m A_{11})(V_f + V_m A_{22})} \quad \text{and} \quad \sigma_{11}^m = \frac{V_f A_{12} \sigma_{22}}{(V_f + V_m A_{11})(V_f + V_m A_{22})} \tag{3.53}$$

From $\varepsilon_{22} = \frac{\sigma_{22}}{E_{22}} = V_f \varepsilon_{22}^f + V_m \varepsilon_{22}^m = V_f (S_{21}^f \sigma_{11}^f + S_{22}^f \sigma_{22}^f) + V_m (S_{21}^m \sigma_{11}^m + S_{22}^m \sigma_{22}^m)$

$$= \left(\frac{V_f [S_{22}^f (V_f + V_m A_{11}) - V_m A_{12} S_{21}^f] + V_m [S_{22}^m (V_f + V_m A_{11}) A_{22} + V_f A_{12} S_{21}^m]}{(V_f + V_m A_{11})(V_f + V_m A_{22})} \right) \sigma_{22}$$

we get

$$E_{22} = \left(\frac{V_f [S_{22}^f (V_f + V_m A_{11}) - V_m A_{12} S_{21}^f] + V_m [S_{22}^m (V_f + V_m A_{11}) A_{22} + V_f A_{12} S_{21}^m]}{(V_f + V_m A_{11})(V_f + V_m A_{22})} \right)^{-1} \tag{3.54}$$

In summary, the five engineering elastic moduli of the UD composite are expressed as

$$E_{11} = V_f E_{11}^f + V_m E^m \tag{3.55.1}$$

$$\nu_{12} = V_f \nu_{12}^f + V_m \nu^m \tag{3.55.2}$$

$$E_{22} = E_{33} = \frac{(V_f + V_m A_{11})(V_f + V_m A_{22})}{(V_f + V_m A_{11})(V_f S_{22}^f + A_{22} V_m S_{22}^m) + V_f V_m (S_{21}^m - S_{21}^f) A_{12}} \tag{3.55.3}$$

$$G_{12} = G_{13} = \frac{V_f + V_m A_{66}}{V_f / G_{12}^f + V_m A_{66} / G^m} \tag{3.55.4}$$

$$G_{23} = \frac{V_f + V_m A_{22}}{V_f / G_{23}^f + V_m A_{22} / G^m} \tag{3.55.5}$$

It is seen that the longitudinal Young's modulus and Poisson's ratio formulae, Eqs. (3.55.1) and (3.55.2), obtained by virtue of Eq. (3.38.1), are exactly the same as those of the rule of mixtures, which are known to be accurate enough. The formula (3.55.4), obtained by using Eq. (3.38.3), is the result of a precise elastic solution for the longitudinal shear modulus (Hyer, 1997). Only the accuracy of Eq. (3.38.2), based on which the transverse Young's modulus and out-plane shear modulus Eqs. (3.55.3) and (3.55.5) are obtained, needs to be verified. This can be done by comparing predicted results of the present model with experiments and with other micromechanical model formulae including Chamis's formulae (Eq. (1.46)). Fig. 3.1 shows the comparison of predicted results using the bridging (with $\beta = 0.5$) and Chamis's models with experiments (Tsai & Hahn, 1980) for the transverse modulus of a glass/epoxy composite. Fig. 3.2 gives a detailed comparison for the predicted transverse moduli between using Eq. (3.55.3), in which the bridging parameter β was taken as β = 0.5, and using Chamis's transverse modulus formula, Eq. (1.46.3). A high correlation between the theoretical and experimental data clearly indicates that the independent bridging elements defined by Eq. (3.38) are correct and accurate.

Remark 3.2

When both the constituent materials are isotropic, the resulting UD composite is still transversely isotropic, and hence has five independent material constants in general. It can be easily verified that the shear modulus given by Eq. (3.55.4) is generally not equal to that given by Eq. (3.55.5), even if both the fiber and the matrix are isotropic. However, several other simple micromechanics models, such as the rule of mixture model and the Chamis model (Eqs. (1.38.5) and (1.46.5)), cannot reveal such a difference (Christensen, 1991).

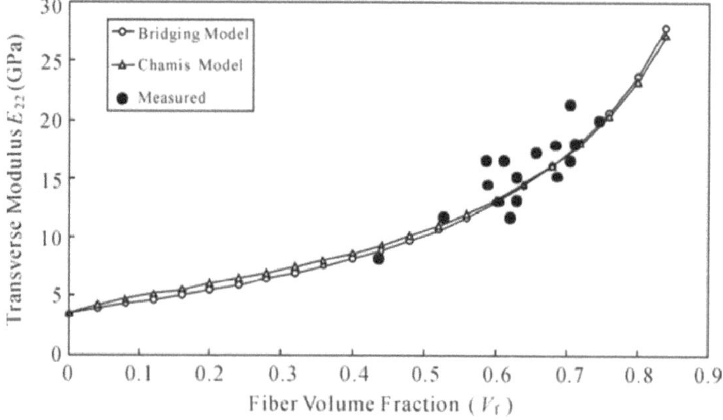

Fig. 3.1 Predicted and measured (Tsai & Hahn, 1980) transverse modulus of a glass/epoxy UD composite (E^f=73.1 GPa, E^m=3.45 GPa, ν^f=0.22 and ν^m=0.35) (from Huang et al., 1999)

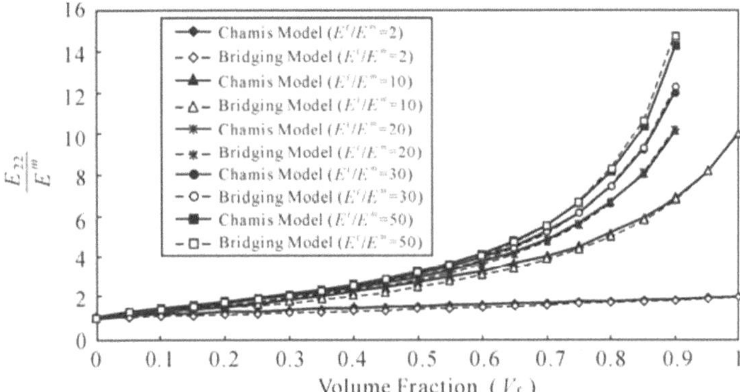

Fig. 3.2 Comparison between Chamis's and bridging models for the overall transverse modulus (E_{22}) of UD composites having different fiber and matrix moduli (Poisson's ratios of the materials have little effect on the predicted results of the two models)

Remark 3.3

In Eqs. (3.55.3) – (3.55.5), the bridging parameters β and α have been involved in the bridging elements A_{22} and A_{66} respectively. It can be expected that a different choice of these parameters will result in different predictions for transverse and shear moduli. Fig. 3.3 shows the influence of different bridging parameters of β on the predictions for the transverse modulus E_{22}. Three predictions from other micromechanics models as well as the experimental data (Tsai & Hahn, 1980) are also graphed in the figure for comparison. It is seen that the smaller the bridging parameter β used, the stiffer the predicted transverse modulus will be. For the glass/epoxy UD composite under consideration, the predictions from the bridging model based on $0.4 \leq \beta \leq 0.5$ have the best agreement with the experimental data. Fig. 3.4 shows the effect of different bridging parameters of α on the predictions for the longitudinal shear modulus G_{12}. A similar comparison has been made and a similar feature as the predictions for the transverse modulus has also been found. The figure indicates that, for the considered composite, all of the predictions using the bridging model with $0.3<\alpha<0.6$ are acceptable.

Further comparisons by choosing different bridging parameters of β and α are shown in Figs. 3.5(a) – 3.5(d) and 3.6(a) – 3.6(d). A main feature of these figures is that the predicted moduli increase with a decrease in the bridging parameters. However, if the moduli of the fiber and matrix are comparable, e.g., when the modulus ratio, E^f/E^m, equals 2.0, the influence of different bridging parameters is insignificant.

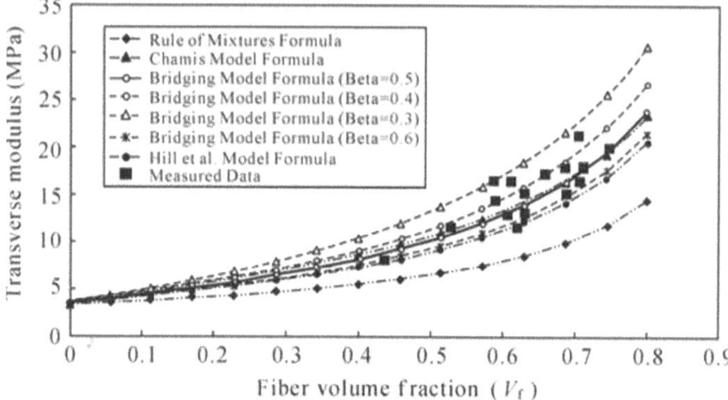

Fig. 3.3 Predicted and measured (Tsai & Hahn, 1980) transverse moduli of a glass/epoxy UD composite versus fiber volume fraction. The used material parameters are: E^f=73.1 GPa, ν^f=0.22, E^m=3.45 GPa and ν^m=0.35

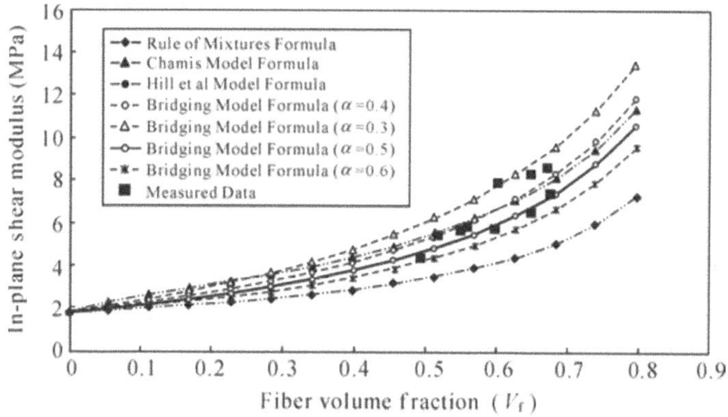

Fig. 3.4 Predicted and measured (Tsai & Hahn, 1980) in-plane shear modulus of a glass/epoxy UD composite versus fiber volume fraction. The material parameters used are: G^f=30.2 GPa and G^m=1.8 GPa

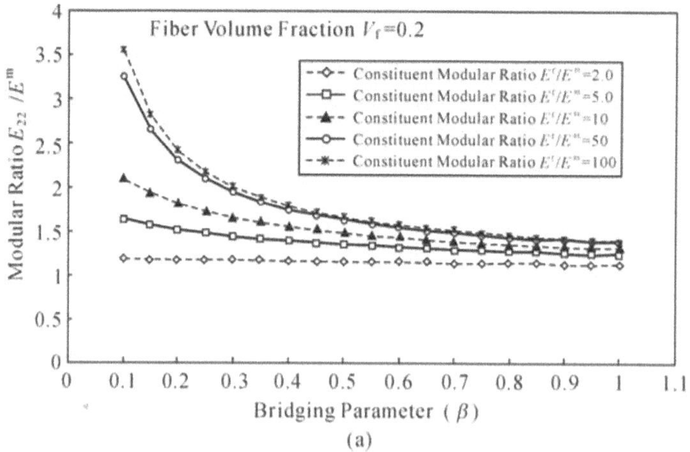

(a)

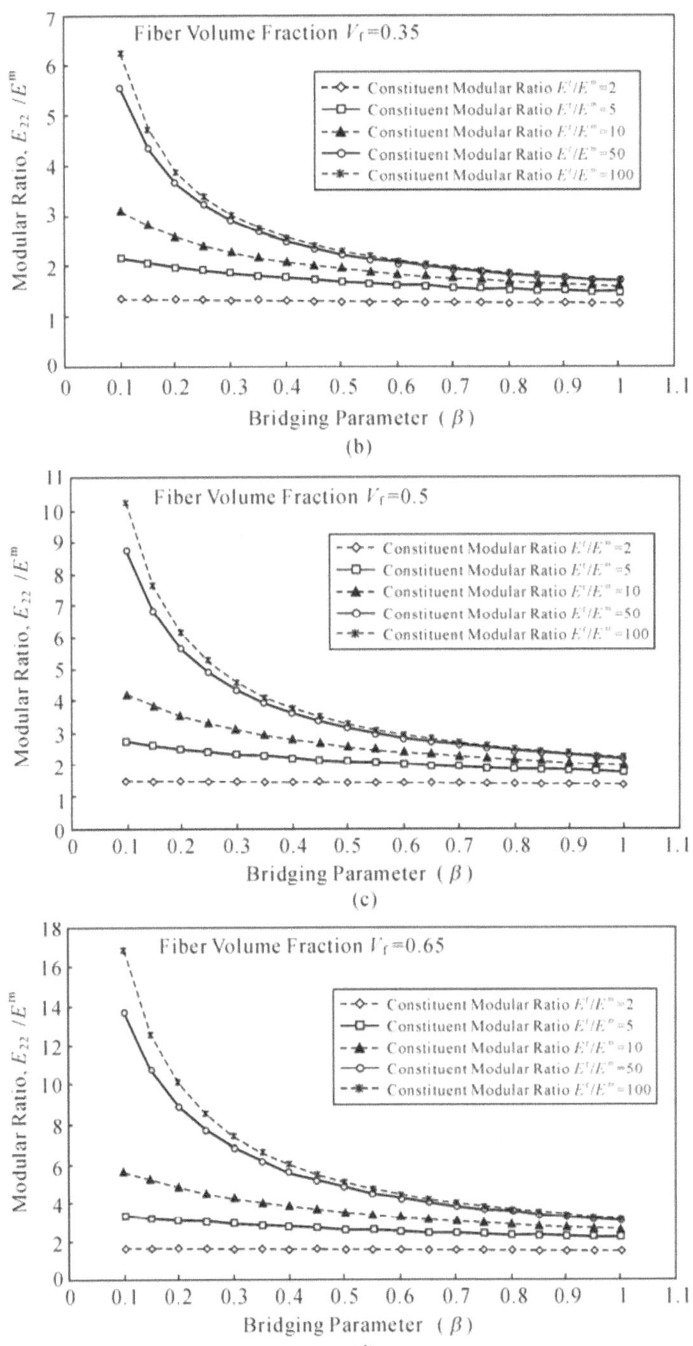

Fig. 3.5 Influence of bridging parameter β on the overall transverse modulus of UD composites. Poisson's ratios of ν^f=0.2 and ν^m=0.33 have been used

Fiber Volume Fraction V_f=0.2
Constituent Modular Ratio G^f/G^m=2
Constituent Modular Ratio G^f/G^m=5
Constituent Modular Ratio G^f/G^m=10
Constituent Modular Ratio G^f/G^m=50
Constituent Modular Ratio G^f/G^m=100
Modular Ratio, G_{12}/G^m
Bridging Parameter (α)
0
0.5
1
1.5
2
2.5
3
3.5
0.1
0.2
0.3
0.4
0.5
0.6
0.7
0.8
0.9
1
1.1
(a)

Fiber Volume Fraction V_f=0.35
Constituent Modular Ratio G^f/G^m=2
Constituent Modular Ratio G^f/G^m=5
Constituent Modular Ratio G^f/G^m=10
Constituent Modular Ratio G^f/G^m=50
Constituent Modular Ratio G^f/G^m=100
Modular Ratio, G_{12}/G^m
Bridging Parameter (α)
0
1
2
3
4
5
6
0.1
0.2
0.3
0.4
0.5
0.6
0.7
0.8
0.9
1
1.1
(b)

Fiber Volume Fraction V_f=0.5
Constituent Modular Ratio G^f/G^m=2
Constituent Modular Ratio G^f/G^m=5
Constituent Modular Ratio G^f/G^m=10
Constituent Modular Ratio G^f/G^m=50
Constituent Modular Ratio G^f/G^m=100
Modular Ratio, G_{12}/G^m
Bridging Parameter (α)
0
1
2
3
4
5
6
7
8
9
10
0.1
0.2
0.3
0.4
0.5
0.6
0.7
0.8
0.9
1
1.1
(c)

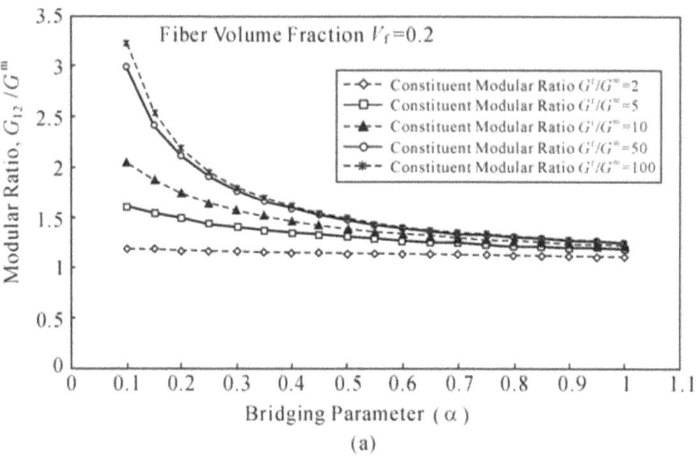

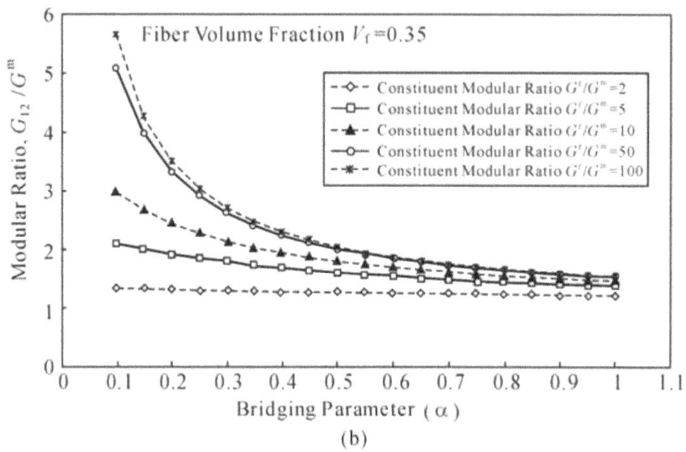

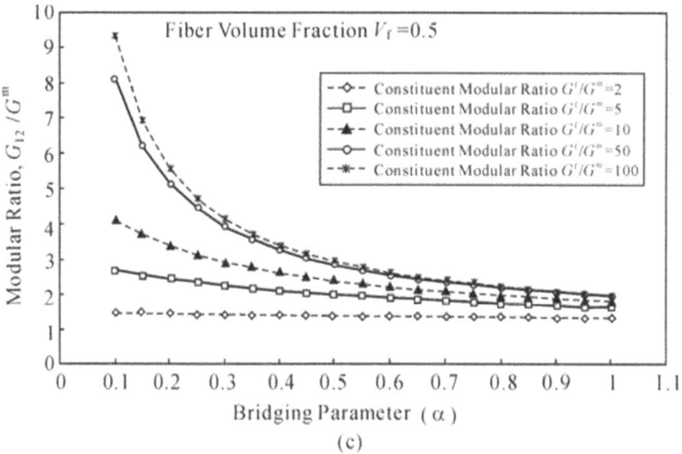

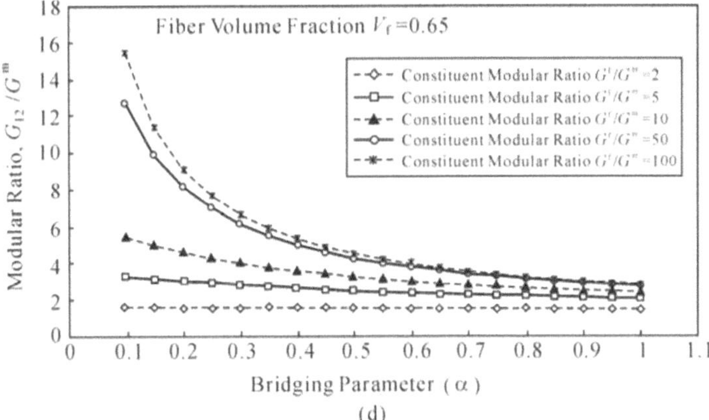

Fig. 3.6 Influence of bridging parameter α on the overall in-plane shear modulus of UD composites

3.7 Unified Formulae for Elastic Moduli

Using a bridging matrix, it is possible to obtain unified expressions for the effective elastic moduli of a UD composite. Let us start with the general form of a 2D bridging matrix, i.e.,

$$[A_{ij}] = \begin{bmatrix} a_{11} & a_{12} & 0 \\ a_{21} & a_{22} & 0 \\ 0 & 0 & a_{33} \end{bmatrix} \tag{3.32}$$

Similar to what was done previously, we first apply a uniaxial tension load, σ_{11}, along the longitudinal direction of the UD composite, as schematically shown in Fig. 3.7(a).

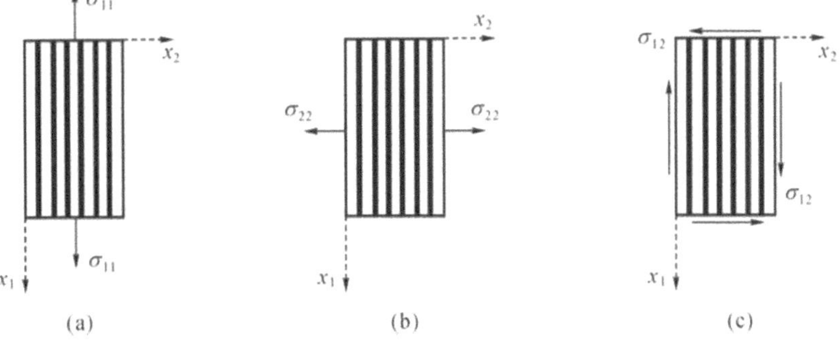

Fig. 3.7 Schematic of a UD composite subjected to: (a) longitudinal, (b) transverse and (c) in-plane shear load, respectively

To determine internal stresses generated in the fiber and matrix materials, the following equations are applicable:

$$\sigma_{11} = V_f \sigma_{11}^f + V_m \sigma_{11}^m \tag{3.56.1}$$

$$\sigma_{22} = V_f \sigma_{22}^f + V_m \sigma_{22}^m = 0 \tag{3.56.2}$$

$$\sigma_{11}^m = a_{11} \sigma_{11}^f + a_{12} \sigma_{22}^f \tag{3.56.3}$$

$$\sigma_{22}^m = a_{21} \sigma_{11}^f + a_{22} \sigma_{22}^f \tag{3.56.4}$$

Eqs. (3.56.1) and (3.56.2) are from the fundamental equation, Eq. (3.1), whereas Eqs. (3.56.3) and (3.56.4) are due to Eq. (3.7) together with Eq. (3.32). From these equations, it is found that

$$\sigma_{11}^f = \left[V_f + V_m \left(a_{11} - \frac{V_m a_{21} a_{12}}{V_f + V_m a_{22}} \right) \right]^{-1} \sigma_{11} = e_{11}^f \sigma_{11} \tag{3.57.1}$$

$$\sigma_{11}^m = \left[a_{11} - \frac{V_m a_{21} a_{12}}{(V_f + V_m a_{22})} \right] \sigma_{11}^f = e_{11}^m \sigma_{11} \tag{3.57.2}$$

$$\sigma_{22}^f = -\frac{V_m a_{21} \sigma_{11}^f}{(V_f + V_m a_{22})} = e_{21}^f \sigma_{11} \tag{3.57.3}$$

$$\sigma_{22}^m = \frac{V_f a_{21} \sigma_{11}^f}{(V_f + V_m a_{22})} = e_{21}^m \sigma_{11} \tag{3.57.4}$$

where

$$e_{11}^f = \left[V_f + V_m \left(a_{11} - \frac{V_m a_{21} a_{12}}{V_f + V_m a_{22}} \right) \right]^{-1} \tag{3.58.1}$$

$$e_{11}^m = \left[a_{11} - \frac{V_m a_{21} a_{12}}{(V_f + V_m a_{22})} \right] e_{11}^f \tag{3.58.2}$$

$$e_{21}^f = -\frac{V_m a_{21} e_{11}^f}{(V_f + V_m a_{22})} \tag{3.58.3}$$

$$e_{21}^m = \frac{V_f a_{21} e_{11}^f}{(V_f + V_m a_{22})} \tag{3.58.4}$$

As the composite is subjected to a uniaxial stress state, under which the constituents are bi-axially loaded, it is obtained by virtue of Eq. (3.2) that

$$\varepsilon_{11} = \frac{\sigma_{11}}{E_{11}} = V_f \varepsilon_{11}^f + V_m \varepsilon_{11}^m = V_f (S_{11}^f \sigma_{11}^f + S_{12}^f \sigma_{22}^f) + V_m (S_{11}^m \sigma_{11}^m + S_{12}^m \sigma_{22}^m)$$
$$= [V_f (S_{11}^f e_{11}^f + S_{12}^f e_{21}^f) + V_m (S_{11}^m e_{11}^m + S_{12}^m e_{21}^m)] \sigma_{11}$$

from which the longitudinal modulus, E_{11}, is found to be

$$E_{11} = [V_f (S_{11}^f e_{11}^f + S_{12}^f e_{21}^f) + V_m (S_{11}^m e_{11}^m + S_{12}^m e_{21}^m)]^{-1} \tag{3.59}$$

Furthermore, we have

$$\varepsilon_{22} = -\nu_{12} \varepsilon_{11} = V_f \varepsilon_{22}^f + V_m \varepsilon_{22}^m = V_f (S_{21}^f \sigma_{11}^f + S_{22}^f \sigma_{22}^f) + V_m (S_{21}^m \sigma_{11}^m + S_{22}^m \sigma_{22}^m)$$
$$= [V_f (S_{21}^f e_{11}^f + S_{22}^f e_{21}^f) + V_m (S_{21}^m e_{11}^m + S_{22}^m e_{21}^m)] E_{11} \varepsilon_{11}$$

Therefore, the longitudinal Poisson's ratio is given by

$$\nu_{12} = -\frac{[V_f (S_{21}^f e_{11}^f + S_{22}^f e_{21}^f) + V_m (S_{21}^m e_{11}^m + S_{22}^m e_{21}^m)]}{[V_f (S_{11}^f e_{11}^f + S_{12}^f e_{21}^f) + V_m (S_{11}^m e_{11}^m + S_{12}^m e_{21}^m)]} \tag{3.60}$$

Similarly, by only applying a transverse load to the UD composite, as shown in Fig. 3.7(b), we have the following simultaneous equations

$$\sigma_{11} = V_f \sigma_{11}^f + V_m \sigma_{11}^m = 0 \tag{3.61.1}$$

$$\sigma_{22} = V_f \sigma_{22}^f + V_m \sigma_{22}^m \tag{3.61.2}$$

$$\sigma_{11}^m = a_{11} \sigma_{11}^f + a_{12} \sigma_{22}^f \tag{3.61.3}$$

$$\sigma_{22}^m = a_{21} \sigma_{11}^f + a_{22} \sigma_{22}^f \tag{3.61.4}$$

which result in the following solutions

$$\sigma_{11}^f = -\frac{V_m a_{12} \sigma_{22}^f}{(V_f + V_m a_{11})} = e_{12}^f \sigma_{22} \tag{3.62.1}$$

$$\sigma_{11}^m = \frac{V_f a_{12} \sigma_{22}^f}{(V_f + V_m a_{11})} = e_{12}^m \sigma_{22} \tag{3.62.2}$$

$$\sigma_{22}^{f}=\left[V_f+V_m\left(a_{22}-\frac{V_m a_{21}a_{12}}{V_f+V_m a_{11}}\right)\right]^{-1}\sigma_{22}=e_{22}^{f}\sigma_{22} \tag{3.62.3}$$

$$\sigma_{22}^{m}=\left[a_{22}-\frac{V_m a_{21}a_{12}}{(V_f+V_m a_{11})}\right]\sigma_{22}^{f}=e_{22}^{m}\sigma_{22} \tag{3.62.4}$$

where
$$e_{22}^{f}=\left[V_f+V_m\left(a_{22}-\frac{V_m a_{21}a_{12}}{V_f+V_m a_{11}}\right)\right]^{-1} \tag{3.58.5}$$

$$e_{12}^{f}=-\frac{V_m a_{12}e_{22}^{f}}{(V_f+V_m a_{11})} \tag{3.58.6}$$

$$e_{12}^{m}=\frac{V_f a_{12}e_{22}^{f}}{(V_f+V_m a_{11})} \tag{3.58.7}$$

$$e_{22}^{m}=\left[a_{22}-\frac{V_m a_{21}a_{12}}{(V_f+V_m a_{11})}\right]e_{22}^{f} \tag{3.58.8}$$

Now, by using the stress-strain relationship in the transverse direction, i.e.,

$$\begin{aligned}\varepsilon_{22}&=\frac{\sigma_{22}}{E_{22}}=V_f\varepsilon_{22}^{f}+V_m\varepsilon_{22}^{m}=V_f(S_{21}^{f}\sigma_{11}^{f}+S_{22}^{f}\sigma_{22}^{f})+V_m(S_{21}^{m}\sigma_{11}^{m}+S_{22}^{m}\sigma_{22}^{m})\\&=[V_f(S_{21}^{f}e_{12}^{f}+S_{22}^{f}e_{22}^{f})+V_m(S_{21}^{m}e_{12}^{m}+S_{22}^{m}e_{22}^{m})]\sigma_{22}\end{aligned}$$

we obtain the transverse modulus as

$$E_{22}=[V_f(S_{21}^{f}e_{12}^{f}+S_{22}^{f}e_{22}^{f})+V_m(S_{21}^{m}e_{12}^{m}+S_{22}^{m}e_{22}^{m})]^{-1} \tag{3.63}$$

Finally, by applying an in-plane shear load to the UD composite, as indicated in Fig. 3.7(c), it is seen that the in-plane shear modulus has the same expression as Eq. (3.51) or

$$G_{12}=\frac{V_f+V_m a_{33}}{V_f/G_{12}^{f}+V_m a_{33}/G^{m}} \tag{3.64}$$

As any micromechanics model for unidirectional composites must correspond to a specific form of bridging matrix, Eq. (3.32), the expressions of Eqs. (3.59), (3.60), (3.63) and (3.64) are unified formulae for the four effective in-plane elastic moduli of a UD composite. For instance, by setting $a_{11}=A_{11}$, $a_{12}=A_{12}$, $a_{21}=A_{21}$,

$a_{22}=A_{22}$ and $a_{33}=A_{66}$, where A_{11}, A_{12}, A_{21}, A_{22} and A_{66} are defined by Eqs. (3.30.1), (3.30.2), (3.30.3), (3.30.4) and (3.30.10), we obtain the Mori-Tanaka formulae.

Let us consider two UD composites made of isotropic glass fibers and epoxy matrix, whose elastic properties are summarized in Table 3.1. Three micromechanics models are used to calculate in-plane effective elastic moduli of the two composites. They are the bridging model, represented by Eqs. (3.55.1) – (3.55.4) with $\alpha=\beta=0.4$ and $\alpha=\beta=0.5$, the Mori-Tanaka model and Mori-Tanaka model with symmetry constraint. Here, the Mori-Tanaka model, denoted by the M-T model, refers to Eqs. (3.59), (3.60), (3.63) and (3.64) with $a_{11}=A_{11}$, $a_{12}=A_{12}$, $a_{21}=A_{21}$, $a_{22}=A_{22}$ and $a_{33}=A_{66}$ defined by Eqs. (3.30.1), (3.30.2), (3.30.3), (3.30.4) and (3.30.10) respectively. Furthermore, the Mori-Tanaka model with symmetry constraint, abbreviated to M-T-S model, implies that the four independent elements $a_{11}=A_{11}$, $a_{21}=A_{21}$, $a_{22}=A_{22}$ and $a_{33}=A_{66}$ are specified by Eqs. (3.30.1), (3.30.3), (3.30.4) and (3.30.10) but $a_{12}=A_{12}$ is determined by the symmetric condition, i.e., (Eq. (3.33)),

$$A_{12}=\frac{(S_{12}^{f}-S_{12}^{m})(A_{22}-A_{11})+(S_{22}^{m}-S_{22}^{f})A_{21}}{S_{11}^{m}-S_{11}^{f}} \tag{3.65}$$

The bridging matrix elements for the two composites calculated using Mori-Tanaka approach formulae, Eqs. (3.30.1), (3.30.3), (3.30.4), (3.30.10) and Eq. (3.30.2) or Eq. (3.65), are listed in Table 3.2. The predicted results for the composite 1 using the three models are given in Table 3.3, whereas those for the composite 2 are shown in Table 3.4. Experimental data for the two composites, provided by Soden et al. (1998), are also summarized in Tables 3.3 and 3.4, respectively, for comparison purposes. Although the predictions using the three models for the effective moduli of the two composites under consideration are not very much different, the bridging model exhibits the highest accuracy. Moreover, the Mori-Tanaka models with and without the symmetry constraint give almost the same predictions for these two composites. However, there does exist a difference between A_{12}'s with and without the symmetry constraint, see Table 3.2. This clearly shows that the compliance matrix of a UD composite defined exactly by the Mori-Tanaka's bridging matrix may not in general be symmetric.

Table 3.1 Constituent properties of two UD composites (Soden et al., 1998)

Composite	V_f	E^f (GPa)	ν^f	G^f (GPa)	E^m (GPa)	ν^m	G^m (GPa)
1	0.62	80	0.2	33.333	3.35	0.35	1.241
2	0.6	74	0.2	30.833	3.35	0.35	1.241

Table 3.2 Bridging matrix elements calculated from Mori-Tanaka approach

Composite	A_{11}	A_{21}	A_{22}	A_{66}	A_{12} (M-T)	A_{12} (M-T-S)
1	0.04438	0.00358	0.70734	0.51861	0.26637	0.23996
2	0.04798	0.00387	0.70856	0.52012	0.26614	0.23977

Table 3.3 Effective elastic moduli of UD composite 1 by different methods

Model	E_{11} (GPa)	ν_{12}	E_{22} (GPa)	G_{12} (GPa)
Bridging (α=β=0.4)	50.873	0.257	15.382	5.275
Bridging (α=β=0.5)	50.873	0.257	13.545	4.605
M-T-S	50.351	0.239	11.266	4.605
M-T	50.368	0.239	11.411	4.605
Experiment (Soden, 1998)	53.48	0.278	17.7	5.83

Table 3.4 Effective elastic moduli of UD composite 2 by different methods

Model	E_{11} (GPa)	ν_{12}	E_{22} (GPa)	G_{12} (GPa)
Bridging (α=β=0.4)	45.74	0.26	14.349	4.927
Bridging (α=β=0.5)	45.74	0.26	12.686	4.318
M-T-S	45.251	0.242	10.617	4.318
M-T	45.268	0.242	10.753	4.318
Experiment (Soden, 1998)	45.6	0.278	16.2	5.83

3.8 Plastic Theory

It has already been recognized that most UD composites can display significant inelastic deformation before failure when subjected to a transverse or an in-plane shear load. Such deformation will affect the load carrying capacity of the composites, and will play a key role in the analysis of laminated composites (Huang, 2000b, 2000c, 2000d, 2000e, 2001c). In a laminated composite, due to different lay-up configuration, each lamina ply in the laminate may carry a different load share. Some lamina plies will fail first before others, and a progressive failure process is developed. The resulting transverse and in-plane shear stresses occur in the lamina even if a uniaxial load is applied to the laminate (Zhang et al., 2000). Therefore, determination of the instantaneous stiffness/ compliance matrix of each lamina is crucial to the understanding of the laminate progressive failure process. This is because the laminate, with respect to the laminas involved, is a statically indeterminate structure. Such determination cannot be accomplished without a thorough understanding of the inelastic deformation characteristic of a UD composite.

The following assumptions will be made in the development of a plastic theory for a UD composite:

(1) A perfect bonding exists between the fiber and matrix interface during the whole load sequence.

(2) Both the fiber and the matrix materials undergo isotropic hardening. In other words, the fiber can either be transversely isotropic and linearly elastic until rupture or be isotropically elastic-plastic. This implies that the same plastic flow theory of isotropic materials can be used to define the instantaneous compliance matrices of both the constituents.

(3) The geometrical deformation of the composite is so small that the volume

fractions of the fiber and the matrix, V_f and V_m, remain unchanged during the entire load sequence.

(4) Each load increment is so small that whenever the instantaneous compliance matrix of a constituent, $[S_{ij}^f]$ or $[S_{ij}^m]$, is evaluated at the beginning of the load increment, it will remain unchanged during the whole load increment.

To obtain an instantaneous stiffness/compliance matrix, or to establish a plastic constitutive relationship for a UD composite, we will need a theorem stated below.

***Theorem* 3.1** *A necessary and sufficient condition for the establishment of a plastic constitutive relationship for a unidirectional composite is to know the internal stresses in its constituent fiber and matrix materials.*

Proof. First, let us prove necessity, i.e., if a plastic constitutive relationship for a UD composite is established, the internal stresses in the fiber and matrix materials must be known in advance.

Let $\{\mathrm{d}\sigma_i^f\}$ and $\{\mathrm{d}\varepsilon_i^f\}$, and $\{\mathrm{d}\sigma_i^m\}$ and $\{\mathrm{d}\varepsilon_i^m\}$, respectively, represent the internal stress and strain increments in the fiber and matrix materials of the composite. $\{\mathrm{d}\sigma_i\}$ and $\{\mathrm{d}\varepsilon_i\}$ denote the overall incremental stresses and strains of the composite. Furthermore, $[S_{ij}^f]$, $[S_{ij}^m]$ and $[S_{ij}]$ denote the elastic-plastic instantaneous compliance matrices of the fiber, matrix and the composite, respectively. At each load level, according to the assumptions (3) and (4), we have the following volume averaged relationships (refer to Subsection 1.3.2):

$$\{\mathrm{d}\varepsilon_i\}=V_f\{\mathrm{d}\varepsilon_i^f\}+V_m\{\mathrm{d}\varepsilon_i^m\} \tag{3.66.1}$$

$$\{\mathrm{d}\sigma_i\}=V_f\{\mathrm{d}\sigma_i^f\}+V_m\{\mathrm{d}\sigma_i^m\} \tag{3.66.2}$$

$$\{\mathrm{d}\varepsilon_i^f\}=[S_{ij}^f]\{\mathrm{d}\sigma_j^f\} \tag{3.66.3}$$

$$\{\mathrm{d}\varepsilon_i^m\}=[S_{ij}^m]\{\mathrm{d}\sigma_j^m\} \tag{3.66.4}$$

$$\{\mathrm{d}\varepsilon_i\}=[S_{ij}]\{\mathrm{d}\sigma_j\} \tag{3.66.5}$$

Similarly, as in a linear elastic analysis, let us assume that there is a bridging matrix such that

$$\{\mathrm{d}\sigma_i^m\}=[A_{ij}]\{d\sigma_j^f\} \tag{3.67}$$

The bridging matrix, $[A_{ij}]$, is also an instantaneous quantity. With the bridging matrix, the following equations are obtained as well

$$\{\mathrm{d}\sigma_i^f\}=(V_f[I]+V_m[A_{ij}])^{-1}\{\mathrm{d}\sigma_j\} \tag{3.68.1}$$

$$\{\mathrm{d}\sigma_i^m\}=[A_{ij}](V_f[\mathrm{I}]+V_m[A_{ij}])^{-1}\{\mathrm{d}\sigma_j\} \tag{3.68.2}$$

and

$$[S_{ij}]=(V_f[S_{ij}^f]+V_m[S_{ij}^m][A_{ij}])(V_f[I]+V_m[A_{ij}])^{-1} \tag{3.69}$$

From Eq. (3.69), it is apparent that the instantaneous compliance matrices, $[S_{ij}^{f}]$ and $[S_{ij}^{m}]$, of the constituent fiber and matrix materials must be determined before the composite compliance matrix, $[S_{ij}]$, can be obtained. However, the determination of an instantaneous compliance matrix of a constituent material under a plastic deformation requires current stress components, as indicated by Eq. (2.28) or Eq. (2.34). This completes the proof for necessity.

Next, let us prove sufficiency, which says that when all of the stress states in the constituent fiber and matrix materials are known, the instantaneous compliance matrix of the composite at every load level can be obtained. In fact, according to the given conditions, the instantaneous compliance matrices, $[S_{ij}^{f}]$ and $[S_{ij}^{m}]$, of the fiber and matrix materials together with stress increments $\{d\sigma_i^{f}\}$ and $\{d\sigma_i^{m}\}$ at each load level are all known. By using Eqs. (3.66.3) and (3.66.4), the strain increments in the fiber and matrix, $\{d\varepsilon_i^{f}\}$ and $\{d\varepsilon_i^{m}\}$, are obtained. Furthermore, from Eqs. (3.66.1) and (3.66.2), all of the strain and stress increments of the composite, $\{d\varepsilon_i\}$ and $\{d\sigma_i\}$, are available. Letting these quantities be interconnected with Eq. (3.66.5), we are always able to determine a compliance matrix $[S_{ij}]$. The proof for the sufficiency is completed.

The theorem just proven releases two kinds of information. Firstly, an elastic-plastic constitutive relationship for a UD composite can be established only micromechanically. Secondly, the only way to approach the establishment of such a constitutive relationship is to evaluate the stresses in the constituent fiber and matrix materials at every load level. From Eqs. (3.68.1) and (3.68.2), it is only necessary to determine the corresponding instantaneous bridging matrix.

The most significant feature of the bridging model is that it can be easily extended to the determination of a bridging matrix in a plastic deformation region. The extension is based on a logical consideration. As mentioned in Section 3.3, the bridging matrix correlating the stress-state in the matrix with that in the fiber of the composite should depend only on the material properties of the constituents and on the fiber packing geometries, which include the fiber volume fraction, the fiber arrangement pattern in the matrix, the fiber cross-sectional shape and the interface bonding between the fiber and the matrix. As long as a bridging matrix has been determined using an elastic deformation condition, only the material parameters involved vary when any constituent material undergoes a plastic deformation, because the fiber packing geometries only change or vary by a negligibly small amount during the plastic deformation. Since the bridging matrix in the elastic region has already been defined, as per Eqs. (3.37) and (3.38), a generalization of it to the plastic region is straightforward.

According to the aforementioned consideration, the bridging matrix, $[A_{ij}]$, should have the same structure as that used in a linear elastic analysis. Namely, it should contain 5 independent elements and 16 other dependent elements. Due to plastic coupling, a general form of the bridging matrix is expressed as (Huang, 2000b, 2001c, 2004)

$$[A_{ij}] = \begin{bmatrix} A_{11} & A_{12} & A_{13} & A_{14} & A_{15} & A_{16} \\ & A_{22} & A_{23} & A_{24} & A_{25} & A_{26} \\ & & A_{33} & A_{34} & A_{35} & A_{36} \\ & & & A_{44} & A_{45} & A_{46} \\ & & & & A_{55} & A_{56} \\ \text{zero} & & & & & A_{66} \end{bmatrix} \quad (3.70)$$

All of the upper off-diagonal elements are dependent. A_{44} is also a dependent element due to Eq. (3.15). Determination of these dependent elements is standard. For instance, the off-diagonal elements are obtained by substituting Eq. (3.70) into Eq. (3.69) and by requiring the resulting overall compliance matrix to be symmetric, i.e.,

$$S_{ji}=S_{ij}, \quad i,j=1, 2, \ldots, 6 \quad (3.71)$$

The equations to define the independent elements should be the same in form as Eq. (3.39). Only the elastic constants of a constituent material involved should be replaced by its plastic counterparts when the material undergoes a plastic deformation. This is due to the fact that the expansion coefficients λ_{ij} given by Eq. (3.40) remain unchanged. More precisely, the non-zero independent elements of the bridging matrix should take the following forms

$$A_{11} = E_m / E_{f1} \quad (3.72.1)$$

$$A_{22} = A_{33} = A_{44} = \beta + (1-\beta)\frac{E_m}{E_{f2}}, \ 0<\beta<1 \quad (3.72.2)$$

$$A_{55} = A_{66} = \alpha + (1-\alpha)\frac{G_m}{G_f}, \ 0<\alpha<1 \quad (3.72.3)$$

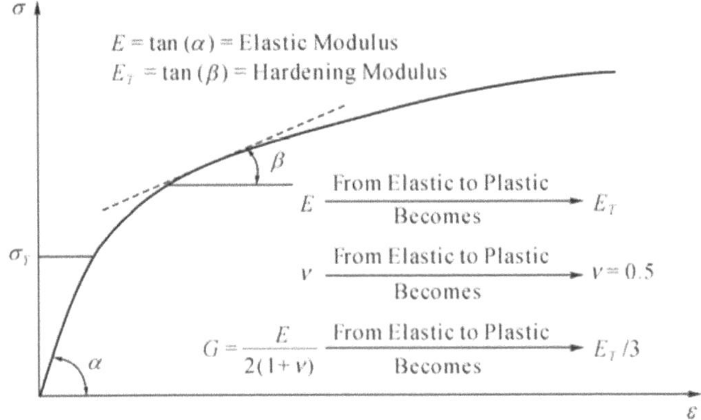

Fig. 3.8 Schematic variation in mechanical properties of an elastic-plastic material

where E_m, G_m, E_{f1}, E_{f2}, and G_f are called effective moduli and are defined as (refer to Fig. 3.8)

$$E_{f1} = \begin{cases} E_{11}^f, \text{when} \quad \tau_0^f \le \sqrt{2}\sigma_Y^f / 3 \\ E_T^f, \text{when} \quad \tau_0^f > \sqrt{2}\sigma_Y^f / 3 \end{cases} \tag{3.73.1}$$

$$E_{f2} = \begin{cases} E_{22}^f, \text{when} \quad \tau_0^f \le \sqrt{2}\sigma_Y^f / 3 \\ E_T^f, \text{when} \quad \tau_0^f > \sqrt{2}\sigma_Y^f / 3 \end{cases} \tag{3.73.2}$$

$$G_f = \begin{cases} G_{12}^f, \text{when} \quad \tau_0^f \le \sqrt{2}\sigma_Y^f / 3 \\ E_T^f / 3, \text{when} \quad \tau_0^f > \sqrt{2}\sigma_Y^f / 3 \end{cases} \tag{3.73.3}$$

$$E_m = \begin{cases} E^m, \text{when} \quad \tau_0^m \le \sqrt{2}\sigma_Y^m / 3 \\ E_T^m, \text{when} \quad \tau_0^m > \sqrt{2}\sigma_Y^m / 3 \end{cases} \tag{3.73.4}$$

$$G_m = \begin{cases} 0.5E^m / (1+\nu^m), \text{when} \quad \tau_0^m \le \sqrt{2}\sigma_Y^m / 3 \\ E_T^m / 3, \text{when} \quad \tau_0^m > \sqrt{2}\sigma_Y^m / 3 \end{cases} \tag{3.73.5}$$

E_T^f and E_T^m are the hardening moduli of the fiber and the matrix in their plastic region, τ_0^f and τ_0^m are the octahedral shear stresses (Eq. (2.13)) and σ_Y^f and σ_Y^m are the uniaxial yield strengths of the fiber and the matrix, respectively.

In order to solve the dependent elements from Eq. (3.71), the instantaneous elastic-plastic compliance matrices of the constituents, $[S_{ij}^f]$ and $[S_{ij}^m]$, must be specified. This can be accomplished by using any established plastic flow theory for isotropic materials. Let us employ the general Prandtl-Reuss theory to do this. For the matrix material, for instance, we have (Section 2.2)

$$\left[S_{ij}^m\right] = \begin{cases} \left[S_{ij}^m\right]^e, \text{when} \quad \tau_0^m \le \dfrac{\sqrt{2}}{3}\sigma_Y^m \\ \left[S_{ij}^m\right]^e + \left[S_{ij}^m\right]^p, \text{when} \quad \tau_0^m > \dfrac{\sqrt{2}}{3}\sigma_Y^m \end{cases} \tag{3.74.1}$$

$$[S_{ij}^m]^p = \frac{1}{2M_T\tau_0^2}\begin{bmatrix} \sigma'_{11}\sigma'_{11} & \sigma'_{22}\sigma'_{11} & \sigma'_{33}\sigma'_{11} & 2\sigma'_{23}\sigma'_{11} & 2\sigma'_{13}\sigma'_{11} & 2\sigma'_{12}\sigma'_{11} \\ & \sigma'_{22}\sigma'_{22} & \sigma'_{33}\sigma'_{22} & 2\sigma'_{23}\sigma'_{22} & 2\sigma'_{13}\sigma'_{22} & 2\sigma'_{12}\sigma'_{22} \\ & & \sigma'_{33}\sigma'_{33} & 2\sigma'_{23}\sigma'_{33} & 2\sigma'_{13}\sigma'_{33} & 2\sigma'_{12}\sigma'_{33} \\ & & & 4\sigma'_{23}\sigma'_{23} & 4\sigma'_{13}\sigma'_{23} & 4\sigma'_{12}\sigma'_{23} \\ & & & & 4\sigma'_{13}\sigma'_{13} & 4\sigma'_{12}\sigma'_{13} \\ & \text{symmetry} & & & & 4\sigma'_{12}\sigma'_{12} \end{bmatrix}_{\sigma_{ij}=\sigma_{ij}^m}$$

(3.74.2)

$$M_T^m = \frac{E^m E_T^m}{E^m - E_T^m} \tag{3.74.3}$$

$[S_{ij}^m]^e$ is the elastic component of the compliance matrix, $[S_{ij}^m]$, specified using Hooke's law.

Having obtained the internal stress increments from Eqs. (3.68.1) and (3.68.2), the total stresses at the current load level are simply updated from the following formulae

$$[\sigma_{ij}^f]^{(K+1)} = [\sigma_{ij}^f]^{(K)} + [\mathrm{d}\sigma_{ij}^f],\ K=0,\ 1\ \ldots \tag{3.75.1}$$

$$[\sigma_{ij}^m]^{(K+1)} = [\sigma_{ij}^m]^{(K)} + [\mathrm{d}\sigma_{ij}^m],\ K=0,\ 1\ \ldots \tag{3.75.2}$$

K=0 corresponds to the initial load. If there are no initial stresses such as no residual thermal stresses, we simply have $[\sigma_{ij}^f]^{(0)} = [\sigma_{ij}^m]^{(0)} = [0]$. The updated overall stresses on the composite are given by

$$[\sigma_{ij}]^{(K+1)} = [\sigma_{ij}]^{(K)} + [\mathrm{d}\sigma_{ij}],\ K=0,\ 1\ \ldots \tag{3.75.3}$$

Similarly, the internal strain and the overall strain increments can be calculated from Eqs. (3.66.3) – (3.66.5), whereas the total strains can be updated accordingly.

Remark 3.4

(1) The weakest assumption involved in developing the plastic theory is the first one, i.e., the perfect bonding assumption. In reality, bonding defects or debonding before failure can occur in a composite. Even worse, some less successful fabrication may leave an apparent separation between the fibers and the matrix. For example, when the fibers are placed in between matrix sheets/films and a hot press is applied to make the resulting composite, fabrication defects (voids) may probably occur. In such cases, the bridging

model can give an inaccurate prediction if the constituent properties, namely the uniaxial stress-strain curve up to failure, are obtained based on monolithic material tests. To resolve such a problem, two different approaches can be employed. The first approach is to redefine the constituent *in situ* properties based on some measured overall response of the composite. Namely, the constituent properties should be retrieved using the measured data of the composite responses. We will illustrate such retrieval in the next chapter. In the second approach, the bridging matrix can be redefined, by using micro-structural information between the fiber and matrix interface. The bridging matrix (independent elements) may even be variable during the whole loading sequence, to account for damage evolution. This second approach is not our concern in the present book.

(2) In describing the constituent plastic behavior, the Prandtl-Reuss theory was incorporated. The reason for this incorporation is that this theory can be easily expressed in an incremental form, which is suitable for the determination of the composite instantaneous compliance matrix (Eq. (3.69)). For some materials, however, other plastic flow theories may be more powerful to achieve an accurate description of their stress-strain relationships. For instance, silicon carbide fibers/titanium matrix composites can be used in very sophisticated environments. The titanium matrix is most commonly described using the unified Bodner-Partom (B-P) flow law (Section 2.3). The B-P theory, however, is described using total stress and total strain. In order to incorporate the B-P flow law into the bridging model, a possible treatment (Section 2.4) is to first plot the uniaxial stress-strain curve of the matrix at given conditions (temperature, strain rate, etc.), based on the B-P theory. Then, the matrix parameters involved in the Prandtl-Reuss theory are obtainable, and the composite properties can be simulated using the computer program included in this book.

(3) In most cases, the reinforcing fibers used in a composite can be regarded as linearly elastic until rupture. In fact, all of the composites considered in this book have been assumed to be fabricated with reinforcing fibers of linear elasticity until rupture.

Remark 3.5

Strictly speaking, the dependent bridging element A_{44} should be determined as per Eq. (3.15), i.e., through $S_{44}=2(S_{22}-S_{23})$. However, compared with the much simpler expression, Eq. (3.72.2), used for A_{44}, no significant difference in predictions has been found.

3.9 Planar Stress Formulae

In the most general case, a three-dimensional (3D) instantaneous compliance matrix of the composite is obtainable once the off-diagonal bridging elements have been solved from Eq. (3.71). However, they are a set of (fifteen) simultaneous, nonlinear algebraic equations, to which iterative solutions are generally required. In reality, the majority of composite components are either subjected to a planar load condition or composed of plate/shell structures. In the analysis of a laminated plate/shell structure, only the in-plane stress and strain components are retained. Therefore, most composites only involve a plane stress analysis, which can be accomplished using a number of explicit and closed form formulae given below.

Thus, let $\{d\sigma_i\}^T=\{d\sigma_{11}, d\sigma_{22}, d\sigma_{12}\}$ be the overall applied stress increments. The incremental stresses in the matrix can be correlated with those in the fiber through the bridging matrix via

$$\begin{Bmatrix} d\sigma_{11}^m \\ d\sigma_{22}^m \\ d\sigma_{12}^m \end{Bmatrix} = \begin{bmatrix} a_{11} & a_{12} & a_{13} \\ 0 & a_{22} & a_{23} \\ 0 & 0 & a_{33} \end{bmatrix} \begin{Bmatrix} d\sigma_{11}^f \\ d\sigma_{22}^f \\ d\sigma_{12}^f \end{Bmatrix} = [A_{ij}] \begin{Bmatrix} d\sigma_{11}^f \\ d\sigma_{22}^f \\ d\sigma_{12}^f \end{Bmatrix} \tag{3.76}$$

where

$$[A_{ij}] = \begin{bmatrix} a_{11} & a_{12} & a_{13} \\ 0 & a_{22} & a_{23} \\ 0 & 0 & a_{33} \end{bmatrix} \tag{3.77}$$

is a 2D instantaneous bridging matrix, whose dependent elements can be determined as done similarly for a 3D bridging matrix. Due to Eq. (3.34.2) or Eq. (3.36.2), the independent element a_{21} has been set to zero in Eq. (3.77). It is noted that compared with a 2D elastic bridging matrix, Eq. (3.32), two other dependent elements, a_{13} and a_{23}, have been introduced to account for coupling of elastic-plastic behavior. The non-zero independent elements are given by,

$$a_{11} = E_m / E_{f1} \tag{3.78.1}$$

$$a_{22} = \beta + (1-\beta)E_m / E_{f2} \tag{3.78.2}$$

$$a_{33} = \alpha + (1-\alpha)G_m / G_f \tag{3.78.3}$$

Definitions for E_m, G_m, E_{f1}, E_{f2} and G_f are the same as those given by Eqs. (3.73.1) – (3.73.5). However, due to a planar stress state, the von Mises equivalent stresses, σ_e^f and σ_e^m, which are much easier to be calculated through Eq. (2.33), are used instead of an octahedral shear stress to express a yield condition. For instance, a matrix yield condition in terms of σ_e^m is expressed as

$$\sigma_e^m = \sqrt{(\sigma_{11}^m)^2 + (\sigma_{22}^m)^2 - (\sigma_{11}^m)(\sigma_{22}^m) + 3(\sigma_{12}^m)^2} \geq \sigma_Y^m \tag{3.79}$$

Substituting the bridging matrix, $[A_{ij}]$, defined in Eq. (3.77) into the overall compliance matrix of the composite, i.e.,

$$[S_{ij}]=(V_f[S_{ij}^f]+V_m[S_{ij}^m][A_{ij}])(V_f[I]+V_m[A_{ij}])^{-1}=(V_f[S_{ij}^f]+V_m[S_{ij}^m][A_{ij}])[B_{ij}]$$

and by requiring the resulting compliance matrix to be symmetric, it is found that

$$a_{12}=(S_{12}^f - S_{12}^m)(a_{11}-a_{22})/(S_{11}^f - S_{11}^m) \tag{3.80.1}$$

$$a_{13} = \frac{d_2\beta_{11} - d_1\beta_{21}}{\beta_{11}\beta_{22} - \beta_{12}\beta_{21}} \tag{3.80.2}$$

$$a_{23} = \frac{d_1\beta_{22} - d_2\beta_{12}}{\beta_{11}\beta_{22} - \beta_{12}\beta_{21}} \tag{3.80.3}$$

$$d_1 = (S_{13}^m - S_{13}^f)(a_{11} - a_{33}) \tag{3.80.4}$$

$$d_2 = (S_{23}^m - S_{23}^f)(V_f + V_m a_{11})(a_{22} - a_{33}) + (S_{13}^m - S_{13}^f)(V_f + V_m a_{33})a_{12} \tag{3.80.5}$$

$$\beta_{11} = S_{12}^m - S_{12}^f,\ \beta_{12} = S_{11}^m - S_{11}^f,\ \ \beta_{22} = (V_f + V_m a_{22})(S_{12}^m - S_{12}^f) \tag{3.80.6}$$

$$\beta_{21} = V_m(S_{12}^f - S_{12}^m)a_{12} - (V_f + V_m a_{11})(S_{22}^f - S_{22}^m) \tag{3.80.7}$$

It should be noted that the constituent instantaneous compliance matrices, $[S_{ij}^f]$ and $[S_{ij}^m]$, must be also two-dimensional (Eq. (2.31)). For the matrix material for instance, its instantaneous compliance matrix, according to the Prandtl-Reuss theory, is defined as

$$\left[S_{ij}^m\right] = \begin{cases} \left[S_{ij}^m\right]^e, \text{when } \ \sigma_e^m \leq \sigma_Y^m \\ \left[S_{ij}^m\right]^e + \left[S_{ij}^m\right]^p, \text{when } \ \sigma_e^m > \sigma_Y^m \end{cases} \tag{3.81.1}$$

where

$$[S_{ij}^m]^e = \begin{bmatrix} \frac{1}{E^m} & -\frac{\nu^m}{E^m} & 0 \\ & \frac{1}{E^m} & 0 \\ & \text{symmetry} & \frac{1}{G^m} \end{bmatrix} \tag{3.81.2}$$

$$[S_{ij}^m]^p = \frac{9}{4M_T^m(\sigma_e^m)^2}\begin{bmatrix} \sigma'_{11}\sigma'_{11} & \sigma'_{22}\sigma'_{11} & 2\sigma'_{12}\sigma'_{11} \\ & \sigma'_{22}\sigma'_{22} & 2\sigma'_{12}\sigma'_{22} \\ & \text{symmetry} & 4\sigma'_{12}\sigma'_{12} \end{bmatrix}_{\sigma_{ij}=\sigma_{ij}^m} \quad (3.81.3)$$

$$M_T^m = \frac{E^m E_T^m}{E^m - E_T^m} \quad (3.81.4)$$

$$\sigma'_{ij} = \sigma_{ij} - \frac{1}{3}(\sigma_{11}+\sigma_{22})\delta_{ij},\ i,j=1,2 \quad (3.81.5)$$

The internal stress increments can be related to the overall applied incremental stresses through

$$\begin{Bmatrix} \mathrm{d}\sigma_{11}^f \\ \mathrm{d}\sigma_{22}^f \\ \mathrm{d}\sigma_{12}^f \end{Bmatrix} = \begin{bmatrix} b_{11} & b_{12} & b_{13} \\ 0 & b_{22} & b_{23} \\ 0 & 0 & b_{33} \end{bmatrix}\begin{Bmatrix} \mathrm{d}\sigma_{11} \\ \mathrm{d}\sigma_{22} \\ \mathrm{d}\sigma_{12} \end{Bmatrix} = [B_{ij}]\begin{Bmatrix} \mathrm{d}\sigma_{11} \\ \mathrm{d}\sigma_{22} \\ \mathrm{d}\sigma_{12} \end{Bmatrix} \quad (3.82.1)$$

$$\begin{Bmatrix} \mathrm{d}\sigma_{11}^m \\ \mathrm{d}\sigma_{22}^m \\ \mathrm{d}\sigma_{12}^m \end{Bmatrix} = \begin{bmatrix} a_{11} & a_{12} & a_{13} \\ 0 & a_{22} & a_{23} \\ 0 & 0 & a_{33} \end{bmatrix}\begin{bmatrix} b_{11} & b_{12} & b_{13} \\ 0 & b_{22} & b_{23} \\ 0 & 0 & b_{33} \end{bmatrix}\begin{Bmatrix} \mathrm{d}\sigma_{11} \\ \mathrm{d}\sigma_{22} \\ \mathrm{d}\sigma_{12} \end{Bmatrix} = [A_{ij}][B_{ij}]\begin{Bmatrix} \mathrm{d}\sigma_{11} \\ \mathrm{d}\sigma_{22} \\ \mathrm{d}\sigma_{12} \end{Bmatrix} \quad (3.82.2)$$

where

$$b_{11}=(V_f+V_m a_{22})(V_f+V_m a_{33})/c \quad (3.82.3)$$

$$b_{12}=-(V_m a_{12})(V_f+V_m a_{33})/c \quad (3.82.4)$$

$$b_{13}=[(V_m a_{12})(V_m a_{23})-(V_f+V_m a_{22})(V_m a_{13})]/c \quad (3.82.5)$$

$$b_{22}=(V_f+V_m a_{11})(V_f+V_m a_{33})/c \quad (3.82.6)$$

$$b_{23}=-(V_m a_{23})(V_f+V_m a_{11})/c \quad (3.82.7)$$

$$b_{33}=(V_f+V_m a_{22})(V_f+V_m a_{11})/c \quad (3.82.8)$$

$$c=(V_f+V_m a_{11})(V_f+V_m a_{22})(V_f+V_m a_{33}) \quad (3.82.9)$$

Apparently, when both the fiber and the matrix are in elastic deformation, one has $S_{13}^m = S_{23}^m = S_{13}^f = S_{23}^f = 0$. Substituting them into Eqs. (3.80.4) and (3.80.5) and then the resulting d_1 and d_2 into Eqs. (3.80.2) and (3.80.3), one obtains $a_{13}=a_{23}=0$. Thus, an elastic-plastic bridging matrix given by Eq. (3.77) together with Eq. (3.78) becomes the same as an elastic one specified using Eqs. (3.32), (3.33) and (3.36).

3.10 Thermal Analysis

In general, the thermal expansion coefficients of the fiber material in a composite do not match those of the matrix material. As a matter of fact, internal stresses will be generated in both the fibers and the matrix whenever the resulting composite is subjected to a temperature variation, even if no external load is applied to the composite. This is because the fibers in the composite are completely constrained by the enclosing matrix, having no ability of free thermal expansion or contraction. The resulting stresses are called thermal stresses. In reality, composites may have a surviving temperature more or less different from their processing temperature in fabrication. This is especially true for metal matrix and thermoplastic polymer matrix composites, which are generally fabricated at temperatures higher than room temperature. Furthermore, many composites such as metal and ceramic matrix composites may have a rather high working temperature, and even sustain a temperature cycling condition. Thus, thermal stresses can generally occur in composites. These thermal stresses can influence the mechanical performance, especially the inelastic and strength behavior of the composites. In order to simulate the mechanical properties of the composites accurately, the thermal stresses in the composites must be fully identified.

Let us suppose that the working temperature of a UD composite, T_1, is different from a reference temperature, T_0, at which the internal stresses in both the fibers and the matrix are already known. For example, T_0 can be the fabrication temperature at which all of the internal stresses are free (of zero). Due to the different thermal expansion coefficients of the fibers and matrix, thermal stresses will be generated in the constituent materials during the temperature variation, $\mathrm{d}T=T_1-T_0$. The general constitutive equations of the fibers, matrix and the composite are then modified to (Huang, 2000f, 2000g, 2001d)

$$\{\mathrm{d}\varepsilon_i^f\} = [S_{ij}^f]\{\mathrm{d}\sigma_j^f\} + \{\alpha_i^f\}\mathrm{d}T \tag{3.83.1}$$

$$\{\mathrm{d}\varepsilon_i^m\} = [S_{ij}^m]\{\mathrm{d}\sigma_j^m\} + \{\alpha_i^m\}\mathrm{d}T \tag{3.83.2}$$

and

$$\{\mathrm{d}\varepsilon_i\} = [S_{ij}]\{\mathrm{d}\sigma_j\} + \{\alpha_i\}\mathrm{d}T \tag{3.83.3}$$

where α_i^f, α_i^m and α_i, respectively, are the thermal expansion coefficients of the fibers, matrix and the composite at the initial temperature T_0. Furthermore, it should be noted that the compliance matrices in Eqs. (3.83.1) and (3.83.2), $[S_{ij}^f]$ and $[S_{ij}^m]$, are also defined at the initial temperature T_0.

Now, let us consider the internal thermal stresses in the fiber and matrix materials. The internal stresses only due to this temperature variation can be represented as

$$\{d\sigma_i^f\}^{(T)} = \{b_i^f\}dT \tag{3.84.1}$$

$$\{d\sigma_i^m\}^{(T)} = \{b_i^m\}dT \tag{3.84.2}$$

where $\{b_i^f\}$ and $\{b_i^m\}$ are called thermal stress concentration factors of the fiber and the matrix materials, respectively. As the UD composite is free of an external stress, it follows that

$$V_f\{b_i^f\} + V_m\{b_i^m\} = \{0\} \tag{3.85}$$

Using the bridging matrix, $[A_{ij}]$, the internal strains in the constituents due to combined thermal-mechanical loads can be related to the overall strains of the composite through

$$\begin{aligned}\{d\varepsilon_i^f\} &= [S_{ij}^f](V_f[S_{ij}^f] + V_m[S_{ij}^m][A_{ij}])^{-1}\{d\sigma_j\} + ([S_{ij}^f]\{b_j^f\} + \{\alpha_i^f\})dT \\ &= [A_{ij}^f]\{d\sigma_j\} + ([S_{ij}^f]\{b_j^f\} + \{\alpha_i^f\})dT\end{aligned} \tag{3.86.1}$$

$$\begin{aligned}\{d\varepsilon_i^m\} &= [S_{ij}^m][A_{ij}](V_f[S_{ij}^f] + V_m[S_{ij}^m][A_{ij}])^{-1}\{d\sigma_j\} + ([S_{ij}^m]\{b_j^m\} + \{\alpha^m\})dT \\ &= [A_{ij}^m]\{d\sigma_j\} + ([S_{ij}^m]\{b_j^m\} + \{\alpha_i^m\})dT\end{aligned} \tag{3.86.2}$$

where $[A_{ij}^f]$ and $[A_{ij}^m]$ are called the strain concentration matrices of the fiber and matrix materials, respectively. Substituting Eqs. (3.84.1) and (3.84.2) into Eqs. (3.83.1) and (3.83.2), respectively, and then taking a volume average from the resulting equations, the thermal expansion coefficients of the composite, $\{\alpha_i\}$, can be obtained as

$$\{\alpha_i\} = V_f([S_{ij}^f]\{b_j^f\} + \{\alpha_i^f\}) + V_m([S_{ij}^m]\{b_j^m\} + \{\alpha_i^m\})$$

Here, the condition of free overall stresses, i.e., $\{d\varepsilon_i\} = \{\alpha_j\}dT$, has been employed. By making use of Eq. (3.85), the last equation becomes

$$\{\alpha_i\} = V_f([S_{ij}^f] - [S_{ij}^m])\{b_j^f\} + V_f\{\alpha_i^f\} + V_m\{\alpha_i^m\} \tag{3.87}$$

In Eqs. (3.84) – (3.87) there is still one set of variables, $\{b_i^f\}$, to be specified. To completely address the problem, we need another set of equations which can be obtained based on some thermal dynamics consideration. Fortunately, the problem has already been completely solved in the literature. Levin (1967) and Benveniste and Dvorak (1990) independently derived rigorous expressions for the overall thermal expansion coefficients of the composites by making use of the strain concentration matrices, $[A_{ij}^f]$ and $[A_{ij}^m]$ (Eqs. (3.86.1) and (3.86.2)), both of which are applicable with the bridging model. Let us make use of Benveniste and

Dvorak's exact expression for a thermal stress concentration factor. In terms of the bridging matrix, Benveniste and Dvorak's expression is given by (Huang, 2001d)

$$\{b_i^m\} = ([I]-[A_{ij}][B_{ij}])([S_{ij}^f]-[S_{ij}^m])^{-1}(\{\alpha_j^m\}-\{\alpha_j^f\}) \tag{3.88.1}$$

or

$$\{b_i^f\} = \frac{V_m}{V_f}\left([I]-[A_{ij}](V_f[I]+V_m[A_{ij}])^{-1}\right)([S_{ij}^f]-[S_{ij}^m])^{-1}(\{\alpha_j^f\}-\{\alpha_j^m\}) \tag{3.88.2}$$

If the temperature variation, $[T_0, T_1]$, is large, a subdivision may be necessary. The final thermal stresses, $\{\sigma_i^f\}^{(T)}$ and $\{\sigma_i^m\}^{(T)}$, are obtained by adding up all those stress increments from each sub-interval. These quantities will serve as the initial stresses in Eqs. (3.75.1) and (3.75.2), if a further mechanical load is applied to the composite.

The total stresses in the constituent materials are obtained by the summation of, respectively, the mechanical, thermal, as well as residual, if any, stress components. Let us assume that the final temperature is applied to the composite before applying any mechanical load and that the thermal stresses, $\{\sigma_i^f\}^{(T)}$ and $\{\sigma_i^m\}^{(T)}$, have been obtained. The total stresses in the fibers and the matrix at any mechanical load level are given by

$$\{\sigma_i^f\}^{K+1} = \{\sigma_i^f\}^{(T)} + \{\sigma_i^f\}^{(R)} + \{\sigma_i^f\}^{(M),\ K+1} \tag{3.89.1}$$

$$\{\sigma_i^m\}^{K+1} = \{\sigma_i^m\}^{(T)} + \{\sigma_i^m\}^{(R)} + \{\sigma_i^m\}^{(M),\ K+1} \tag{3.89.2}$$

where

$$\{\sigma_i^f\}^{(M),\ K+1} = \{\sigma_i^f\}^{(M),\ K} + \{d\sigma_i^f\}^{(M)},\ K=0,1\ldots \text{ with } \{\sigma_i^f\}^{(M),\ 0} = \{0\} \tag{3.89.3}$$

$$\{\sigma_i^m\}^{(M),\ K+1} = \{\sigma_i^m\}^{(M),\ K} + \{d\sigma_i^m\}^{(M)},\ K=0,1\ldots \text{ with } \{\sigma_i^m\}^{(M),\ 0} = \{0\} \tag{3.89.4}$$

In Eqs. (3.89.1) and (3.89.2), the superscripts T, R and M refer to the temperature, residual and externally applied mechanical loads, respectively. If no other but only the thermal residual stresses exist, $\{\sigma_i^f\}^{(R)}$ and $\{\sigma_i^m\}^{(R)}$ are also calculated based on Eqs. (3.84.1) and (3.84.2) (note that $\{b_i^f\}=-V_m\{b_i^m\}/V_f$ because of Eq. (3.85)). For instance, if a stress-free temperature (such as fabrication temperature) is T_0 which is higher than an initial working temperature (e.g., room temperature) T_L, thermal residual stresses $\{\sigma_i^f\}^{(R)}$ and $\{\sigma_i^m\}^{(R)}$ will be generated when the composite cools down from T_0 to T_L. Furthermore, when a current working temperature T_U is higher than T_L, thermal stresses, $\{\sigma_i^f\}^{(T)}$ and

$\{\sigma_i^m\}^{(T)}$, will be induced. Supposing T_U=T_0 and both of the constituents are in elastic deformations, we would have $\{\sigma_i^f\}^{(R)}$ + $\{\sigma_i^f\}^{(T)}$ ={0} and $\{\sigma_i^m\}^{(R)}$ + $\{\sigma_i^m\}^{(T)}$ ={0}. It should be noted that the mechanical load increments, $\{d\sigma_i^f\}^{(M)}$ and $\{d\sigma_i^m\}^{(M)}$, are calculated from Eqs. (3.8.1) and (3.8.2) or Eqs. (3.68.1) and (3.68.2) with constituent properties specified at the final temperature, T_1.

Eqs. (3.89.1) and (3.89.2) are uncoupled thermo-mechanical stress formulae. Namely, the thermal stresses have been evaluated before applying any mechanical load, whereas the mechanical stresses are later calculated without changing any temperature. On the other hand, we can also obtain coupled formulae for the total internal stresses in the constituent materials, which may be useful in the analysis of some specific problems such as an in-phase or out-of-phase fatigue. This can be accomplished by using the following equations (refer to Eqs. (3.68.1) and (3.68.2))

$$\{d\sigma_i^f\} = (V_f[I] + V_m[A_{ij}])^{-1}\{d\sigma_j\} + \{b_i^f\}dT \tag{3.90.1}$$

$$\{d\sigma_i^m\} = [A_{ij}](V_f[I] + V_m[A_{ij}])^{-1}\{d\sigma_j\} + \{b_i^m\}dT \tag{3.90.2}$$

Namely, the current total stresses are given by

$$\left\{\sigma_i^f\right\}^{K+1} = \left\{\sigma_i^f\right\}^{(R)} + \left\{\sigma_i^f\right\}^{(M-T),K+1} \tag{3.91.1}$$

$$\left\{\sigma_i^m\right\}^{K+1} = \left\{\sigma_i^m\right\}^{(R)} + \left\{\sigma_i^m\right\}^{(M-T),K+1} \tag{3.91.2}$$

where

$$\left\{\sigma_i^f\right\}^{(M-T),K+1} = \left\{\sigma_i^f\right\}^{(M-T),K} + \left\{d\sigma_i^f\right\}^{(M-T)}, K=0,1,\ldots \text{ with } \left\{\sigma_i^f\right\}^{(M-T),0} = \{0\} \tag{3.91.3}$$

$$\left\{\sigma_i^m\right\}^{(M-T),K+1} = \left\{\sigma_i^m\right\}^{(M-T),K} + \left\{d\sigma_i^m\right\}^{(M-T)}, K=0,1,\ldots, \text{ with } \left\{\sigma_i^m\right\}^{(M-T),0} = \{0\} \tag{3.91.4}$$

The superscript "*M–T*" indicates that the quantity involved results from the coupled thermo-mechanical effect and the stress increments, $\{d\sigma_i^f\}^{(M-T)}$ and $\{d\sigma_i^m\}^{(M-T)}$, are calculated from Eqs. (3.90.1) and (3.90.2), rather than from Eqs. (3.8.1) and (3.8.2) or Eqs. (3.68.1) and (3.68.2), with constituent properties specified at the current temperature, T. It must be pointed out that no matter whether Eqs. (3.8.1) and (3.8.2) (or Eqs. (3.68.1) and (3.68.2)) or Eqs. (3.90.1) and (3.90.2) are employed, the overall stress increments involved, $\{d\sigma_j\}$, are the externally applied mechanical loads on the composite, which are the same for both sets of equations. Without them, the composite is said to be stress-free. However, the constituents of the composite are not necessarily stress-free. Furthermore, the internal stresses calculated from Eq. (3.89.1) or Eq. (3.89.2) can be different from those calculated from Eq. (3.91.1) or Eq. (3.91.2) in general, even though the same amount of mechanical load has been applied. This is because Eqs. (3.89.1) and (3.89.2) give the internal stresses always pertaining to the final temperature, whereas Eqs. (3.91.1) and (3.91.2) give the stresses corresponding to the current temperature.

References

Benveniste, Y. & Dvorak, G.J. (1990) On a Correspondence between Mechanical and Thermal Effects in Two-Phase Composites, in The Toshio Muta Anniversary Volume: Micromechanics and Inhomogeneity (pp. 65-81). G.J. Weng, M. Taya, & H. Abe (eds.). New York: Springer.

Benveniste, Y. (1987) A new approach to the application of Mori-Tanaka's Theory in Composite Materials. Mechanics of Materials 6, 147-157.

Bogdanovich, A.E. & Pastore, C.M. (1996) Mechanics of Textile and Laminated Composites with Applications to Structural Analysis, Chapter 3: Elastic Properties of Fiber Reinforced Composites. London: Chapman & Hall.

Bogdanovich, A.E. & Sierakowski, R.L. (1999) Composite materials and structures: Science, technology and applications: a compendium of books, review papers and other sources of information. Applied Mechanics Reviews 52(12), 351-365.

Chamis, C.C. & Sendeckyj G.P. (1968) Critique on theories predicting thermoelastic properties of fibrous composites. J. Composite Materials 2, 332-358.

Christensen, R.M. (1991) Mechanics of composite materials. Malabar: Krieger Pub. Co.

Halpin, J.C. (1992) Primer on composite materials analysis (2nd ed.). Lancaster, Basel: Technomic Publishing Co., Inc.153-192.

Hashin, Z. (1964) Theory of mechanical behavior of heterogeneous media. Applied Mechanics Reviews 17, 1-39.

Hashin, Z. (1983) Analysis of composite materials—a survey. J. Applied Mechanics 50, 481-505.

Huang, Z. M., Ramakrishna, S., & Tay, A. O. A. (1999) A micromechanical approach to the tensile strength of a knitted fabric composite. J. Comp. Mater 33, 1758-1791.

Huang, Z.M. (2000a) A unified micromechanical model for the mechanical properties of two constituent composite materials, Part I: Elastic behavior. J. Thermoplastic Comp. Mater. 13(4), 252-271.

Huang, Z.M. (2000b) A unified micromechanical model for the mechanical properties of two constituent composite materials, Part II: Plastic behavior. J. Thermoplastic Comp. Mater. 13(5), 344-362.

Huang, Z.M. (2000c) A unified micromechanical model for the mechanical properties of two constituent composite materials, Part IV: Rubber-Elastic behavior. J. of Thermoplastic Comp. Mater. 13(2), 119-139.

Huang, Z.M. (2000d) A unified micromechanical model for the mechanical properties of two constituent composite materials, Part V: Laminate strength. J. Thermoplastic Comp. Mat. 13(3), 190-206.

Huang, Z.M. (2000e) Simulation of inelastic response of multidirectional laminates based on stress failure criteria. Mater. Sci. Tech. 16(6), 692-698.

Huang, Z.M. (2000f) Tensile strength of fibrous composites at elevated temperature. Mater. Sci. Tech. 16(1), 81-94.

Huang, Z.M. (2000g) Strength formulae of unidirectional composites including thermal residual stresses. Materials Letters 43(1-2), 36-42.

Huang, Z.M. (2001a) A unified micromechanical model for the mechanical properties of two constituent composite materials, Part III: Strength behavior. J. Thermoplastic. Comp. Mater. 14(1), 54-69.

Huang, Z.M. (2001b) Micromechanical prediction of ultimate tensile strength of transversely isotropic fibrous composites. Int. J. Solids & Struct. 38(22-23), 4147-4172.

Huang, Z.M. (2001c) Simulation of the mechanical properties of fibrous composites by the bridging micromechanics model. Composites Part A, 32(2), 143-172.

Huang, Z.M. (2001d) Modeling strength of multidirectional laminates under thermo-mechanical loads. J. Comp. Mater. 35(4), 281-315.

Huang, Z.M., (2004) A bridging model prediction of the ultimate strength of composite laminates subjected to biaxial loads. Comp. Sci. & Tech. 64, 395-448.

Huang, Z.M. (2007) Inelastic and failure analysis of laminate structures by ABAQUS incorporated with a general constitutive relationship. Journal of Reinforced Plastics and Composites 26(11), 1135-1181.

Hyer, M.W. (1997) Stress analysis of fiber-reinforced composite materials (pp.120-124). Boston: WCB, McGraw-Hill.

Levin, V.M. (1967) On the coefficients of thermal expansion of heterogeneous materials. Mekhanika Tverdovo Tela, 1, 88-94 (in Russian).

McCullough, R.L. (1990) Micro-models for composite materials- continuous fiber composites, micromechanical materials modeling, Delaware Composites Design Encyclopedia (Vol. 2), M. Whitney James & Roy L. McCullough (eds.). Lancaster: Technomic Publishing Co., Inc.

Mori, T. & Tanaka, K. (1973) Average stress in Matrix and average energy of materials with misfitting Inclusion. Acta Metall 21, 571-574.

Soden, P.D., Hinton, M.J., & Kaddour, A.S. (1998) Lamina properties, Lay-up configurations and loading conditions for a range of fiber- reinforced composite laminates. Comp. Sci. Tech 58, 1011-1022.

Tsai, S.W. & Hahn, H.T. (1980) Introduction to Composite Materials. Lancaster, Basel: Technomic Publishing Co., Inc.

Wang, Y.M. & Weng, G.J. (1992) The influence of inclusion shape on the over viscoelastic behavior of composites. ASME J Appl Mech 59, 510-518.

Weng, G.J. (1984) Some elastic properties of reinforced solids with special reference to isotropic ones containing spherical inclusion. Int J Eng Sci 22(7), 845-856.

Zhang, H.S., Huang, Z.M. (2008) Micromechanics study on elastic-plastic behavior of fiber reinforced composites. Acta Materiae Compositae Sinica 25(5), 157-162 (in Chinese).

Zhang, Y.Z., Bini, T.B., Huang, Z.M., & Ramakrishna, S. (2000) Fracture characteristics of nnitted fabric composites under tensile load. Advanced Composites Letters 9(2), 133-137.

4

Strength of Unidirectional Composites

4.1 Introduction

It has been recognized that composite materials are being used widely and successfully in primary load bearing structures in many modern industrial fields. Thus, the strength of composites is an important issue to be addressed (Hinton & Soden, 1998). One must be confident in their load carrying capacity before making efficient use of them. The purpose of any stress analysis is to predict the conditions under which the materials will fail, and to determine the allowable external loads for a desired margin of safety.

Great efforts have been made over the last fifty years on the development of reliable methods to analyze the failure strength of composite materials. The majority of these methods or theories are based on a phenomenological approach to a UD lamina (Hinton & Soden, 1998). A failure criterion expressed with respect to the stresses applied on the lamina is formulated. In general, extensive experiments on the composite lamina are necessary in order to determine the critical strength parameters involved in the phenomenological or macromechanical strength theory. Such experiments may be difficult or expensive, and even impossible in some circumstances. Even with the same constituent materials, a different composite only having a different fiber volume fraction generally requires repeated tests (Rowlands, 1985). Thus, large databases, generally at great expense, have to be established for a composite design.

On the contrary, a micromechanical strength theory only makes use of mechanical properties of the constituent materials to predict the composite strength. It is developed based on an assumption that any composite failure is caused by the failure of a constituent material. Compared with its counterpart, a macromechanical strength theory, several advantages are apparent. First, fewer experiments and hence less cost will be incurred, since a repeated test on a composite will not be necessary if it is made of the same constituents whose properties are already known. Second, the failure mechanism (i.e., whether the failure is caused by the fiber or by the matrix) is clearly known, which will be useful for the development of a stiffness discount scheme used for a laminate strength prediction (see the next chapter for details). Furthermore, an optimal design for a composite load carrying capacity can be achieved. For instance, under a given load condition, such an optimal design may be obtained if the fiber and the matrix of the composite attain their failure simultaneously.

In this chapter, the bridging model introduced in Chapter 3 is applied to the development of micromechanics strength theory and is used to estimate the strength of UD composites subjected to various combinations of load conditions. If the composites are under a uniaxial load (longitudinal, transverse or in-plane shear load), explicit and closed-form formulae for the composite strengths are obtainable (Huang, 1999a, 2000d, 2001). As the determination of the internal stresses in the constituent fiber and matrix materials has already been shown, the present development begins with the introduction of stress failure criteria.

4.2 Failure Criteria

Failure occurs in a material when the applied load reaches a threshold that is the limit of its load carrying capacity. "Failure" can be defined in different ways according to the mechanical and deformation characteristics of the subject material and the surviving load requirements. For metals, "failure" is defined in most cases either as "yield" (for ductile materials) or "fracture" (for brittle materials). In any case, however, "failure" is directly related to the corresponding strength of the materials. The "yield failure" is controlled by the yield stress (yield strength) of the materials whereas the "fracture failure" is usually controlled by the ultimate stress (ultimate strength) of the material (an unstable crack-induced fracture problem is beyond the scope of our present concern). The occurrence of failure depends on the material properties and also on the loading configuration (uniaxial vs. multiaxial), loading function (e.g., quasi-static, cyclic), loading rate, temperature, etc. On the other hand, the strength is an intrinsic material property. For a composite, its strength depends on, among others, the strengths of the constituent materials as well as on the internal micro-structural geometry of the constituents.

4.2.1 Strength Theories for Isotropic Materials

When an isotropic material is subjected to uniaxial tensile loading, only one principal stress component exists and failure can be defined experimentally by a simple tensile test, which provides a characteristic stress-strain curve for the material. Failure is predicted to occur when the maximum normal stress reaches the yield or the ultimate point on the stress-strain curve. When the material is subjected to a multiaxial state of stress, however, the interaction among the various stress components makes prediction of failure much more difficult. A large number of multiaxial tests are required. Accounting for the entire range of all stress components, the cost and complexity of such testing prohibits a definitive experimental characterization of failure. In some cases, experimental duplication of actual loading conditions is even impossible. Therefore, strength theories have to be developed to predict the failure of materials under a multiaxial state of stress conditions.

For isotropic materials, all of the failure theories are applied in the form of material principal stresses (σ^1, σ^2 and σ^3, with $\sigma^1 \geq \sigma^2 \geq \sigma^3$). Given a general three-dimensional stress state (σ_{xx}, σ_{yy}, σ_{zz}, σ_{yz}, σ_{xz}, σ_{xy}) generated in a material, the principal stresses σ^1, σ^2 and σ^3 are the solutions to the following eigen-value equation

$$\det\left(\begin{bmatrix} \sigma_{xx} & \sigma_{xy} & \sigma_{xz} \\ \sigma_{yx} & \sigma_{yy} & \sigma_{yz} \\ \sigma_{zx} & \sigma_{zy} & \sigma_{zz} \end{bmatrix} - \sigma[I]\right) = 0 \tag{4.1}$$

If only the in-plane stress components, (σ_{xx}, σ_{yy}, σ_{xy}), appear in the material, one of the three principal stresses is negligible and the two others are calculated from

$$\sigma^{\mathrm{I}} = \frac{\sigma_{xx} + \sigma_{yy}}{2} + \frac{1}{2}\sqrt{(\sigma_{xx} - \sigma_{yy})^2 + 4(\sigma_{xy})^2} \tag{4.2.1}$$

$$\sigma^{\mathrm{II}} = \frac{\sigma_{xx} + \sigma_{yy}}{2} - \frac{1}{2}\sqrt{(\sigma_{xx} - \sigma_{yy})^2 + 4(\sigma_{xy})^2} \tag{4.2.2}$$

Comparing the stress values among σ^{I}, σ^{II} and zero, the three principal stresses σ^1, σ^2 and σ^3 can be arranged. With these principal stresses, the two most widely used strength theories for isotropic materials are expressed below.

Maximum Normal Stress Theory

This theory postulates that a material fails as soon as the maximum normal stress, σ^1, generated in the material reaches its ultimate value, no matter whether the material is under a uniaxial or multiaxial state of stress. The failure criterion can be expressed as

$$\sigma^1 \geq \sigma_{\mathrm{u}} \tag{4.3}$$

where σ_{u} is the ultimate tensile stress (tensile strength) of the material obtained from a uniaxial test.

Historically, this theory was developed for governing the strength of brittle materials (Timoshenko, 1953). For this kind of materials, there is no noticeable plastic deformation. Under a uniaxial tensile load, the stress-strain curve of a brittle material is monotonically increased to a peak point that corresponds to the ultimate stress and ultimate strain of the material. If we agree to accept that the failure occurs when $\frac{\mathrm{d}P}{\mathrm{d}\varepsilon} = 0$ (Cooper, 1974), where P is the load applied on the material and ε is the resulting strain, i.e., when the material is unable to take any more load, we can easily appreciate the background to this theory.

However, even for ductile materials for which $\frac{\mathrm{d}P}{\mathrm{d}\varepsilon} = 0$ may occur at another point, the yield point on the stress-strain curve, we can still use Eq. (4.3) to predict the ultimate strength of the materials after yield failure, since the σ^1 involved does not depend on material properties. Therefore, Eq. (4.3) can be used to govern the ultimate tensile strength of any isotropic material.

Maximum Distorted Energy Theory (von Mises-Hencky Theory)

This theory was proposed for governing the yield failure of ductile materials such as ductile metals. When such materials yield, a distinguishable phenomenon, called the plastic incompressibility condition, is involved, i.e., the volumetric strain of the materials in the plastic regime is nearly zero. Hence, no plastic work can be done by the hydrostatic component (mean normal stresses) of the applied stress field (Section 2.2). If we separate the total strain energy into two parts, the volume strain energy part and the distorted strain energy part, we can say that only the distorted strain energy part results in the material yield.

Subtracting the strain energy caused by the change in volume from the total energy expression, we obtain an equivalent stress to characterize the distorted energy (in terms of principal stresses) as

$$\sigma_{\text{eq}} = \sqrt{\frac{1}{2}\left[(\sigma^1 - \sigma^2)^2 + (\sigma^2 - \sigma^3)^2 + (\sigma^3 - \sigma^1)^2\right]} \tag{4.4.1}$$

where the subscript "eq" stands for "equivalent". The maximum distorted energy theory postulates that no matter whether a ductile material is under a uniaxial or multiaxial state of stress, the yield failure of the material occurs if its equivalent stress, defined by Eq. (4.4.1), attains a limit value σ_{Y}. The failure criterion is thus

$$\sigma_{\text{eq}} \geq \sigma_{\text{Y}} \tag{4.4.2}$$

where σ_{Y} is the yield strength of the material corresponding to a uniaxial loading test.

It is seen that the von Mises-Hencky theory has been already incorporated in the development of composite plastic theory.

Generalized Maximum Normal Stress Theory

It should be pointed out that the maximum normal stress failure criterion is efficient when the three principal stresses of the material differ from each other distinctly. If, however, two or three of them become equal, the accuracy of this criterion is certainly questionable. Otherwise, a material would be able to sustain the same amount of equibiaxial or equitriaxial tension as it did uniaxial tension. Phenomenologically, this might be impossible. The load carrying capacity of an isotropic material when subjected to an equibiaxial or equitriaxial tension should be lowered to some extent compared to a uniaxial tension. Recognizing this fact, a generalized criterion (Huang, 2000a, 2001, 2004) to detect tensile failure of the material is expressed as

$$\sigma_{\mathrm{eq}} \geq \sigma_{\mathrm{u}} \tag{4.5.1}$$

where
$$\sigma_{\mathrm{eq}} = \begin{cases} \sigma^{1}, \ \text{when} \ \sigma^{3} < 0 \\ [(\sigma^{1})^{q} + (\sigma^{2})^{q}]^{\frac{1}{q}}, \ \text{when} \ \sigma^{3} = 0 \\ [(\sigma^{1})^{q} + (\sigma^{2})^{q} + (\sigma^{3})^{q}]^{\frac{1}{q}} \ \text{when} \ \sigma^{3} > 0, \ 1 < q \leq \infty \end{cases} \tag{4.5.2}$$

The power-index q, a real number greater than 1, can be determined through experiment. It is seen that when the power-index $q=\infty$, Eq. (4.5.1) together with Eq. (4.5.2) is equivalent to the classical maximum normal stress criterion, Eq. (4.3). For this reason, the condition represented by Eq. (4.5) is named as a generalized maximum normal stress criterion. In fact, the difference between the generalized and the classical maximum normal stress criteria is distinct only when the second or the third principal stress of the material is close to its first principal stress.

In contrast to a multiaxial tension which results in a reduction of the material load carrying ability, a tri-axial compression will increase that ability to a certain level. One evident example is that no failure will occur if an isotropic material is subjected to a tri-axially equal compression. To compensate for this increased ability, a generalized maximum compressive stress failure criterion to govern the compressive failure of the material under any compressive load condition is simply given by

$$\sigma_{\mathrm{eq}}^{c} \leq (-\sigma_{\mathrm{u},c}) \tag{4.6.1}$$

where
$$\sigma_{\mathrm{eq}}^{c} = \begin{cases} \sigma^{3}, \quad \text{when} \quad \sigma^{1} > 0, \\ \sigma^{3} - \sigma^{1}, \text{when} \quad \sigma^{1} \leq 0 \end{cases} \tag{4.6.2}$$

where $\sigma_{\mathrm{u},c}$ is the ultimate compressive strength of the material under a uniaxial load. Eq. (4.6.2) is used to take into account an enhancement in the material load-carrying capacity due to a tri-axial compression. It must be pointed out that if only in-plane (or two-dimensional) loads are considered, the first principal stress σ^1 is always larger than, or equal to, 0, i.e., σ^c_{eq} is always equal to the stress σ^3 in Eq. (4.6.2). It should be further noted that no material buckling is assumed in the compression concerned here.

4.2.2 Composite Strength Theories

As for composite materials, strength theories are much more abundant. Most of them are developed phenomenologically. These theories treat composites as general anisotropic materials and can be considered more or less as generalizations from the corresponding failure theories of isotropic materials. In general, these theories are directly applied to the stress components of the composite laminae, but in their local (or material) coordinate system. Namely, the coordinate x_1 is always along the fiber axial direction. Several comprehensive surveys of these phenomenological strength theories exist in the literature (Rowlands, 1985; Nahas, 1986; Labossiere & Neal, 1987; Echaabi et al., 1996; Soden et al., 1998). Only two such typical theories are cited here. They are Hashin-Rotem and Tsai-Wu criteria. As most composites are subjected to plane stress states, only two-dimensional forms of these theories are presented here.

The Hashin-Rotem (Hashin & Rotem, 1973) criterion is expressed as

$$\max\left\{\left(\frac{\sigma_{11}}{X}\right)^2, \left(\frac{\sigma_{11}}{X'}\right)^2, \left(\frac{\sigma_{22}}{Y}\right)^2 + \left(\frac{\sigma_{12}}{S}\right)^2, \left(\frac{\sigma_{22}}{Y'}\right)^2 + \left(\frac{\sigma_{12}}{S}\right)^2\right\} \geq 1 \tag{4.7}$$

where X, X', Y, Y' and S are the longitudinal (in the fiber axial direction) tension, longitudinal compression, transverse (in the direction perpendicular to the fiber axes) tension, transverse compression and in-plane shear strengths of the UD lamina, respectively.

Furthermore, the Tsai-Wu (Tsai & Wu, 1971) criterion is expressed as

$$F_1(\sigma_{11})^2 + F_2(\sigma_{22})^2 + F_3\sigma_{11}\sigma_{22} + F_4(\sigma_{12})^2 + F_5\sigma_{11} + F_6\sigma_{22} \geq 1 \tag{4.8}$$

where $F_1 = \frac{1}{XX'}$, $F_2 = \frac{1}{YY'}$, $F_3 = -\sqrt{F_1F_2}$, $F_4 = \frac{1}{S^2}$, $F_5 = \frac{X'-X}{XX'}$ and $F_6 = \frac{Y'-Y}{YY'}$ (Tsai & Hahn, 1980).

It is seen that many more experiments will have to be performed in order to determine the critical strength parameters involved in the Hashin-Rotem and Tsai-Wu criteria. Even worse, different laminae with the same constituent materials but only having different fiber content (i.e., different fiber volume fraction) still require the same set of repeated tests, as the composite critical strength parameters also depend significantly on the fiber volume fraction.

Since both the internal stresses in the constituent materials and the overall stresses on the composite lamina are explicitly known by employing the bridging model, any strength theory either applied to the constituents, such as the maximum normal stress criterion, or applied to the composite lamina, such as the Hashin-Rotem and Tsai-Wu criteria, can be incorporated to determine the maximum allowable load. The composite strength is thus defined accordingly.

4.3 Strength Formulae under Uniaxial Loads

In Chapters 1 and 3, we have seen that the elastic properties (stiffness) of a UD composite can be easily calculated using explicit and concise formulae, such as Eqs. (1.38.1) – (1.38.5), (1.46.1) – (1.46.5), or Eqs. (3.55.1) – (3.55.5). In this section, we will present a set of similarly explicit and concise formulae for uniaxial strengths of the composite. The simplest one is the rule of mixture formula for the composite longitudinal strength (Bushby, 1998), which states that

$$\sigma_{11}^{u} = V_f \sigma_u^f + V_m \sigma_u^m \tag{4.9}$$

where σ_u^f is the fiber tensile strength (in the axial direction) and σ_u^m the matrix tensile strength. In Eq. (4.9), the composite strength only depends on the constituent strengths and the constituent contents. No relation is linked to the constituent stiffness (i.e., deformation ability). This may not be the case in general. Phenomenologically, the composite can be regarded as an indeterminate structural system between the fibers and the matrix. The load shared by each constituent is highly dependent on the deformability of both the constituents. Therefore, the composite strength should also depend on the constituent stiffness as well as on the constituent plastic parameters. The strength formulae presented in the following are mainly adapted from Huang (1999a, 2000a, 2000b).

For simplicity, let us assume that the fibers used are linearly elastic until rupture and the matrix is a bilinearly elastic-plastic material (Fig. 2.1). If both the constituent materials are in linear elastic deformation, the internal stresses in the fiber and matrix materials are explicitly related to the overall applied stresses on the composite through Eqs. (3.44.1) – (3.44.12). When the matrix has undergone a plastic deformation, the internal stress formulae, Eqs. (3.44.1) – (3.44.12), are generally no longer valid for an arbitrarily applied overall load. If, however, the composite is only subjected to a uniaxial load (e.g., a longitudinal, a transverse, or an in-plane shear load), the stress formulae are still applicable, although they should be in an incremental form. This can be confirmed by noticing Eqs. (2.31), (2.32) and (2.34) for an instantaneous compliance matrix of the matrix material. If there is no coupling between the normal and shear stresses, the plastic component, Eq. (2.34), will have the same structure as the elastic component, Eq. (2.32). Furthermore, Eqs. (3.80.4) and (3.80.5) show that $d_1=d_2=0$. Hence $a_{13}=a_{23}=0$, which means that the constituent internal stress formulae in the plastic region should be the same in form as Eqs. (3.44.1) – (3.44.12).

Thus, let the composite be subjected to only a longitudinal load, i.e., $\sigma_{11}\neq 0$ and all the other $\sigma_{ij}=0$. From Eqs. (3.44.11) and (3.44.12) and using an incremental form, we obtain

$$\mathrm{d}\sigma_{11}^{f}=\frac{\mathrm{d}\sigma_{11}}{V_f+V_m a_{11}} \quad \text{and} \quad \mathrm{d}\sigma_{11}^{m}=\frac{a_{11}\mathrm{d}\sigma_{11}}{V_f+V_m a_{11}} \tag{4.10}$$

We assume that a composite failure occurs whenever the fiber or the matrix attains its ultimate strength. Accordingly, the overall applied stress, σ_{11}, is defined as the composite longitudinal strength. Suppose, for example, σ_{11} is in tension and no residual stress occurs in the fiber and the matrix. From Eqs. (4.10) and (3.36.1) and supposing both the fiber and the matrix are in elastic deformation, it follows that

$$\sigma_{11}^{f}=\frac{\sigma_{11}}{V_f+V_m a_{11}}=\alpha_{e1}^{f}\sigma_{11}\le\sigma_{u}^{f} \quad \text{and} \quad \sigma_{11}^{m}=\frac{a_{11}\sigma_{11}}{V_f+V_m a_{11}}\mp\alpha_{t}^{m}\sigma_{1}\le\sigma_{Y}^{m}$$

which leads to

$$\sigma_{11}=\sigma_{11}^{0}=\min\left\{\frac{\sigma_{Y}^{m}}{\alpha_{e1}^{m}},\frac{\sigma_{u}^{f}}{\alpha_{e1}^{f}}\right\} \tag{4.11}$$

where
$$\alpha_{e1}^{f}=\frac{E_{11}^{f}}{V_f E_{11}^{f}+V_m E^{m}} \tag{4.12.1}$$

$$\alpha_{e1}^{m}=\frac{E^{m}}{V_f E_{11}^{f}+V_m E^{m}} \tag{4.12.2}$$

In most cases, the σ_{11}^{0} is not given by $\sigma_{u}^{f}/\alpha_{e1}^{f}$ and the composite can still sustain an additional load. The resulting internal stresses in the constituents should then fulfill the following conditions

$$\sigma_{11}^{f}=\alpha_{e1}^{f}\sigma_{11}^{0}+\alpha_{p1}^{f}(\sigma_{11}-\sigma_{11}^{0})\le\sigma_{u}^{f} \quad \text{and} \quad \sigma_{11}^{m}=\alpha_{e1}^{m}\sigma_{11}^{0}+\alpha_{p1}^{m}(\sigma_{11}-\sigma_{11}^{0})\le\sigma_{u}^{m} \tag{4.13}$$

where
$$\alpha_{p1}^{f}=\frac{E_{11}^{f}}{V_f E_{11}^{f}+V_m E_{T}^{m}} \tag{4.14.1}$$

$$\alpha_{p1}^{m} = \frac{E_T^m}{V_f E_{11}^f + V_m E_T^m} \tag{4.14.2}$$

From Eq. (4.13), the composite longitudinal tensile strength is derived as

$$\sigma_{11}^{u} = \min\left\{\frac{\sigma_u^f - (\alpha_{e1}^f - \alpha_{p1}^f)\sigma_{11}^0}{\alpha_{p1}^f}, \frac{\sigma_u^m - (\alpha_{e1}^m - \alpha_{p1}^m)\sigma_{11}^0}{\alpha_{p1}^m}\right\} \tag{4.15}$$

Now, let us only apply a transverse stress, $d\sigma_{22}$, to the composite. From Eqs. (3.44.1), (3.44.2), (3.44.11) and (3.44.12), the constituent internal stress components are obtained as

$$d\sigma_{22}^{f} = \frac{d\sigma_{22}}{V_f + V_m a_{22}}, \quad d\sigma_{11}^{f} = -\frac{V_m a_{12} d\sigma_{22}}{(V_f + V_m a_{11})(V_f + V_m a_{22})} \tag{4.16.1}$$

$$d\sigma_{22}^{m} = \frac{a_{22} d\sigma_{22}}{V_f + V_m a_{22}} \text{ and } d\sigma_{11}^{m} = \frac{V_f a_{12} d\sigma_{22}}{(V_f + V_m a_{11})(V_f + V_m a_{22})} \tag{4.16.2}$$

Evidently, the transverse stress components, i.e., $d\sigma_{22}^f$ and $d\sigma_{22}^m$ in both the constituents should be dominant. By neglecting the effect of the longitudinal stress components, we obtain the following formula for the composite transverse strength (subjected to only the transverse load).

$$\sigma_{22}^{u} = \min\left\{\frac{\sigma_u^f - (\alpha_{e2}^f - \alpha_{p2}^f)\sigma_{22}^0}{\alpha_{p2}^f}, \frac{\sigma_u^m - (\alpha_{e2}^m - \alpha_{p2}^m)\sigma_{22}^0}{\alpha_{p2}^m}\right\} \tag{4.17}$$

where

$$\sigma_{22}^{0} = \min\left\{\frac{\sigma_Y^m}{\alpha_{e2}^m}, \frac{\sigma_u^f}{\alpha_{e2}^f}\right\} \tag{4.18.1}$$

$$\alpha_{e2}^{f} = \frac{E_{22}^f}{V_f E_{22}^f + (1-V_f)[(1-\beta)E^m + \beta E_{22}^f]} \tag{4.18.2}$$

$$\alpha_{e2}^{m} = \frac{(1-\beta)E^m + \beta E_{22}^f}{V_f E_{22}^f + (1-V_f)[(1-\beta)E^m + \beta E_{22}^f]} \tag{4.18.3}$$

$$\alpha_{p2}^{f} = \frac{E_{22}^f}{V_f E_{22}^f + (1-V_f)[(1-\beta)E_T^m + \beta E_{22}^f]} \tag{4.18.4}$$

$$\alpha_{p2}^{m} = \frac{(1-\beta)E_T^m + \beta E_{22}^f}{V_f E_{22}^f + (1-V_f)[(1-\beta)E_T^m + \beta E_{22}^f]} \tag{4.18.5}$$

It should be noticed that the fiber ultimate strength, σ_u^f, occurring in Eqs. (4.17) and (4.18.1), should be understood to be that along the fiber transverse direction. However, in reality, the fiber transverse strength is difficult to obtain experimentally. Thus, the fiber longitudinal strength may have to be used instead. Fortunately, the composite failure under a transverse load is generally controlled by a matrix failure. So the fiber strength is actually immaterial here.

Similarly, if we apply only an in-plane shear stress, σ_{12}, to the composite, we can estimate the composite in-plane shear strength by making use of internal stress Eqs. (3.44.7) and (3.44.8). In such a case, however, the composite shear strength formula will depend on the failure criterion chosen to control the composite failure. If the maximum normal strength criterion, Eq. (4.3), is used and a von Mises equivalent stress, Eq. (2.33), of the matrix is checked with its yield stress under uniaxial tension, we obtain the following in-plane shear strength formula

$$\sigma_{12}^u = \min\left\{ \frac{\sigma_u^f - (\alpha_{e3}^f - \alpha_{p3}^f)\sigma_{12}^0}{\alpha_{p3}^f}, \frac{\sigma_u^m - (\alpha_{e3}^m - \alpha_{p3}^m)\sigma_{12}^0}{\alpha_{p3}^m} \right\} \tag{4.19}$$

where

$$\sigma_{12}^0 = \min\left\{ \frac{\sigma_Y^m}{\sqrt{3}\alpha_{e3}^m}, \frac{\sigma_u^f}{\alpha_{e3}^f} \right\} \tag{4.20.1}$$

$$\alpha_{e3}^f = \frac{G_{12}^f}{V_f G_{12}^f + (1-V_f)[(1-\alpha)G^m + \alpha G_{12}^f]} \tag{4.20.2}$$

$$\alpha_{e3}^m = \frac{(1-\alpha)G^m + \alpha G_{12}^f}{V_f G_{12}^f + (1-V_f)[(1-\alpha)G^m + \alpha G_{12}^f]} \tag{4.20.3}$$

$$\alpha_{p3}^f = \frac{G_{12}^f}{V_f G_{12}^f + (1-V_f)[(1-\alpha)G_T^m + \alpha G_{12}^f]} \tag{4.20.4}$$

$$\alpha_{p3}^m = \frac{(1-\alpha)G_T^m + \alpha G_{12}^f}{V_f G_{12}^f + (1-V_f)[(1-\alpha)G_T^m + \alpha G_{12}^f]} \tag{4.20.5}$$

On the other hand, if we use the constituent shear strengths to control constituent shear failure, we get

$$\sigma_{12}^u = \min\left\{ \frac{\tau_u^f - (\alpha_{e3}^f - \alpha_{p3}^f)\tau^0}{\alpha_{p3}^f}, \frac{\tau_u^m - (\alpha_{e3}^m - \alpha_{p3}^m)\tau^0}{\alpha_{p3}^m} \right\} \tag{4.21.1}$$

where

$$\tau^0 = \min\left\{ \frac{\tau_Y^m}{\alpha_{e3}^m}, \frac{\tau_u^f}{\alpha_{e3}^f} \right\} \tag{4.21.2}$$

τ_u^f is the fiber shear strength, τ_Y^m and τ_u^m are the yield and ultimate strengths of the matrix under pure shear load, respectively. G_T^m is the hardening modulus of the matrix under pure shear, which can be set to $E_T^m/3$ if only a tensile stress-strain curve is available. Strength formulae for other kinds of uniaxial as well as biaxial (not coupled with shear) loads can be derived in the same way.

The above strength formulae have not incorporated any thermal residual stress effect. However, if the composite is fabricated at a higher temperature than working temperature, the composite strength, especially the transverse strength at room temperature, can be apparently affected. Employing the Benrniste & Dvorak model (Section 3.10), the composite uniaxial strength formulae with thermal residual stress effect can be amended. Thus, let T_I and T_F be the initial and final temperatures of the composite respectively. At T_I, the constituents of the composite are free of stress. Let the interval $[T_I, T_F]$ be divided into N sub-intervals, in each of which the constituent thermal-mechanical properties remain unchanged. The total thermal stress components in the fiber and matrix are thus given by

$$\sigma_{11}^{f,0} = \sum_{k=1}^{N} (b_1^f)_k (\mathrm{d}T)_k \,, \quad \sigma_{22}^{f,0} = \sum_{k=1}^{N} (b_2^f)_k (\mathrm{d}T)_k \tag{4.22.1}$$

$$\sigma_{11}^{m,0} = \sum_{k=1}^{N} (b_1^m)_k (\mathrm{d}T)_k \quad \text{and} \quad \sigma_{22}^{m,0} = \sum_{k=1}^{N} (b_2^m)_k (\mathrm{d}T)_k \tag{4.22.2}$$

where b_i^f and b_i^m are obtained from Eqs. (3.88.2) and (3.88.1), respectively. If all of the temperature increments are the same, we have $(\mathrm{d}T)_k = \mathrm{d}T = const.$ Supposing that the thermal residual stresses, given by Eqs. (4.22), are below yield or failure strength of a constituent material, the ultimate tensile strengths of the composite under different uniaxial loads are amended as in the following.

Longitudinal tensile strength due to a longitudinal tensile load (σ_{11}) only

$$\sigma_{11}^u = \min\left\{ \frac{(\sigma_u^f - \sigma_{11}^{f,0}) - (\alpha_{e1}^f - \alpha_{p1}^f)\sigma_{11}^0}{\alpha_{p1}^f}, \frac{(\sigma_u^m - \sigma_{11}^{m,0}) - (\alpha_{e1}^m - \alpha_{p1}^m)\sigma_{11}^0}{\alpha_{p1}^m} \right\} \tag{4.23.1}$$

where $$\sigma_{11}^0 = \min\left\{ \frac{(\sigma_{22}^{m,0} - 2\sigma_{11}^{m,0}) + \sqrt{4(\sigma_Y^m)^2 - 3(\sigma_{22}^{m,0})^2}}{2\alpha_{e1}^m}, \frac{(\sigma_u^f - \sigma_{11}^{f,0})}{\alpha_{e1}^f} \right\} \tag{4.23.2}$$

Transverse tensile strength due to a transverse tensile load (σ_{22}) only

$$\sigma_{22}^{u} = \min\left\{\frac{(\sigma_{u}^{f} - \sigma_{22}^{f,0}) - (\alpha_{e2}^{f} - \alpha_{p2}^{f})\sigma_{22}^{0}}{\alpha_{p2}^{f}}, \frac{(\sigma_{u}^{m} - \sigma_{22}^{m,0}) - (\alpha_{e2}^{m} - \alpha_{p2}^{m})\sigma_{22}^{0}}{\alpha_{p2}^{m}}\right\} \quad (4.24.1)$$

where $$\sigma_{22}^{0} = \min\left\{\frac{(\sigma_{11}^{m,0} - 2\sigma_{22}^{m,0}) + \sqrt{4(\sigma_{Y}^{m})^{2} - 3(\sigma_{11}^{m,0})^{2}}}{2\alpha_{e2}^{m}}, \frac{(\sigma_{u}^{f} - \sigma_{22}^{f,0})}{\alpha_{e2}^{f}}\right\} \quad (4.24.2)$$

In-plane shear strength due to an in-plane shear load (σ_{12}) only

$$\sigma_{12}^{u} = \min\left\{\frac{\sigma_{u}^{f} - (\alpha_{e3}^{f} - \alpha_{p3}^{f})\sigma_{12}^{0}}{\alpha_{p3}^{f}}, \frac{\sigma_{u}^{m} - (\alpha_{e3}^{m} - \alpha_{p3}^{m})\sigma_{12}^{0}}{\alpha_{p3}^{m}}\right\} \quad (4.25.1)$$

where $$\sigma_{12}^{0} = \min\left\{\sqrt{\frac{(\sigma_{Y}^{m})^{2} + (\sigma_{11}^{m,0})(\sigma_{22}^{m,0}) - (\sigma_{11}^{m,0})^{2} - (\sigma_{22}^{m,0})^{2}}{3(\sigma_{e3}^{m})^{2}}}, \frac{\sigma_{u}^{f}}{\alpha_{e3}^{f}}\right\} \quad (4.25.2)$$

In all of the above strength formulae, the bridging parameters, β and α, have been entered as variables. As can be expected, a different choice of these parameters will result in different predictions. An overall feature of their effect is helpful. Figs. 4.1(a) – 4.1(d) and Figs. 4.2(a) – 4.2(d) show the influence of the bridging parameters β and α on the predicted transverse and in-plane shear strengths of UD composites, respectively. These results have been obtained with no thermal residual stress assumption. In the predictions, all of the matrix materials have been assumed to be bilinear in a stress-strain curve, with a yield strength of $\sigma_{Y}^{m} = 0.65\sigma_{u}^{m}$ and a hardening modulus of E_{T}^{m}=0.25E^{m}. Furthermore, the in-plane shear strength has been predicted using Eq. (4.19) rather than Eq. (4.21.1), and the hardening shear modulus, G_{T}^{m}, has been set to $E_{T}^{m}/3$.

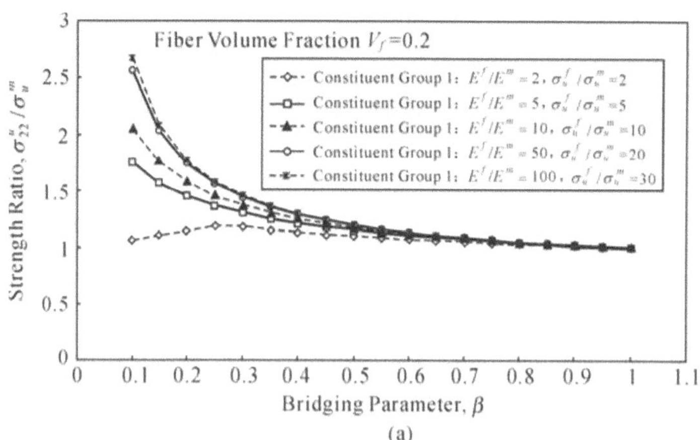

(a)

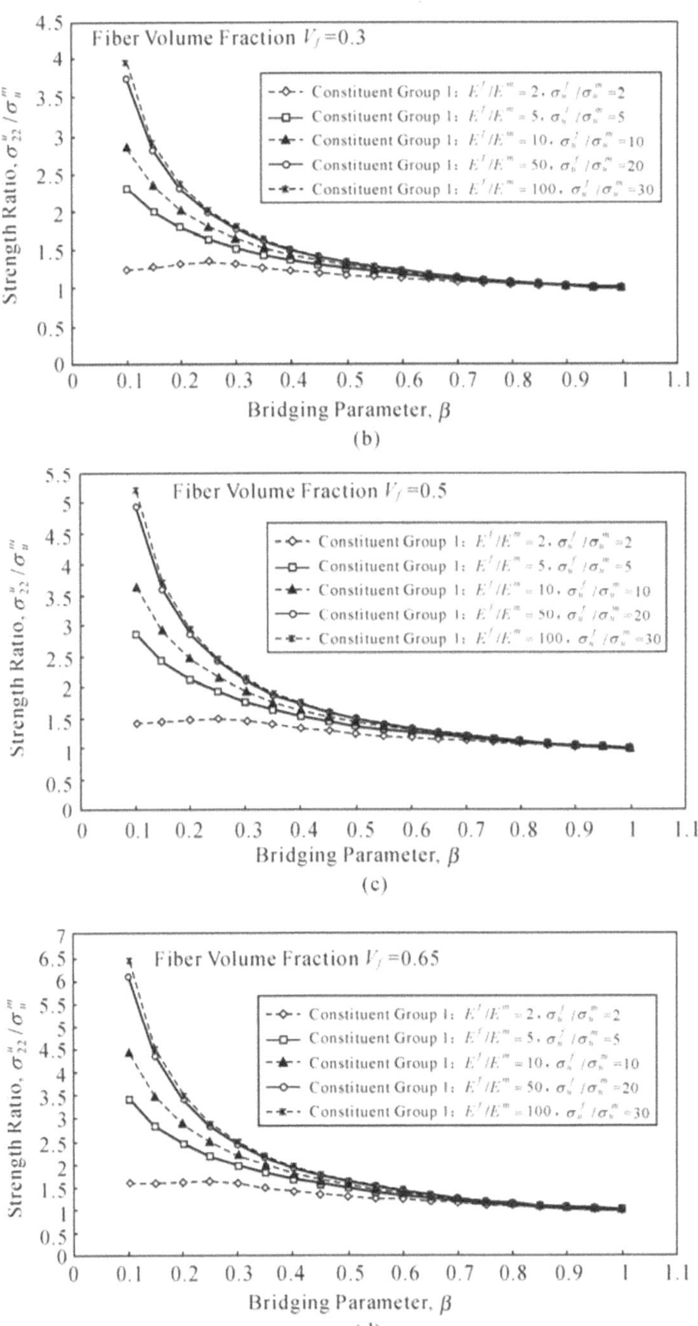

Fig. 4.1 Influence of bridging parameter, β, on the predicted transverse strengths of UD composites. Poisson's ratios of ν^f=0.2 and ν^m=0.33 have been used

Fiber Volume Fraction V_f=0.2
Strength Ratio, σ_{12}^u/σ_u^m
Bridging Parameter, α
Constituent Group 1: $E^f/E^m=2$, $\sigma_u^f/\sigma_u^m=2$
Constituent Group 1: $E^f/E^m=5$, $\sigma_u^f/\sigma_u^m=5$
Constituent Group 1: $E^f/E^m=10$, $\sigma_u^f/\sigma_u^m=10$
Constituent Group 1: $E^f/E^m=50$, $\sigma_u^f/\sigma_u^m=20$
Constituent Group 1: $E^f/E^m=100$, $\sigma_u^f/\sigma_u^m=30$
(a)

Fiber Volume Fraction V_f=0.3
Strength Ratio, σ_{12}^u/σ_u^m
Bridging Parameter, α
Constituent Group 1: $E^f/E^m=2$, $\sigma_u^f/\sigma_u^m=2$
Constituent Group 1: $E^f/E^m=5$, $\sigma_u^f/\sigma_u^m=5$
Constituent Group 1: $E^f/E^m=10$, $\sigma_u^f/\sigma_u^m=10$
Constituent Group 1: $E^f/E^m=50$, $\sigma_u^f/\sigma_u^m=20$
Constituent Group 1: $E^f/E^m=100$, $\sigma_u^f/\sigma_u^m=30$
(b)

Fiber Volume Fraction V_f=0.5
Strength Ratio, σ_{12}^u/σ_u^m
Bridging Parameter, α
Constituent Group 1: $E^f/E^m=2$, $\sigma_u^f/\sigma_u^m=2$
Constituent Group 1: $E^f/E^m=5$, $\sigma_u^f/\sigma_u^m=5$
Constituent Group 1: $E^f/E^m=10$, $\sigma_u^f/\sigma_u^m=10$
Constituent Group 1: $E^f/E^m=50$, $\sigma_u^f/\sigma_u^m=20$
Constituent Group 1: $E^f/E^m=100$, $\sigma_u^f/\sigma_u^m=30$
(c)

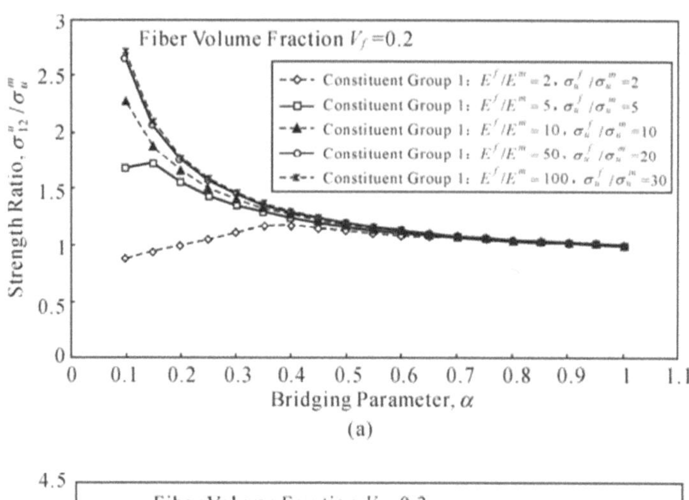

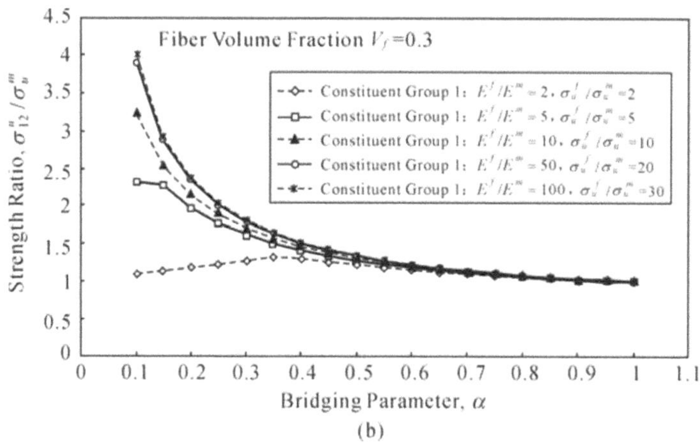

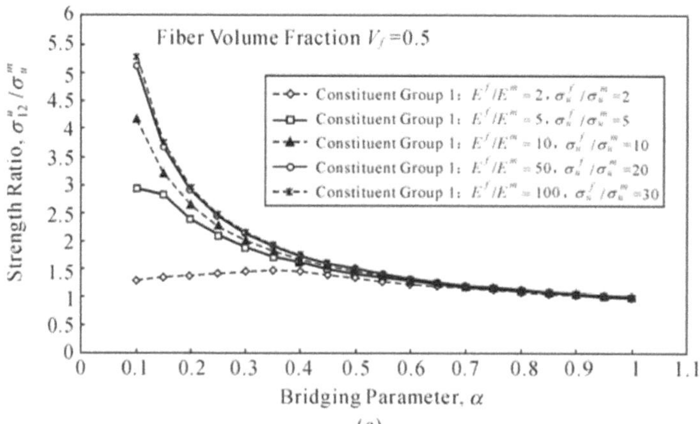

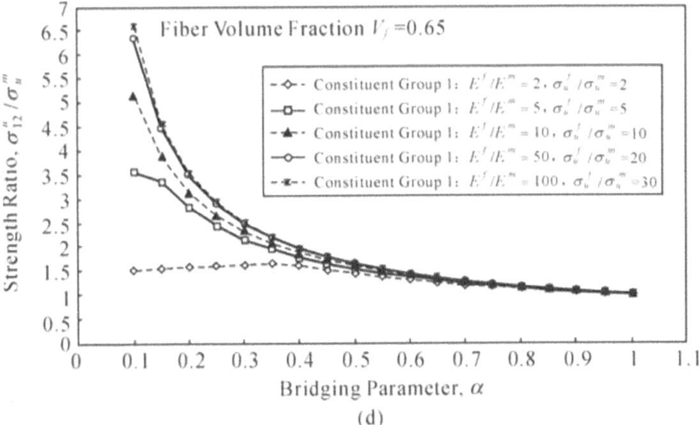

Fig. 4.2 Influence of bridging parameter, α, on the predicted in-plane shear strengths of UD composites. Poisson's ratios of ν^f=0.2 and ν^m=0.33 have been used

On the whole, the influence of the bridging parameters on the composite strengths is similar to that on the composite moduli (Figs. 3.5 and 3.6), i.e, the predicted strengths increase with a decrease in the bridging parameters. However, when the constituent modular and strength ratios are not significantly large (less than 10), there is a threshold value for the corresponding bridging parameter. Below that value, a further decrease in the bridging parameter will reduce the predicted strength. Furthermore, the quantitative effect of the bridging parameters on the predicted strengths is slightly heavier than that on the predicted moduli. For example, when E^f/E^m=50, σ_u^f/σ_u^m =20 and V_f=0.65, the difference in the predicted transverse strengths between using β=0.5 and β=0.4 is 19.4%, whereas that in the predicted in-plane shear strengths between using α=0.5 and α=0.35 is 33.3%, relative to 17% for the transverse moduli and 23% for the in-plane shear moduli. In light of the results shown in Figs. 3.3 – 3.6 and Figs. 4.1 – 4.2, it can be concluded that any choice of the bridging parameters with

$$0.35 \leq \beta \leq 0.5 \text{ and } 0.3 \leq \alpha \leq 0.5 \tag{4.26}$$

is pertinent for both stiffness and strength prediction, if no further information is available.

The strength formulae developed above have been used to calculate ultimate strengths of several UD composites under longitudinal loads. The first composite considered is a UD SiC-fiber and titanium (Ti) matrix composite. Gundel and Wawner (1997) carried out an experimental investigation on the longitudinal tensile behavior of this composite with varied fiber reinforcements. According to their report, the SiC-fiber used is isotropically linear elastic until rupture, having a Young's modulus of E^f= 400 GPa and a Poisson's ratio of ν^f=0.25. The measured ultimate tensile strength of the extracted fiber specimens, however, varied from 2,520 MPa to 4,540 MPa. In the present prediction, a fiber ultimate strength of

$\sigma_u^f = 3{,}480$ MPa, which was measured using fiber samples extracted from a composite panel whose tensile stress-strain curve was plotted in Fig. 6 of Gundel and Wawner (1997), is used. No thermal residual stress effect is taken into account in the calculation, as no related material thermal and processing parameters were reported. However, as will be seen in the next example, the influence of thermal residual stresses on the composite longitudinal strength is generally insignificant. In Gundel and Wawner's measurement, the Ti-matrix exhibited a typical bilinear elastic-plastic behavior (Fig. 2 of Gundel & Wawner, 1997). According to the information provided in Gundel and Wawner (1997), the Ti-matrix together with the SiC fiber properties are summarized in Table 4.1. Using these parameters, the ultimate tensile strength of the composite with any volume fraction can be easily calculated from Eqs. (4.11) – (4.15). Fig. 4.3 shows the calculated tensile strength varied with fiber volume fraction, V_f. The experimental data taken from Table 5 of Gundel and Wawner (1997) are also shown in the figure. A good correlation is seen to exist.

Table 4.1 Constituent properties of the SiC-Ti UD composite (β=α=0.5)

	E_{11} (GPa)	E_{22} (GPa)	G_{12} (GPa)	ν_{12}	ν_{23}	E_T (GPa)	σ_Y (MPa)	σ_u (MPa)
Fiber	400	400	160	0.25	0.25	-	-	3,480
Matrix	110	110	41.4	0.33	0.33	2.16	850	1,000

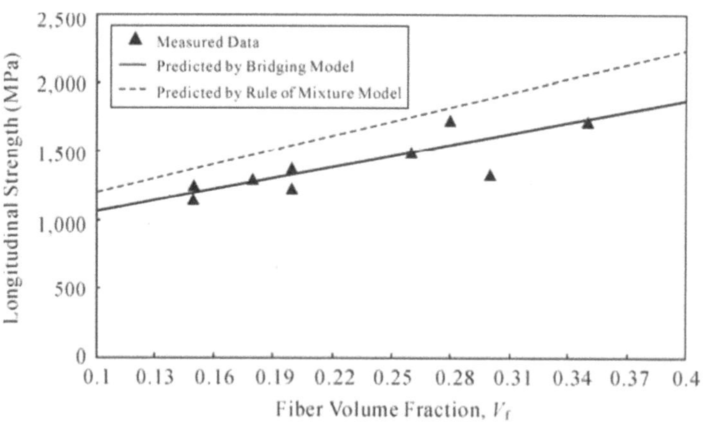

Fig. 4.3 Predicted and measured (Gundel & Wawner, 1997) tensile strengths of SiC-Ti UD composites. The material parameters used are given in Table 4.1

For further comparison, estimation by the rule of mixture formula, Eq. (4.9), has also been made and is plotted in Fig. 4.3. It is seen that the rule of mixture formula gave an over estimation for the ultimate strength of this composite. Relative errors at the two end points in the figure, i.e., at V_f=0.1 and V_f=0.4, between the two formula predictions are

$$e_{0.1}=(1198-1071.3)/(1071.3)=11.8\% \quad \text{and} \quad e_{0.4}=(2242-1874)/(1874)=19.6\%$$

It is noted that the bridging model strength formulae can also indicate the failure mode of the composite clearly. Namely, it can tell whether fiber or matrix failure causes the composite failure. On the other hand, the rule of mixture model formula cannot. For illustration, let us estimate all of the three uniaxial strengths and identify corresponding failure modes of the composite with V_f=0.15.

Longitudinal tensile strength

a) From Eq. (4.9) we get

σ_{11}^{u} =(0.15)(3480)+(0.85)(1000)=1372 (MPa)

The formula did not show whether the fiber or the matrix fracture caused the failure of the composite.

b) From Eqs. (4.11) − (4.15) we have

α_{e1}^{f} =(400)/[(0.15)(400)+(0.85)(110)]=2.606

α_{e1}^{m} =(110)/[(0.15)(400)+(0.85)(110)]=0.717

α_{p1}^{f} =(400)/[(0.15)(400)+(0.85)(2.16)]=6.550

α_{p1}^{m} =(2.16)/[(0.15)(400)+(0.85)(2.16)]=0.0349

σ_{11}^{0} =min{(850)/(0.717),(3480)/(2.606)}=min{1185.5,1335.4}=1185.5

σ_{11}^{u} =min{[3480−(2.606−6.55)(1185.5)]/6.55, [1,000−(0.717−0.0349)(1185.5)]/(0.0349)}=min{1245.1, 5483.4}=1245.1(MPa)

The last expression indicated that it was fiber fracture that caused the composite failure (since under the longitudinal tensile load the fiber failure stress is 1,245.1 MPa whereas the matrix failure stress is 5,483.4 MPa).

Transverse tensile strength

According to Eqs. (4.17) − (4.18) we obtain

α_{e2}^{f} =(400)/[(0.15)(400)+(0.5)(0.85)(110+400)]=1.445

α_{e2}^{m} =(0.5)(110+400)/[(0.15)(400)+(0.5)(0.85)(110+400)]=0.921

α_{p2}^{f} =(400)/[(0.15)(400)+(0.5)(0.85)(2.16+400)]=1.732

α_{p2}^{m} =(0.5)(400+2.16)/[(0.15)(400)+(0.5)(0.85)(402.16)]=0.871

σ_{22}^{0} =min{(850)/(0.921),(3480)/(1.445)}=min{922.9,2408.3}=922.9

σ_{22}^{u} =min{[(3480)−(1.445−1.732)(922.9)]/(1.732),[1000−(0.921−0.871)(922.9)]/(0.871)}=min{2162.2, 1095.1}=1095.1 (MPa)

The last expression indicated that it was matrix fracture that caused the composite failure (since under the transverse tensile load the fiber failure stress is 2,162.2 MPa whereas the matrix failure stress is 1,095.1 MPa).

In-plane shear strength

First we have

G^f=(400)/[(2)(1+0.25)]=160 (GPa), G^m=(110)/[(2)(1+0.33)]=41.4 (GPa)
Then Eqs. (4.19) − (4.20) give

α_{e3}^{f} =(160)/[(0.15)+(0.5)(0.85)(160+41.4)]=1.460

α_{e3}^{m} =(0.5)(201.4)/[(0.15)(160)+(0.5)(0.85)(201.4)]=0.919

α_{p3}^{f} =(3)(160)/[(3)(0.15)(160)+(0.5)(0.85)(2.16+480)]=1.733

α_{p3}^{m} =(0.5)(480+2.16)/[(3)(0.15)(160)+(0.5)(0.85)(2.16+480)]=0.871

σ_{12}^{0} =min{(850)/[(1.732)(0.871)],(3480)/(1.46)}=min{563.4, 2383.6}=563.4

σ_{12}^{u} =min{[3480−(1.46−1.733)(563.4)]/(1.733), [1000−(0.919−0.871)(563.4)]/(0.871)}=min{2096.8, 1118.4}=1118.4 (MPa)

The last expression indicated that it was matrix fracture that caused the composite failure (since under the in-plane shear load the fiber failure stress is 2,096.8 MPa whereas the matrix failure stress is 1,118.4 MPa).

The longitudinal stress-strain curves of the composite with two different volume fractions (V_f=0.2 and 0.35) under longitudinal tensile load until rupture are predicted and are plotted in Figs. 4.4 and 4.5. The prediction is made according to the following formula (refer to Eqs. (3.66.1) and (4.10)):

$$\mathrm{d}\varepsilon_{11} = V_f S_{11}^f \mathrm{d}\sigma_{11}^f + V_f S_{11}^m \mathrm{d}\sigma_{11}^m = \frac{V_f S_{11}^f + V_f a_{11} S_{11}^m}{V_f + V_m a_{11}} \mathrm{d}\sigma_{11}$$

The measured curves (Gundel & Wawner, 1997) are also shown in the figures. Pretty good agreement has been found in both the figures. More examples without thermal residual stress effect can be found in Huang (2001a, 2001b).

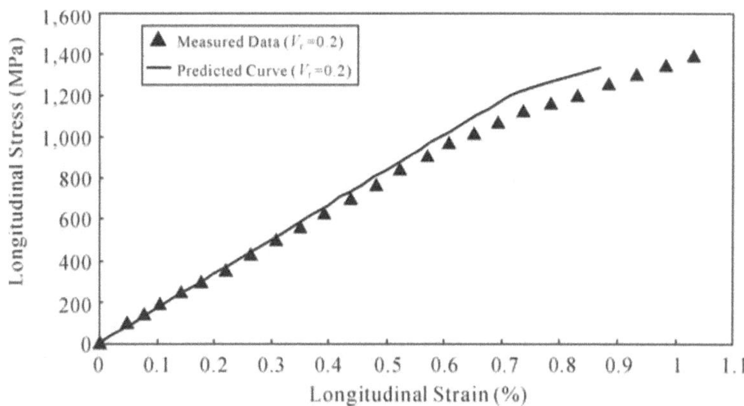

Fig. 4.4 Predicted and measured (Gundel & Wawner, 1997) stress-strain curves of a SiC-Ti UD composite of V_f=0.2. The material parameters used are given in Table 4.1 (from Huang, 2001b)

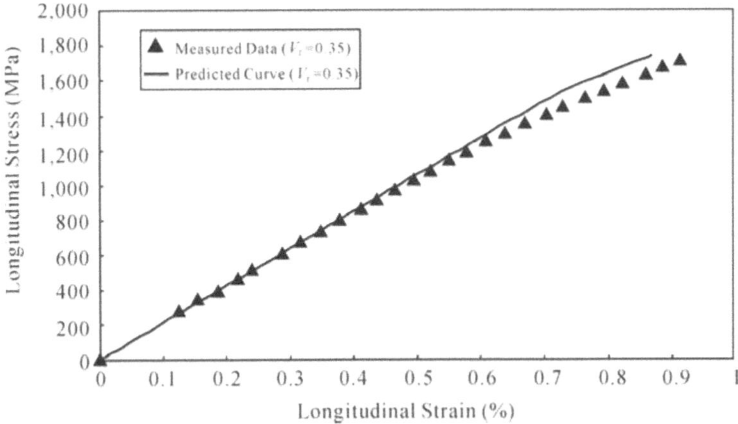

Fig. 4.5 Predicted and measured (Gundel & Wawner, 1997) stress-strain curves of a SiC-Ti UD composite of V_f=0.35. The material parameters used are given in Table 4.1 (from Huang, 2001b)

Next, let us consider a composite in which thermal residual stresses can be determined. Brindley et al. (1992) measured longitudinal tensile strengths of unidirectional SiC/Ti-24-11 composites with different fiber volume fractions at temperatures of 23 °C, 200 °C, 425 °C, 600 °C, 650 °C and 815 °C. A stress-free temperature of 806 °C was employed (Brindley et al., 1992). In their paper, Brindley et al. (1992) also reported elasto-plastic properties of the Ti-24-11 matrix obtained from monolithic material tests. The thermal-mechanical properties of the matrix are summarised in Table 4.2, in which the thermal expansion coefficients have been taken from Robertson and Mall (1997). In contrast, not much information about the SiC fibers was reported. The thermal-elastic properties of the SiC fibers are taken from Robertson and Mall (1997), and are listed in Table 4.3. The ultimate tensile strength of the fibers, σ_u^f =2,900 MPa, is calibrated using an overall composite strength measured at room temperature. The calibration has incorporated the thermal residual stress effect and this strength value is kept unchanged at all the elevated temperatures.

Substituting the material properties of the fibers and the matrix, given in Tables 4.2 and 4.3, into Eq. (4.23.1), the longitudinal tensile strengths of the composites with various fiber volume fractions are calculated. Before the calculations, the thermal residual stresses should be determined first. Corresponding to final (working) temperatures of 23 °C, 200 °C, 425 °C, 600 °C and 650 °C respectively, the composites had to be cooled down from 806 °C, whilst for a working temperature of 815 °C the composites needed to be elevated in temperature. It should be noted that at different working temperatures, the constituents must assume different material properties. The calculated strengths versus V_f at different temperatures are graphed in Fig. 4.6. For comparison, the measured data, taken from Table III of Brindley et al. (1992), are also shown in the figure. It is seen that the correlation between the analytical and the experimental data is reasonably good.

Table 4.2 Material properties of Ti-24-11 matrix (ν^m=0.26)

Temperature (°C)	E^m (GPa)	σ_Y^m (MPa)	E_T^m (GPa)	σ_u^m (MPa)	$\alpha_1^m = \alpha_2^m$ (×10^{-6}/°C)
23	107.9	511.0	3.92	676.6	11.33
200	102.7	423.6	3.00	594.2	11.68
425	87.3	350.8	1.55	577.1	12.10
600	84.1	276.3	0.89	411.0	12.58
650	74.2	252.5	0.58	352.7	12.73
815	44.1	137.8	0.78	220.0	13.53

Table 4.3 Material properties of SiC fibers (ν^f=0.25, σ_u^f =2900 MPa)

Temperature (°C)	E^f (GPa)	$\alpha_1^f = \alpha_2^f$ (×10^{-6}/°C)
23	393.0	3.56
200	386.1	3.62
425	378.1	3.69
600	371.8	3.79
650	370.0	3.83
815	363.0	3.94

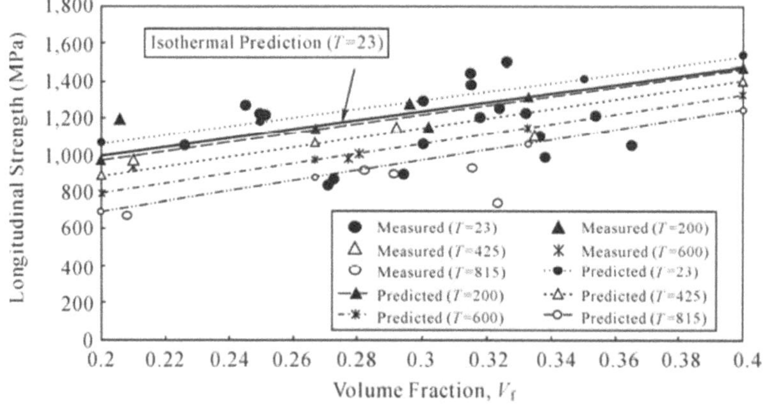

Fig. 4.6 Predicted and measured (Brindley et al., 1992) uniaxial strengths of UD SiC/Ti-24-11 composites at different temperatures T (°C) (from Huang, 2000d)

In order to gain some idea about the difference between the isothermal strength formula, Eq. (4.15), and the non-isothermal formula, Eq. (4.23.1), the composite strengths at room temperature without the influence of thermal residual stresses are calculated and are also plotted in Fig. 4.6. It is seen that for the present composites (thermal expansion coefficients of the fibers do not differ from those of the matrix too much), the isothermal strength formula does not involve a significant error.

Remark 4.1

(1) Equtions. (4.11) – (4.15) and Eqs. (4.17) – (4.18) are applicable for both tensile and compressive strength estimations, as long as the material parameters involved, i.e., the elastic/hardening moduli, yield strength and ultimate strengths are defined accordingly.
(2) In general, the stiffness and strength of the fibers are remarkably higher than those of the matrix. Therefore, the composite transverse and in-plane shear strengths are most probably determined by the matrix strength.
(3) It is seen from Figs. 4.1 and 4.2 that the bridging model always gives a predicted transverse strength higher than the matrix strength if no thermal residual stress is accounted for (providing that the fiber stiffness and strength are higher than those of the matrix). In reality, however, measurements of transverse composite strengths lower than the monolithic matrix strength have been observed. Several reasons may be attributed to this difference. First, the arrangement of isolated fibers in the matrix and especially some fabrication defects such as misalignment, filament breaking, imperfect bounding between the fibers and the matrix, as well as voids/microcracks involved, can cause some stress concentration which may have reduced the matrix *in situ* strength relative to the monolithic matrix strength. Second, the measurement of composite transverse strength is quite sensitive to the testing conditions involved, including, possibly, the effect of specimen dimensions. It has been well recognized that a quite large deviation exists in the measurement of the transverse strength of a UD composite (Zweben, 1990). In fact, people have realised that in some cases the measured transverse strength using an independent (isolated) UD lamina is lower than what it should be when embedded in a multidirectional laminate (Rotem & Hashin, 1975; Bailey et al., 1979; Flaggs & Kural, 1982). For this reason, some researchers have even artificially increased the transverse strength measured from the isolated UD lamina when performing laminate analysis (Rotem, 1998; Sun & Tao, 1998; Hart-Smith, 1998a, 1998b; Soden et al., 1998). However, a possible way to resolve this mismatch problem is to use a measured composite strength (preferably a laminate strength, see the next chapter) to calibrate the matrix *in situ* ultimate strength.

4.4 Off-axial Strength

It must be realized that an off-axial strength of a UD composite cannot generally be obtained from a direct combination of its uniaxial strengths presented in the preceding section. This is because an off-axial load will also generate a shear stress component in the composite. Hence, a general two-dimensional and incremental stress analysis, as described in Section 3.9, is required.

Substituting $\{d\sigma_i\}=\{d\sigma_{11}, d\sigma_{22}, d\sigma_{12}\}^T=d\sigma_\theta\{\cos^2(\theta), \sin^2(\theta), -0.5\sin(2\theta)\}^T$ into Eqs. (3.82.1) and (3.82.2), the internal stress increments, and hence the total stresses in the constituent fibers and matrix, can be determined, where $d\sigma_\theta$ is the overall off-axial stress increment. At each load level, the constituent failure status is checked against the ultimate strength of the constituent using the classical or the generalized maximum normal stress criterion presented in Section 4.2. Throughout this book, the following assumption has been made:

As long as any constituent has attained its ultimate stress state, the resulting composite lamina is considered to have failed.

Accordingly, the overall applied load on the composite, σ_θ, is defined as the composite off-axial strength. Apparently, such an assumption is applicable to most composites. Thus, a strength theory using only constituent properties and volume contents of the constituents is developed. By definition, such a theory is referred to as a micromechanical strength theory.

Let us apply this strength theory to three unidirectional composites of different constituents subjected to off-axial loads with varied off-axial angles. All the composites will be assessed using the classical maximum normal stress criterion, i.e., Eq. (4.3), to detect a constituent failure.

The elastic properties of the fiber and matrix materials as well as the fiber volume fractions of these composites are taken from Aboudi (1988) and are summarised in Table 4.4 through Table 4.6, respectively. However, neither the plastic properties nor ultimate strengths for the constituent materials have been reported. On the other hand, the uniaxial strengths of these composites corresponding to, respectively, longitudinal, transverse and in-plane shear loads are available and are also included in the tables. These uniaxial strengths can be used to recover the plastic parameters of the matrices and the ultimate strengths of the constituents. Let us explain this recovering procedure with the first example, a glass fiber reinforced epoxy matrix composite. The elastic properties of the glass fiber and the epoxy matrix (both are assumed to be isotropic) as well as the longitudinal strength (X), transverse strength (Y) and in-plane shear strength (S) of the composite are given in Table 4.4.

Table 4.4 Constituent properties of a UD Glass/Epoxy composite ($\beta=\alpha=0.5$) (X=1,236 MP*a*, Y=28.45 MPa, S=38 MPa)

Mater.	V	E_{11} (GPa)	E_{22} (GPa)	ν_{12}	ν_{23}	G_{12} (GPa)	E_T (GPa)	σ_Y (MPa)	σ_u (MPa)
Glass	0.6	73	73	0.22	0.22	29.92	-	-	2047
Epoxy	0.4	3.45	3.45	0.35	0.35	1.28	-	-	19

The recovery is begun by assuming that the glass fiber is linearly elastic until rupture and the epoxy matrix is, at most, bilinearly elastic-plastic, so that the uniaxial strength formulae presented in the preceding section can be applied. As the stiffness of the glass fiber is much higher than that of the epoxy matrix, we can imagine that the transverse tensile strength of the composite is governed by the strength of the matrix (Remark 4.1(b)). Therefore, Eq. (4.17) gives

$$\sigma_{22}^{u} = \frac{\sigma_{Y}^{m}}{\alpha_{e2}^{m}} + \frac{\sigma_{u}^{m} - \sigma_{Y}^{m}}{\alpha_{p2}^{m}} = Y \approx \frac{\sigma_{u}^{m}}{\alpha_{e2}^{m}} \tag{4.27}$$

since $\alpha_{p2}^{m} \approx \alpha_{e2}^{m}$. From Eq. (4.27), the matrix strength is recovered to be 18.4 MPa (a slight amendment is made for the recovered parameter used in the final prediction, see Table 4.4, due to incorporation of $\alpha_{p2}^{m} \neq \alpha_{e2}^{m}$). Next, let us use Eq. (4.15) to recover the fiber strength. It is required that

$$\sigma_{11}^{u} = \min\left\{ \frac{\sigma_{u}^{f} - (\alpha_{e1}^{f} - \alpha_{p1}^{f})\sigma_{11}^{0}}{\alpha_{p1}^{f}}, \frac{\sigma_{u}^{m} - (\alpha_{e1}^{m} - \alpha_{p1}^{m})\sigma_{11}^{0}}{\alpha_{p1}^{m}} \right\} = X$$

At this stage, we cannot assume that $\alpha_{p1}^{m} \approx \alpha_{e1}^{m}$. However, the longitudinal strength of the composite is most probably governed by the strength of the fibers. Thus, we can consider $X = \sigma_{11}^{u} \approx \frac{\sigma_{u}^{f}}{\alpha_{e1}^{f}}$, due to $\alpha_{p1}^{f} \approx \alpha_{e1}^{f}$, providing that we can choose the other two parameters of the matrix, E_{T}^{m} and σ_{Y}^{m}, so that

$$\left(\frac{\sigma_{u}^{m} - (\alpha_{e1}^{m} - \alpha_{p1}^{m})\sigma_{11}^{0}}{\alpha_{p1}^{m}} \right)_{\sigma_{u}^{m} = 18.4 \text{ MPa}} = \left(\frac{\sigma_{Y}^{m}}{\alpha_{e1}^{m}} + \frac{\sigma_{u}^{m} - \sigma_{Y}^{m}}{\alpha_{p1}^{m}} \right)_{\sigma_{u}^{m} = 18.4 \text{ MPa}} \geq X \tag{4.28}$$

It is clear that many different combinations of E_{T}^{m} and σ_{Y}^{m}, which satisfy inequality (4.28), exist. Since $\left(\frac{\sigma_{u}^{m}}{\alpha_{e1}^{m}} \right)_{\sigma_{u}^{m} = 18.4 \text{ MPa}} < X$, the epoxy used cannot be considered as linearly elastic until rupture. On the other hand, any combination of E_{T}^{m} and σ_{Y}^{m}, which satisfies Eq. (4.28), is a possible candidate due to no other information. The estimated results for the off-axial tensile strength of the composite with four different combinations of E_{T}^{m} and σ_{Y}^{m} are plotted in Fig. 4.7(a). It is seen that all of the four curves essentially coincide with each other, which are comparable with the experimental data taken from Hashin & Rotem (1973). This is due to the fact that the overall stress components applied on the composite can be calculated without knowing the composite strains, i.e., the composite is statically determinate. The stress sharing parameters of the constituents (Eqs. (4.11) – (4.15), (4.17) – (4.18) and (4.19) – (4.20)) are dominated by the fiber stiffness if this stiffness is significantly higher than the matrix stiffness. Thus, any variation in the matrix hardening modulus (as long as its value is reasonable, e.g., smaller than the matrix Young's modulus) will not bring much significant change in the internal stress shares, and hence will not significantly affect the composite strength prediction. When the lamina is statically indeterminate, such as being arranged in a multidirectional laminate, the overall stresses on it must be determined by using its deformation condition, which can depend heavily on the plastic parameters of the matrix. We will illustrate this phenomenon in the next chapter.

Since the composite tensile strength is estimated from the ultimate stress of either the fiber or the matrix, the failure mode is automatically indicated. Here the failure mode is defined as the source that initiates the composite failure. Under the perfect bonding assumption (Remark 3.4(a)), the mode can be matrix failure, fiber failure, or the failure of both the constituent materials. A constituent's failure status is defined by its maximum normal stress, which can be greater than its tensile stress. Information on the maximum normal stress in a constituent material is therefore important. It shows how much load is sustained by the constituent material, and whether or not the material has attained its ultimate stress state. On the other hand, an applied stress on the composite, corresponding to which one of the constituent materials has failed, is defined as an off-axial strength of the composite. Figure. 4.7(a) shows the composite strengths varied with the off-axial angles, whereas the maximum normal stresses in different materials are plotted in Fig. 4.7(b). It is noted that the maximum normal stress in the composite always corresponds to its tensile strength. Fig. 4.7(b) indicates that, except for the composite with an off-axial angle around 0°, the failure (tensile strength) in any other direction is controlled by the matrix strength. The maximum normal stresses in the fiber and composite with off-axial angles smaller than 10° are not shown due to a scale problem. The figure also indicates that the maximum normal stress or the tensile strength of the matrix is always smaller than the corresponding strength of the composite. It is found that, with a fiber reinforcement of 60% by volume, the strength of the composite is at least 54% higher than that of the neat matrix in any off-axial direction.

The second composite is also made from two isotropic constituent materials, i.e., boron fibers and epoxy matrix, whereas the third composite is made from transversely isotropic graphite fibers and an isotropic polyimide matrix. The recovery procedure for the ultimate and plastic parameters of these two composites is similar to that described above. Both the fibers considered are assumed to be linearly elastic until rupture. The matrices are assumed to be either elastic or bilinearly elastic-plastic, depending on whether the involved ultimate strengths of the constituent materials can be successfully determined or not, with an initial elastic assumption. Namely, if an inequality similar to Eq. (4.28) can be fulfilled by using only the matrix elastic properties, the matrix used is assumed to be linearly elastic until rupture. Otherwise, the matrix must be considered (in the present study) as a bilinear elastic-plastic material. In the case where the bilinear elastic-plasticity is assumed, the plastic parameters for the matrix are determined somewhat arbitrarily (Fig. 4.7(a) and the above comments). The recovered parameters are listed in Tables 4.5 and 4.6, respectively. The predicted off-axial tensile strengths, using these parameters, and the corresponding testing data taken from different sources are plotted in Figs. 4.8(a) and 4.9(a). The longitudinal strength of the composite is used to determine the fiber's strength, whereas the transverse strength of the composite is applied to recover the matrix's strength. Figures. 4.8(a) and 4.9(a) indicate that good agreement exists between the predicted results and available experimental data for both the composites. Figs. 4.8(b) and 4.9(b) plot the corresponding maximum normal stresses generated in the fibers and matrices when the off-axial tensile stresses of the composites attain their ultimate values. The qualitative behavior of these two figures is similar to that of Fig. 4.7(b). More results on off-axial strength prediction can be found in Huang et al. (1999b) and Huang (2000a, 2001a).

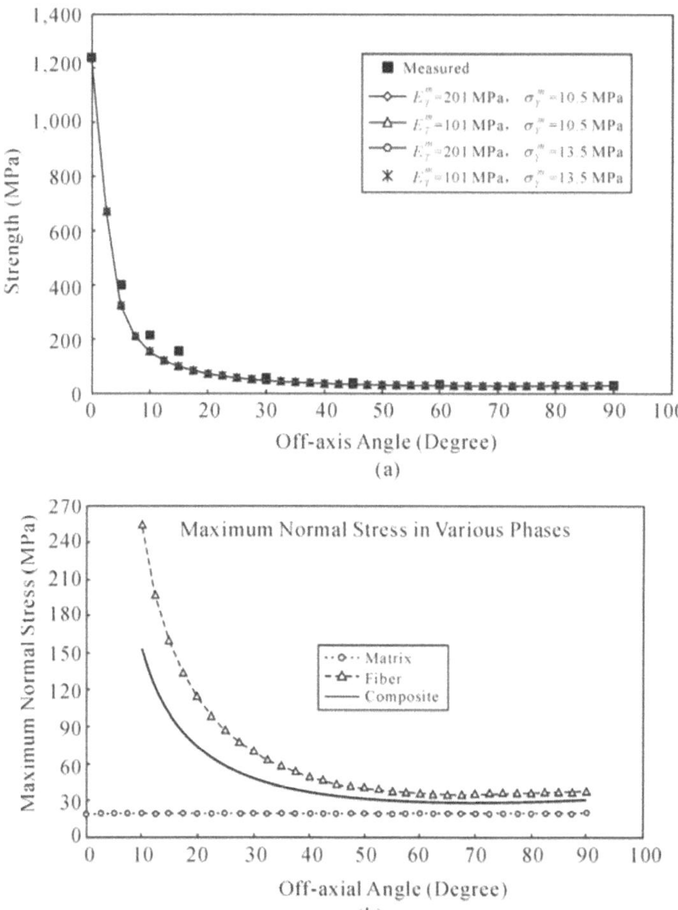

Fig. 4.7 (a) Predicted and measured (Hashin & Rotem, 1973) off-axial strength of a UD glass/epoxy composite. The parameters used are given in Table 4.4; (b) Maximum normal stresses in composite, fiber and matrix *vs* off-axial angle (σ_Y^m =10.5 MPa and E_T^m =201 MPa) (from Huang, 2001a)

Table 4.5 Parameters of UD Boron/Epoxy composite (β=α=0.5) (X=1,296 MPa, Y=62 MPa, S=69 MPa)

Mater.	V	E_{11} (GPa)	E_{22} (GPa)	ν_{12}	ν_{23}	G_{12} (GPa)	E_T (GPa)	σ_Y (MPa)	σ_u (MPa)
Boron	0.5	400	400	0.2	0.2	166.67	-	-	2,570
Epoxy	0.5	3.45	3.45	0.35	0.35	1.28	-	-	41

Table 4.6 Parameters of a UD Graphite/Polyimide composite (β=α=0.5) (X=1,553.6 MPa, Y=52 MPa, S=59 MPa)

Mater.	V	E_{11} (GPa)	E_{22} (GPa)	ν_{12}	ν_{23}	G_{12} (GPa)	E_T (GPa)	σ_Y (MPa)	σ_u (MPa)
Graphite	0.61	222	29.5	0.33	0.73	24.1	-	-	2,530
Polyim.	0.39	3.1	3.1	0.39	0.39	1.12	0.58	20	34.5

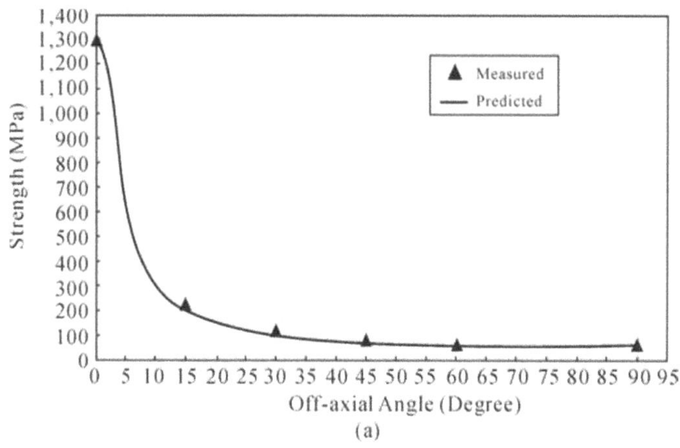

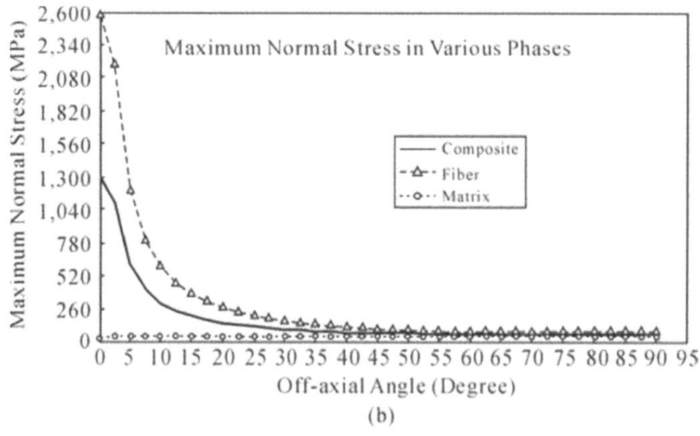

Fig. 4.8 (a) Predicted and measured (Pipes & Cole, 1973) off-axial strength of a UD boron/epoxy composite. The parameters used are given in Table 4.5; (b) Maximum normal stresses in composite, fiber, and matrix vs off-axial angle (from Huang, 2001a)

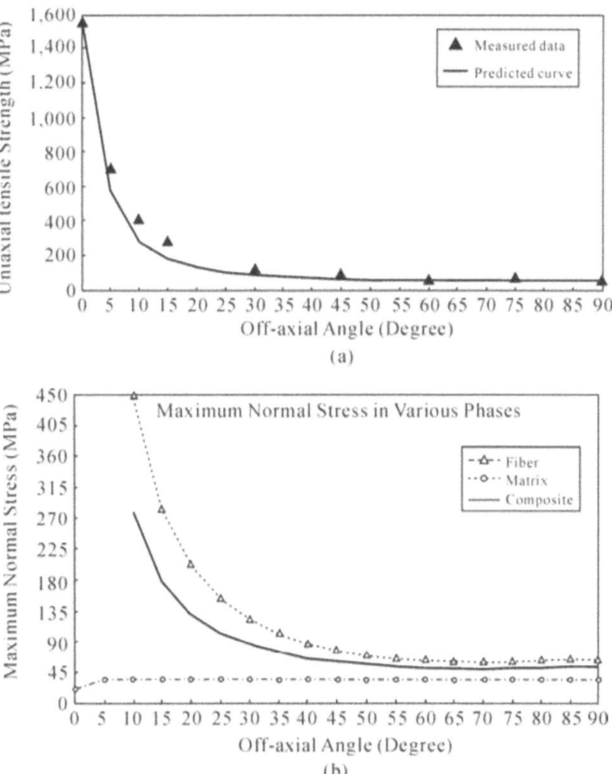

Fig. 4.9 (a) Predicted and measured (Pindera & Herakorvich, 1981) off-axial strength of a UD graphite/polyimide composite. The parameters used are given in Table 4.6; (b) Maximum normal stresses in composite, fiber and matrix vs off-axial angle (from Huang, 2001a)

4.5 Strength Envelope under Combined Loads

One important feature of the bridging model is its simplicity. The model can be used to calculate the internal stress increments in the constituent materials under any multiaxial stress condition, by simply setting the overall stress-increment vector in Eqs. (3.82.1) and (3.82.2) to a prescribed quantity, and estimating the ultimate failure strength of fibrous composites by incorporating simple failure criteria, e.g., maximum normal stress criteria or generalized maximum normal stress criteria as indicated in Section 4.2. Let us consider a graphite/epoxy UD composite subjected to combined transverse tension and in-plane shear (i.e., σ_{22} and σ_{12}) loads. Set

$$\{d\sigma_i\}=\{d\sigma_{11}, d\sigma_{22}, d\sigma_{12}\}^T=d\sigma\{0, \cos\theta, \sin\theta\}^T, 0°\leq\theta\leq 90°$$

Corresponding to a given angle θ, the composite ultimate load, σ, can be determined using the procedure as done in the previous sub-section. In this way, we obtain an ultimate load combination, $\sigma_{22}=\sigma\cos\theta$ and $\sigma_{12}=\sigma\sin\theta$, in the σ_{22}–σ_{12} plane. Varying θ from θ=0° to θ=90°, a continuous curve can be plotted in the σ_{22}–σ_{12} plane, which is called a failure envelope for the composite in that load plane. Thus, composite failure occurs when it is subjected to any combined σ_{22} and σ_{12} load which is in or outside the failure envelope. The composite is safe when the load combination is within the envelope.

The elastic properties of the graphite fibers and the epoxy matrix were taken from Aboudi (1989) and are summarized in Table 4.7. The failure envelope of the composite versus different combinations of σ_{22} and σ_{12} was experimentally measured by Awerbuch and Hahn (1981). The transverse tensile and in-plane shear strengths of the composite were (Awerbuch and Hahn, 1981): Y=56.9 MPa (when σ_{12}=0) and S=86.35 MPa (when σ_{22}=0), respectively. Based on these data, we can back-calculate the ultimate strengths of the constituent materials. From the composite examples studied in the preceding section, it can be expected that these stress levels may hardly cause the failure of the graphite fibers. As the stiffness and strength of the graphite fibers are much higher than the counterparts of the epoxy matrix, the fiber fracture would be mainly caused by excessive load in the longitudinal direction. Hence, the composite failures should result from matrix fracture under the current loads. Using the given transverse tensile strength, Y=56.9 MPa, the recovered matrix strength is 40.8 MPa, no matter whether the matrix is assumed to be linearly elastic or elastic-plastic. Thus, a linear-elastic assumption has been applied to the matrix during prediction. With this matrix strength, the predicted failure envelope of the composite is shown in Fig. 4.10(a) by the broken-line. It can be seen from the figure that, except for σ_{12}=0, the predicted strength of the composite under every other combination of σ_{22} and σ_{12} is lower than a measured datum. It is further noted that the graphite fibers cannot fail first, before the matrix fracture under all of the considered load combinations. More evidence can be gained from the resulting maximum normal stresses in the fiber and matrix, which are plotted in Fig. 4.10(b). These stresses are generated when the composite is loaded to the failure envelope, which is controlled by a matrix strength of 40.8 MPa, and have been plotted versus the *in situ* transverse tensile stress. On the other hand, the recovered ultimate strength of the matrix is 54.5 MPa if the overall shear strength of the composite, S=86.35 MPa, is used. Based on this matrix strength, the predicted failure envelope, as shown by the solid line in Fig. 4.10(a), agrees even better with the measured data for all of the considered load combinations except for σ_{12}=0. It seems that the predicted curve based on σ_u^m=54.5 MPa should be more realistic. The corresponding maximum normal stresses generated in the fiber and matrix phases are indicated in Fig. 4.10(c), which has a similar legend meaning as Fig. 4.10(b). From Fig. 4.10(c), we can see that the maximum normal stress generated in the fiber under every combined transverse tensile and in-plane shear load is below 110 MPa, much lower than an expected graphite fiber strength. Hence, all the failures of the composite must result from matrix fracture, as indicated by the dot-and-dash line in Fig. 4.10(c).

Table 4.7 Parameters of a UD Graphite/Epoxy composite (β=α=0.5) (Y=56.9 MPa, S=86.35 MPa)

Mater.	V	E_{11} (GPa)	E_{22} (GPa)	ν_{12}	ν_{23}	G_{12} (GPa)	E_T (GPa)	σ_Y (MPa)	σ_u (MPa)
Graphite	0.66	213.7	13.8	0.2	0.25	13.8	-	-	-
Epoxy	0.34	3.45	3.45	0.35	0.35	1.28	-	-	40.8*
Epoxy	0.34	3.45	3.45	0.35	0.35	1.28	-	-	54.5**

* recovered using the overall transverse tensile strength of the composite;
** recovered using the overall in-plane shear strength of the composite.

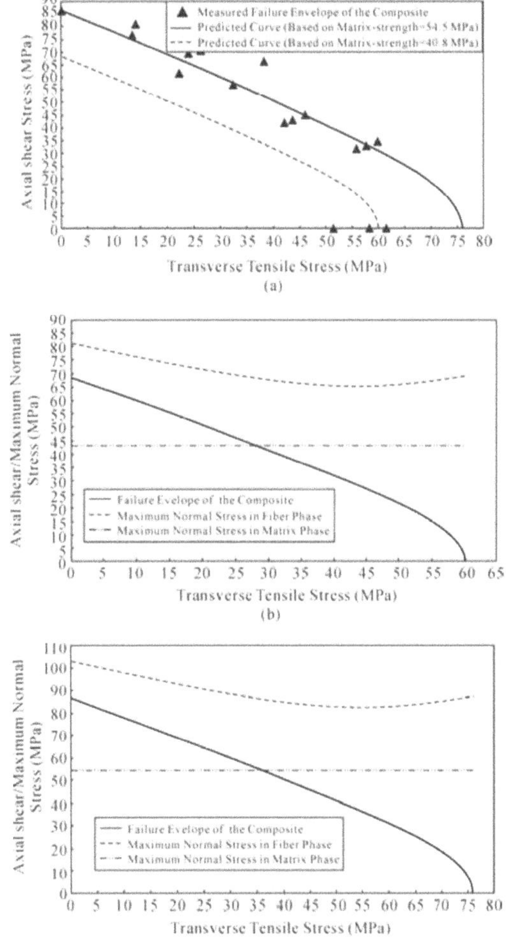

Fig. 4.10 (a) Predicted and measured (Awerbuch & Hahn, 1981) failure envelope of a UD graphite/epoxy composite. The parameters used are given in Table 4.7; (b) Maximum normal stresses in fiber and matrix when the composite is loaded to failure envelope (predicted using $\sigma_{u,c}^{m}$ =43 MPa); (c) Maximum normal stresses in fiber and matrix when the composite is loaded to failure envelope (predicted using $(\sigma_Y^m)_i$ =54.5 MPa) (from Huang, 2001b)

4.6 Strength at Elevated Temperature

Metal matrix composites (MMC) and ceramic matrix composites (CMC) usually have a high service temperature. When these composites are subjected to a mechanical load, the ultimate strength of the composites varies with both the elevated temperature and the external load. It is necessary to understand the composite thermo-mechanical strength for an optimal design and safe use purpose. However, by using the bridging model theory presented in Chapter 3, the work can be accomplished relatively easily: the internal stress increments in the constituent fiber and matrix materials at any thermo-mechanical load level are explicitly determined from Eqs. (3.90.1) and (3.90.2). These internal stresses are then checked against the ultimate strengths of the constituents at the corresponding temperature. If any constituent fails, the overall applied load is defined as the composite ultimate strength at that temperature.

As an example (Huang, 2000d), consider a ceramic alumina fiber reinforced aluminum matrix composite, having a fiber volume fraction of V_f=0.5. The measured off-axial tensile strengths of this composite at different temperatures, from room temperature to 773 K, were obtained experimentally by Matsuda and Matsuura (1997), and are taken for comparison. Unfortunately, they did not report any property of the constituent materials. In the present study, the alumina fibers (taken as linearly elastic until rupture) are considered as temperature independent (ASTM Int., 1994), and their thermal-elastic properties are taken from ASTM Int. (1994) and are listed in Table 4.8. On the other hand, the uniaxial tensile stress-strain curve of the aluminum matrix depends heavily on temperature. The Young's moduli, Poisson's ratios, yield strengths and thermal expansion coefficients at different temperatures are taken from (Chun & Daniel, 1996). These properties are also summarized in Table 4.8. The other three material parameters, i.e., the tensile strength and hardening modulus of the matrix, and the tensile strength of the fiber, have to be retrieved using the overall composite longitudinal and transverse strengths. It is noted that the fiber strength is assumed to be temperature independent. Therefore, except for the reference temperature at which the composite longitudinal strength was employed to retrieve the fiber strength, only the overall transverse strength of the composite at each of the other temperatures is required to extract the matrix strength and hardening modulus.

In the study (Huang, 2000d), the room temperature is taken as the reference temperature and both the fiber and the matrix are assumed to be stress-free (zero residual stresses) at the reference temperature. There are two reasons for doing so. The first one is that no reference temperature was mentioned in (Matsuda & Matsuura, 1997). Although the fabrication temperature of an MMC is usually rather high, post-processing (heat treatment) is generally applied to the composite so that the residual stresses are reduced to a minimum. The second reason comes from the fact that the tensile strength of the matrix at each temperature is extracted using the corresponding overall transverse ultimate stress of the composite. This strength, which may be different from that of the neat matrix at the same temperature, can be considered as some compensation for the influence of the possible residual stresses. Contrary to the "exact" recovery of the fiber (at room temperature) and matrix strengths, the hardening moduli of the matrix are determined somewhat arbitrarily, as indicated in the previous examples (Section 4.4). The recovered hardening moduli at each temperature are also listed in Table 4.8.

Table 4.8 Material properties of the Alumina/Aluminium UD composite (V_f=0.5) (β=α=0.5)

Alumina fiber properties, independent of temperature (ASTM Int., 1994)

Young's modulus (GPa)	Tensile strength (MPa)	Poisson's ratio	Thermal expansion coefficient ($\times 10^{-6}$/K)
300	1380	0.26	6.0

Aluminium matrix properties (Chun & Daniel, 1996)

Temperature (K)	Young's modulus (GPa)	Yield Strength (MPa)	Hardening modulus (MPa)	Tensile strength (MPa)	Poisson's ratio	Thermal expansion coefficient
297	68.9	41.4	6500	78.4	0.33	23.4×10^{-6} K^{-1}
394	63.8	39.3	4500	65	0.33	23.6×10^{-6} K^{-1}
473	59.6	36.5	1150	51	0.33	23.9×10^{-6} K^{-1}
573	54.6	32.5	500	34	0.33	24.8×10^{-6} K^{-1}
673	48.3	15.9	200	21	0.33	24.8×10^{-6} K^{-1}
773	42.0	10.5	80	12.5	0.33	25.7×10^{-6} K^{-1}

Using the data given in Table 4.8, the off-axial tensile strengths of the composite at room temperature, 473 K, 573 K, 673 K and 773 K are predicted. Each temperature variation (such as from room temperature to 473 K) is divided into 100 sub-intervals. In each sub-interval, the resulting thermal stresses were calculated. During prediction, care must be paid to distinguish whether a mechanical load is a loading or an unloading mode after a thermal load has been applied to the composite, when the internal thermal stresses already cause the matrix yielding. Because a free thermal expansion (ΔT>0) results in a compression stress to the matrix, a mechanical tensile load on the composite corresponds to an unloading condition for the matrix at an initial stage. A loading condition is considered to occur again as long as the longitudinal stress component of the matrix (i.e., σ_1^m) becomes positive. The predicted results at different temperatures are plotted in Fig. 4.11 through Fig. 4.15, respectively. The experimental data taken from (Matsuda & Matsuura, 1997) are also shown in the corresponding figures. It is seen that the agreements between the predicted and experimental strengths are satisfactory in all the cases. For more discussion, refer to (Huang, 2000e).

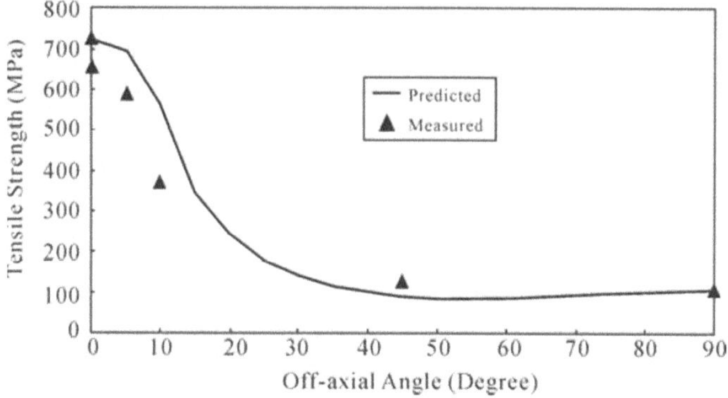

Fig. 4.11 Predicted and measured (Matsuda & Matsuura, 1997) off-axial tensile strengths of a UD alumina/aluminium composite at room temperature. The parameters used are given in Table 4.8 (from Huang, 2000e)

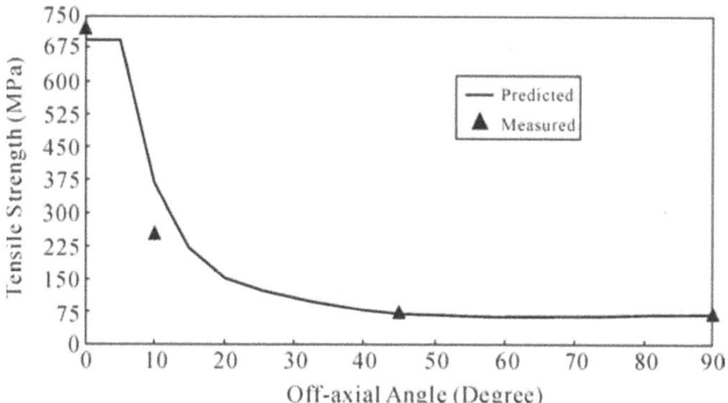

Fig. 4.12 Predicted and measured (Matsuda & Matsuura, 1997) off-axial tensile strengths of a UD alumina/aluminium composite at 473 K. The parameters used are given in Table 4.8 (from Huang, 2000e)

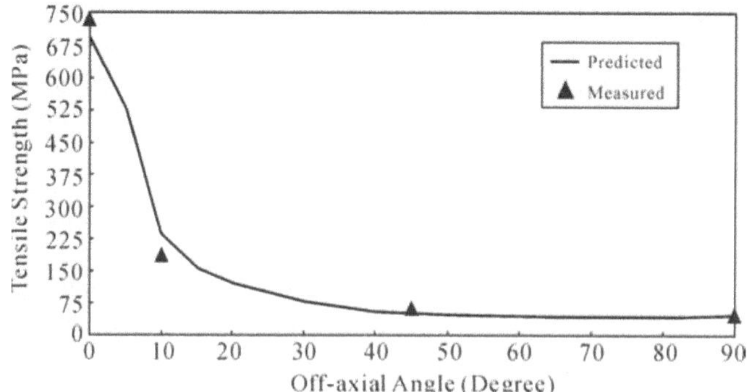

Fig. 4.13 Predicted and measured (Matsuda & Matsuura, 1997) off-axial tensile strengths of a UD alumina/aluminium composite at 573 K. The parameters used are given in Table 4.8 (from Huang, 2000e)

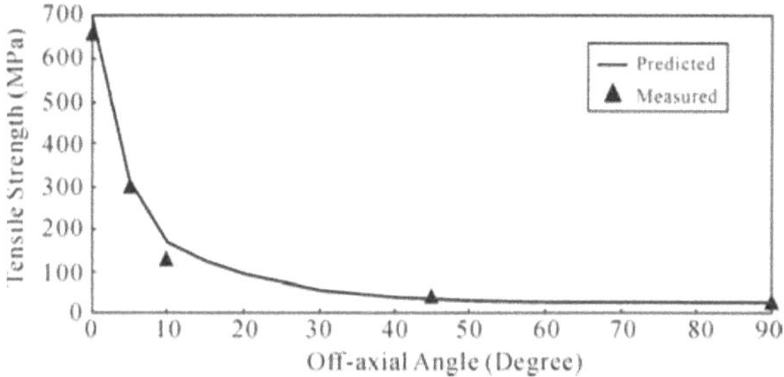

Fig. 4.14 Predicted and measured (Matsuda & Matsuura, 1997) off-axial tensile strengths of a UD alumina/aluminium composite at 673K. The parameters used are given in Table 4.8 (from Huang, 2000e)

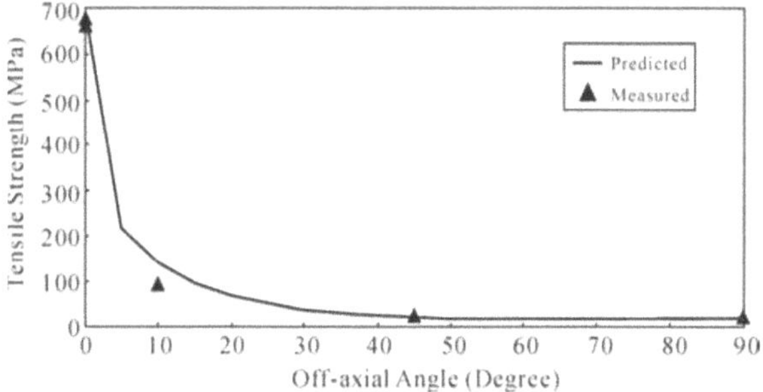

Fig. 4.15 Predicted and measured (Matsuda & Matsuura, 1997) off-axial tensile strengths of a UD alumina/aluminium composite at 773 K. The parameters used are given in Table 4.8 (from Huang, 2000e)

4.7 Fatigue Strength and Life Prediction

It is believed that any failure of a material must result from its internal stresses, no matter what kind of load has been applied to the material. As such, the bridging model can be applied to simulate the fatigue strength of a fibrous composite by using the fatigue behavior of its constituent materials. Again, the maximum normal stress criterion (Section 4.2) is used to control a constituent's fatigue failure. However, as cyclic loads are involved, the criterion must be adapted to incorporate the effect of the fatigue load condition. Namely, the maximum normal stress in the constituent is a function of the stress ratio, R ($=\sigma_{min}/\sigma_{max}$, where σ_{max} and σ_{min} are the maximum and minimum stress amplitudes, respectively), cycle number, N, and cycling frequency, ω. Accordingly, the constituent fatigue strength at the same cyclic load condition must be provided. The modified fatigue failure criterion is expressed as

$$\sigma^1(R, N, \omega) \geq \sigma_u(R, N, \omega) \tag{4.29.1}$$

or

$$\sigma^3(R, N, \omega) \leq -\sigma_{u,c}(R, N, \omega) \tag{4.29.2}$$

where $\sigma^1(R, N, \omega)$ and $\sigma^3(R, N, \omega)$ represent, respectively, the first and the third principal stresses in the constituent, and $\sigma_u(R, N, \omega)$ and $\sigma_{u,c}(R, N, \omega)$ are the constituent tensile and compressive strengths at the same fatigue loading condition, R, N and ω, as applied to the composite.

Two UD fibrous composites subjected to off-axial tensile fatigue loads were analyzed. The first one is a graphite/epoxy (AS/3501-5A) UD lamina. Awerbuch and Hahn (1981) made extensive experiments on this composite with a load ratio of R=0.1 and a cyclic frequency of ω=18 Hz. Their results are taken for comparison. The constituent elastic properties (Table 4.7), together with uniaxial static strengths of the composite, are listed in Table 4.9. From the uniaxial strengths, the fiber and matrix tensile strengths can be retrieved.

Table 4.9 Parameters of a UD graphite/epoxy composite (β=α=0.5)
(Longitudinal tensile strength X=1,836 MPa, transverse tensile strength Y=56.9 MPa)

Mater.	V	E_{11} (GPa)	E_{22} (GPa)	ν_{12}	ν_{23}	G_{12} (GPa)	E_T (GPa)	σ_Y (MPa)	σ_u (MPa)
Graphite	0.66	213.7	13.8	0.2	0.25	13.8	-	-	2,763.7
Epoxy	0.34	3.45	3.45	0.35	0.35	1.28	1.8	25	39.6

The retrieving is performed in the same way as was done in Section 4.4. Using the given elastic properties, the elastic coefficients in the required strength formulae are calculated as follows:

$$\alpha^f_{e1}=1.503,\ \ \alpha^m_{e1}=0.0243,\ \ \alpha^f_{e2}=1.146 \text{ and } \alpha^m_{e2}=0.716$$

Firstly, we get an initial estimation for the matrix strength from

$$\sigma^u_{22}=\frac{\sigma^m_Y}{\alpha^m_{e2}}+\frac{\sigma^m_u-\sigma^m_Y}{\alpha^m_{p2}}=Y\approx\frac{\sigma^m_u}{\alpha^m_{e2}} \quad \text{since } \alpha^m_{p2}\approx\alpha^m_{e2}$$

which gives a value of 40.8 MPa. Next, consider the condition (Eq. (4.28)),

$$\left(\frac{\sigma^m_Y}{\alpha^m_{e1}}+\frac{\sigma^m_u-\sigma^m_Y}{\alpha^m_{p1}}\right)_{\sigma^m_u=40.8\text{ MPa}}\geq X \tag{4.30}$$

Since $\left(\dfrac{\sigma^m_u}{\alpha^m_{e1}}\right)_{\sigma^m_u=40.8\text{ MPa}}$ =1679< X, the epoxy used cannot be considered as linearly elastic until rupture. As aforementioned, any combination of E^m_T and σ^m_Y, which satisfies Eq. (4.30), is possible due to no other information being available. Fortunately, for the UD composites which are statically determinate, it has been shown that as long as the transverse and the longitudinal strengths of the composite are determined, respectively, by the strengths of the matrix and the fiber, the chosen values of σ^m_Y and E^m_T have an insignificant effect on any predicted off-axis strength of the composite (Fig. 4.7(a)). Choosing σ_Y^m =25 MPa and E_T^m =1,800 MPa, we obtain

$$\alpha_{p1}^f = 1.509, \quad \alpha_{p1}^m = 0.0127, \quad \alpha_{p2}^f = 1.173 \text{ and } \alpha_{p2}^m = 0.663$$

The constituent tensile strengths are thus amended to $\sigma_u^m = (\sigma_{22}^u - \sigma_y^m / \alpha_{e2}^m)\alpha_{p2}^m + \sigma_y^m$ = 39.6 MPa and $\sigma_u^f = (\alpha_{e1}^f - \alpha_{p1}^f)\sigma_y^m / \alpha_{e1}^m + \sigma_{11}^u \alpha_{p1}^f$ =2763.7 MPa.

In order to predict an off-axis S-N curve of the composite under a given load ratio, R=0.1 and a cycling frequency, ω=18 Hz, the constituent fatigue strengths under the same condition must be provided. Unfortunately, no information about constituent fatigue strengths was reported. Instead, the measured longitudinal (0°-directional) and transverse (90°-directional) S-N data of the composite are used to retrieve the constituent fatigue strength parameters, similar to what has been done in the retrieving of the corresponding static parameters. However, the original fatigue data scatter was large. As such, a linear interpolation, which is based on a least-square approximation technique, is used to represent the S-N data of each off-axis angle. Namely, the fatigue data of each off-axis angle is firstly least-square interpolated to a linear function (by using, e.g., *Microsoft Excel* "Add Trendlines" selection plus an extrapolation). This is because the overall fatigue data in each off-axial direction is most properly represented by a linear function, as can be seen from Figs. 4.16 and 4.17. Furthermore, for those off-axial tests which did not provide enough data to cover the entire cycling range under consideration (i.e., from N=10^3 to N=10^6), an extrapolation from the corresponding interpolated function has to be made. In such a case, the linear interpolation generally results in the smallest extrapolation error. The original fatigue data and the corresponding linear interpolations for the off-axial tests of 0° and 90° are shown in Figs. 4.16 and 4.17. From the linear interpolation, the fatigue strengths under the off-axial tests of 0°, 10°, 20°, 30°, 45°, 60° and 90° at selected (integer) cycle numbers are obtained, and are summarized in Table 4.10.

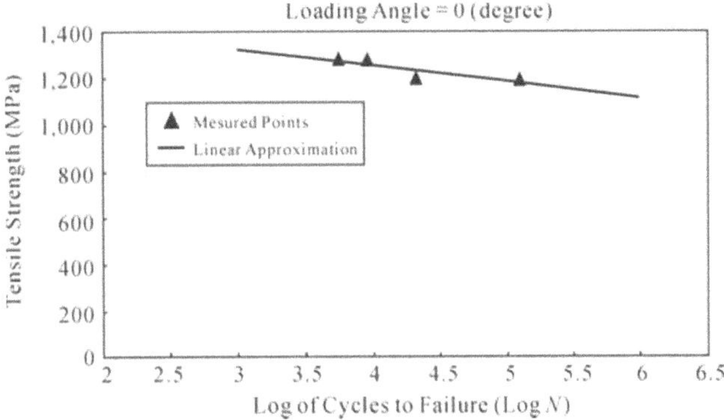

Fig. 4.16 Measured (Awerbuch & Hahn, 1981) S-N data of a UD graphite/epoxy composite loaded in 0° direction (from Huang, 2002)

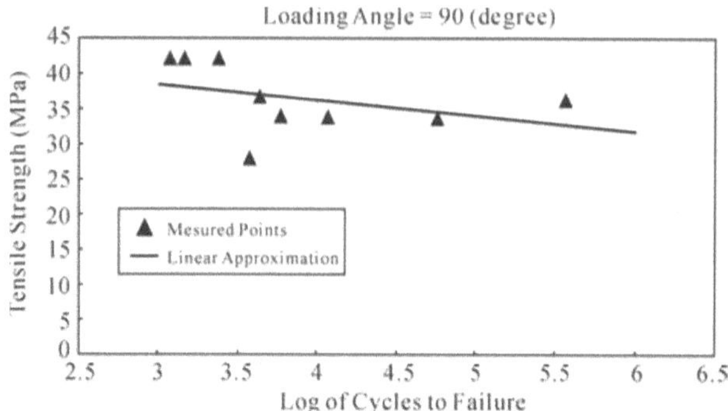

Fig. 4.17 Measured (Awerbuch & Hahn, 1981) S-N data of a UD graphite/epoxy composite loaded in 90° direction (from Huang, 2002)

The longitudinal (0°-directional) and the transverse (90°-directional) fatigue strengths of the composite, i.e., the data on the first and the last lines in Table 4.10, are then used to recover the tensile strength of the fiber and the tensile strength and plastic parameters of the matrix. The constituent compressive fatigue strengths are immaterial, because the overall applied fatigue loads are given in tension. At each cycle number, the recovery is carried out as though the composite had been subjected to an equivalent static tensile load. For example, at the cycle number of N=10^3, the constituent strengths and the matrix plastic parameters are extracted as though the composite had been subjected statically to an overall longitudinal strength of 1,323.7 MPa and an overall transverse strength of 38.5 MPa. The thus obtained constituent fatigue strengths at respective cycle numbers are summarized in Table 4.11. In the retrieving, while the matrix hardening modulus, 1.8 GPa, is kept unchanged, the matrix yield strength at each cycle number has to be adjusted so that the longitudinal fracture of the composite can be governed by the fiber's failure. Namely, an inequality similar to Eq. (4.30) can be fulfilled. Thus, the fatigue failures of the composite in the longitudinal direction at all the cycle numbers are considered to be caused by fiber fracture, whereas all the fatigue failures of the composite in the transverse direction result from the matrix failure.

Table 4.10 Measured off-axis fatigue strengths (MPa) of graphite/epoxy composite (R=0.1 and ω=18 Hz, Awerbuch and Hahn, 1981)

Angle	Cyclic number, N			
	10^3	10^4	10^5	10^6
0°	1,323.7	1,254.6	1,185.5	1,116.4
10°	295.8	257.6	219.5	181.3
20°	156.6	142.7	128.9	115.1
30°	65.5	68.7	71.9	75.2
45°	67.1	59.5	52.0	44.4
60°	47.0	45.3	43.5	41.8
90°	38.5	36.2	33.9	31.7

Table 4.11 Retrieved constituent fatigue properties of graphite/epoxy composite (R=0.1 and ω=18 Hz)

	Cyclic number, N			
	10^3	10^4	10^5	10^6
σ_u^f (MPa)	1,992.5	1,888.5	1,784.8	1,680.8
σ_u^m (MPa)	26.9	25.3	23.6	22.1
σ_Y^m (MPa)	18	17	15	14
E_T^m (GPa)	1.8	1.8	1.8	1.8

Using the constituent properties given in Tables 4.9 and 4.11, the off-axis static and fatigue strengths of the composite are estimated. The predictions indicate that, except for those specimens with off-axial load angles in a very close neighborhood to 0^0, the composite failures in all of the other off-axial directions are caused by the matrix fracture, no matter what kind of (static or fatigue) load has been applied to the composite. The predicted results are graphed in Figs. 4.18 – 4.22, respectively. For comparison, the experimental data taken from Awerbuch & Hahn (1981) (Table 4.10) are also shown in the corresponding figures. It is seen that the agreement between all the predictions and the experimental data is satisfactory.

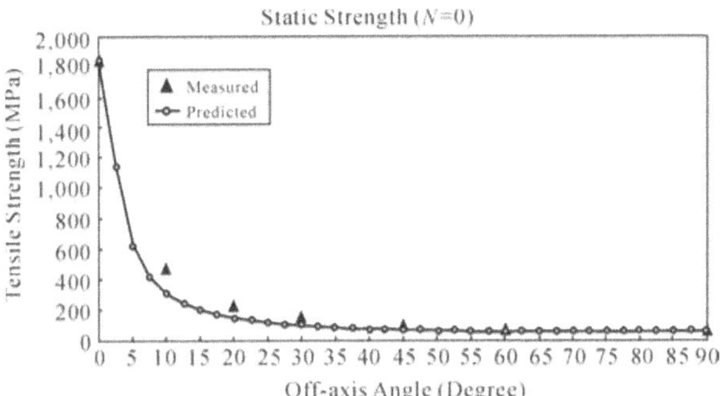

Fig. 4.18 Comparison between measured (Awerbuch & Hahn, 1981) and predicted off-axis static strengths of a UD graphite/epoxy composite (from Huang, 2002)

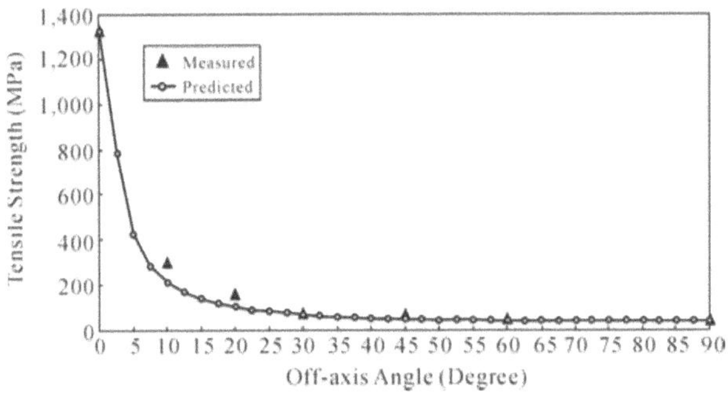

Fig. 4.19 Comparison between measured (Awerbuch & Hahn, 1981) and predicted off-axis fatigue strengths of a UD graphite/epoxy composite at N=10^3 (from Huang, 2002)

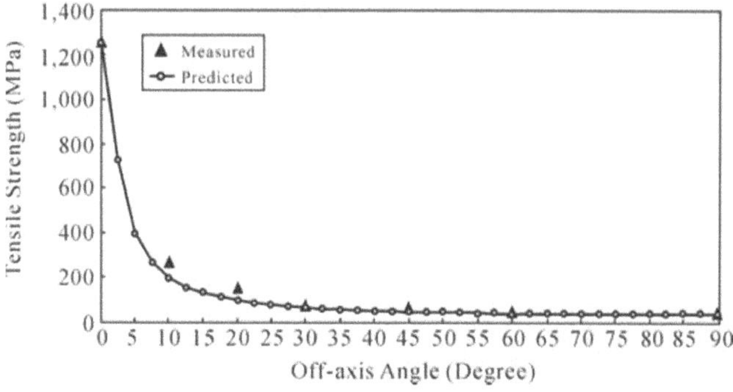

Fig. 4.20 Comparison between measured (Awerbuch & Hahn, 1981) and predicted off-axis fatigue strengths of a UD graphite/epoxy composite at N=10^4 (from Huang, 2002)

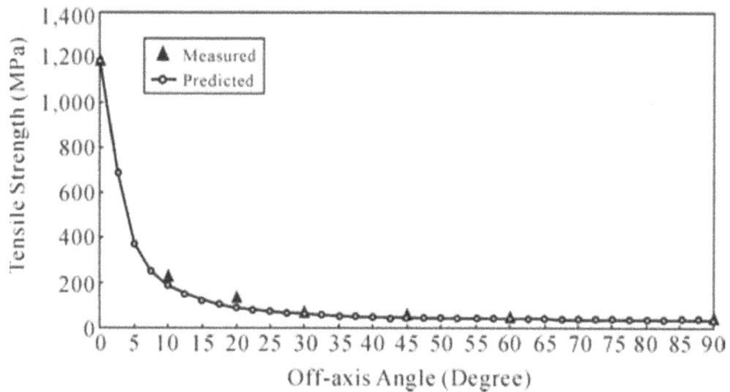

Fig. 4.21 Comparison between measured (Awerbuch & Hahn, 1981) and predicted off-axis fatigue strengths of a UD graphite/epoxy composite at N=10^5 (from Huang, 2002)

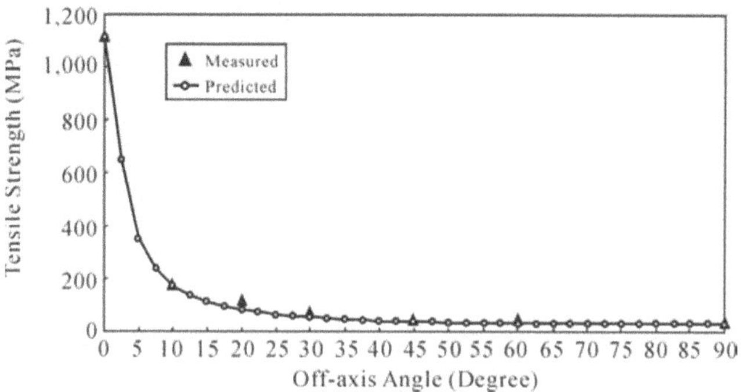

Fig. 4.22 Comparison between measured (Awerbuch & Hahn, 1981) and predicted off-axis fatigue strengths of a UD graphite/epoxy composite at N=10^6 (from Huang, 2002)

The second composite considered is a glass/epoxy UD lamina, having a fiber volume fraction of 0.6. The experimental work for this composite was done by Hashin and Rotem (1973). The load ratio was R=0.1 and the cyclic frequency ω=19 Hz. The elastic properties of the constituents have been listed in Table 4.4. No information on the constituent plastic and strength properties was reported. These data must be retrieved using overall strengths of the composite in two different directions (one should be the longitudinal direction). For the present composite, except for the static loading situation, all the fatigue tests were performed within a maximum off-axis angle of 60°. Therefore, except for the static strengths of the constituents, which are retrieved using the longitudinal and transverse test data, all the fatigue strengths of the fiber and the matrix at various cycle numbers are back-calculated from the corresponding longitudinal (0°) and 60° off-axial test data. These overall fatigue data, defined using the same technique as in the previous composite, are summarized in Table 4.12. The retrieving is similar to that performed for the previous graphite/epoxy composite. Namely, the longitudinal data (the third row in Table 4.12) are used to back-calculate the glass fiber strengths, whereas the 60° off-axial data are employed to recover the matrix strength and plastic parameters. The hardening modulus of the matrix, determined initially using the overall transverse static strength of the composite, has been kept unchanged. On the other hand, the yield strength is adjusted so that the composite failure in the 0° direction is initiated from fiber fracture whereas the composite failure in the 60° direction is caused by matrix failure. In this way, the fiber and matrix tensile strengths at each cycle number are also recovered. The retrieved constituent properties are summarized in Table 4.13.

Table 4.12 Measured off-axis fatigue strengths (MPa) of glass/epoxy composite (R=0.1 and ω=19 Hz, Hashin & Rotem, 1973)

Angle	Cyclic Number, N				
	10^2	10^3	10^4	10^5	10^6
0°	882.8	747.5	614.4	480	347.3
60°	29.8	27.3	25.1	22.9	20.3

Static Strength: X(longitudinal)=1,236 MPa and Y(transverse)=28.45 MPa

Table 4.13 Retrieved constituent fatigue properties of glass/epoxy composite

	Cycling Number, N					
	0	10^2	10^3	10^4	10^5	10^6
σ_u^f (MPa)	2,055	1,460	1,235	1,013	790	570
σ_u^m (MPa)	18.6	19.5	18	16.5	15	13.4
σ_Y^m (MPa)	13	13	13	13	13	12
E_T^m (GPa)	0.21	0.21	0.21	0.21	0.21	0.21

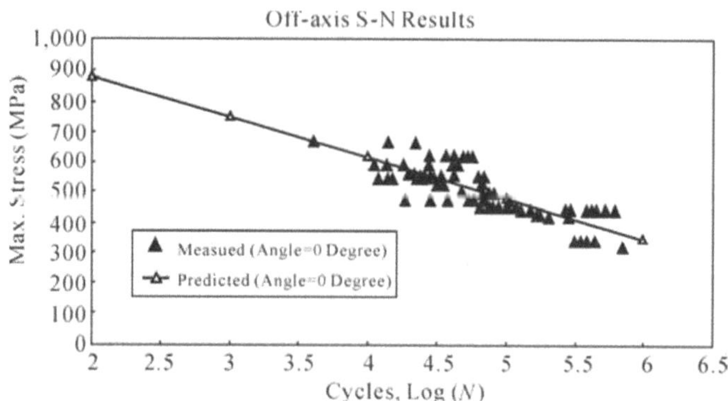

Fig. 4.23 Comparison between measured (Hashin & Rotem, 1973) and predicted S-N data of a glass/epoxy composite in longitudinal direction (from Huang, 2002)

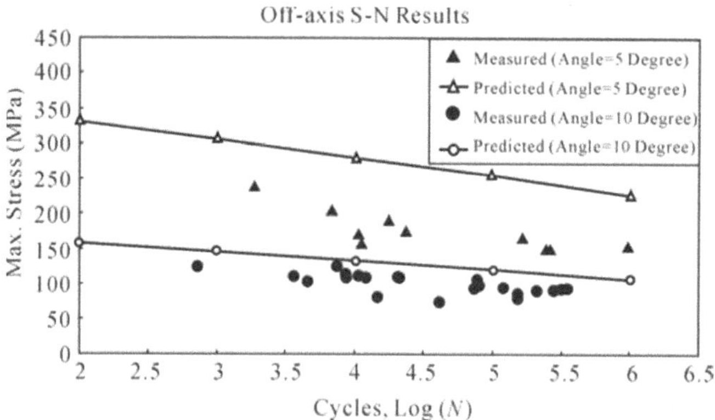

Fig. 4.24 Comparison between measured (Hashin & Rotem, 1973) and predicted off-axial S-N data of glass/epoxy composite with off-axial angles of 5° and 10° (from Huang, 2002)

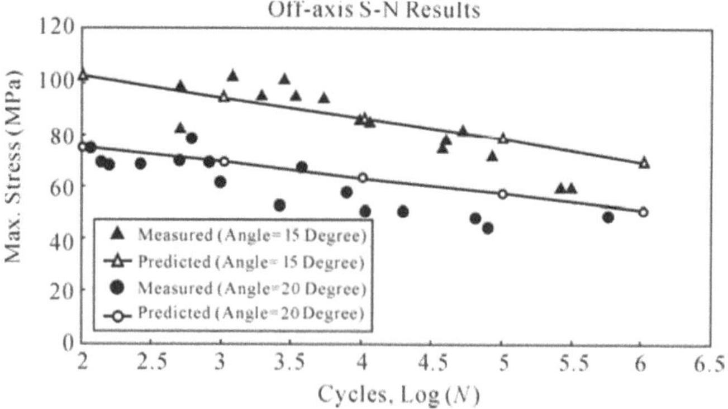

Fig. 4.25 Comparison between measured (Hashin & Rotem, 1973) and predicted off-axial S-N data of glass/epoxy composite with off-axial angles of 15° and 20° (from Huang, 2002)

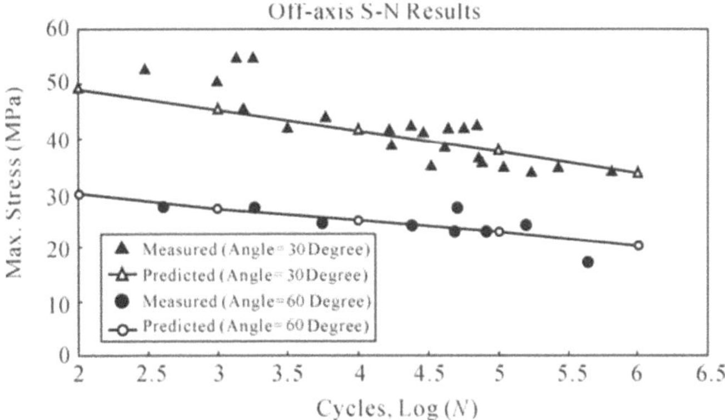

Fig. 4.26 Comparison between measured (Hashin & Rotem, 1973) and predicted off-axial S-N data of glass/epoxy composite with off-axial angles of 30° and 60° (from Huang, 2002)

Given constituent properties as input data, the predicted S-N curves for off-axial angles of 0°, 5°, 10°, 15°, 20°, 30° and 60° are plotted in Figs. 4.23 through 4.26, respectively. Results indicate that, except for the off-axial angle of 5°, the predicted S-N curves in all the other off-axial directions correlate quite well with the experimental data. The discrepancy between the predicted and measured results for the off-axial angle of 5° might be due to some originally inaccurate measurement. From Fig. 4.7(a), one can clearly see that when the off-axial angle is less than 5°, the ultimate strength of the UD lamina is very sensitive to any variation of this angle. Therefore, a very small error in the original preparation for a 5° specimen would give rise to a large degree of inaccuracy in the strength measurement.

Remark 4.2

In this chapter, only in-plane (two-dimensional) load examples have been considered. This is because experiments with a two-dimensional load condition are relatively easy to conduct. However, the theory presented herein is equally applicable to the strength prediction of a unidirectional composite subjected to three-dimensional loads. By using a 3D form of the bridging matrix, Eq.(3.70), the internal stress increments in the constituent fiber and matrix materials due to an applied load increment in three-dimensions on the composite, $\{d\sigma_{11}$ (or dN_{11}), $d\sigma_{22}$ (or dN_{22}), $d\sigma_{33}$ (or dN_{33}), $d\sigma_{23}$, $d\sigma_{13}$, $d\sigma_{12}\}$, can be determined through Eqs. (3.68.1) and (3.68.2). The total stresses at the current load level, $\left\{\sigma^f_{11}, \sigma^f_{22}, \sigma^f_{33}, \sigma^f_{23}, \sigma^f_{13}, \sigma^f_{12}\right\}$ and $\left\{\sigma^m_{11}, \sigma^m_{22}, \sigma^m_{33}, \sigma^m_{23}, \sigma^m_{13}, \sigma^m_{12}\right\}$, are updated as per Eqs. (3.75.1) and (3.75.2). Then the principal stresses of the fiber and matrix are derived from Eq. (4.1), and a maximum normal stress criterion or a generalized maximum normal stress criterion can be employed to detect the failure of the constituent materials. Similarly, as done for a two-dimensional load condition, the failure of the composite is considered to occur as long as any of its constituent materials attains a failure stress state.

Unfortunately, except for limited work on the effects of hydrostatic pressures on a composite strength (Pae & Rhee, 1995; Rhee et al., 2003; Zinoviev et al., 2001; Hine et al., 2005), few experimental data can be found in the open literature. In order to promote the development of three-dimensional strength theories for composite materials, a worldwide failure exercise (called the second worldwide failure exercise, WWFE- II) was initiated by Hinton and Kaddour in 2006. Typical problems were designed by the organizers to check the predictive capacities of the theories for composite materials under some kinds of tri-axial loads. The bridging model analysis for WWFE- II problems will be illustrated in the next chapter.

References

Aboudi, J. (1988) Micromechanical analysis of the strength of unidirectional fiber composites. Comp. Sci. Tech. 33, 79-96.

Aboudi, J. (1989) Micromechanical analysis of composites by the method of cells. Applied Mechanics Reviews 42(7), 193-221.

ASTM Int. (1994) ASM Handbook (OH 440730002, Vol. 6, p.992). The Materials Information Society, Materials Park, USA.

Awerbuch, J. & Hahn, H.T. (1981) Off-Axis Fatigue of Graphite/Epoxy Composite, ASTM STP 723, 243-273.

Bailey, J.E., Curtis, P.T. & Parvisi, A. (1979) On the transverse cracking and longitudinal splitting of glass and carbon fiber reinforced epoxy cross ply laminates and the effect of Poisson and thermally generated strain. Proc. R. Soc. Lond. A366, 599-623.

Benveniste, Y. & Dvorak, G.J. (1990) On a Correspondence between Mechanical and Thermal Effects in Two-Phase Composites, in G. J.Weng, M. Taya & H. Abe (eds.), The Toshio Muta Anniversary Volume: Micromechanics and Inhomogeneity (pp.65-81). New York: Springer.

Brindley, P.K., Draper, S.L., Eldridge, J.I., Nathal, M.V. & Arnold, S.M. (1992) The Effect of Temperature on the Deformation and Fracture of SiC/Ti-24Al-11Nb, Metallurgical Transactions A, 23A, 2527-2540.

Bushby, R.S. (1998) Evaluation of continuous alumina fiber reinforced composites based upon pure aluminium. Mater. Sci. Tech., 14, 877-886.

Chun, H.J. & Daniel, I.M. (1996) Behavior of a unidirectional metal-matrix composite under thermomechanical loading. J. Eng. Mater. Tech. ASME 118, 310-316.

Cooper, G.A. (1974) Micromechanics Aspects of Fracture and Toughness, Composite Materials, Vol. 5, Brontman & Krock (eds.), Academic Press, p. 415.

Echaabi, J., Trochu, F. & Gauvin, R. (1996) Review of failure criteria of fibrous composite materials. Polymer Composites 17(6), 786-798.

Flaggs, D.L. & Kural, M.H. (1982) Experimental determination of the in-situ transverse laminate strength in graphite epoxy laminates. J. Comp. Mater. 16, 103-116.

Gundel, D.B. & Wawner, F.E. (1997) Experimental and theoretical assessment of the longitudinal tensile strength of unidirectional SiC-Fiber/ Titanium-Matrix composites. Comp. Sci. & Tech. 57, 471-481.

Hashin, Z. & Rotem, A. (1973) A fatigue failure criterion for fiber Reinforced Composites. J. Comp. Materials 7, 448-464.

Hart-Smith, L.J. (1998a) Predictions of the original and truncated maximum-strain failure models for certain fibrous composite laminates. Comp. Sci. Tech. 58(7), 1151.

Hart-Smith, L.J. (1998b) Predictions of a generalized maximum-shear-stress failure criterion for certain fibrous composite laminates. Comp. Sci. Tech. 58(7), 1179.

Hine, P.J., Duckett, R.A., Kaddour, A.S., et al. (2005) The effect of hydrostatic pressure on the mechanical properties of glass fiber/epoxy unidirectional composites. Composites: Part A, 36, 279-289.

Hinton M.J., Soden P.D. (1998) Predicting failure in composite laminates: the background to the exercise. Comp. Sci. Tech. 58, 1001-1010.

Huang, Z.M. (1999) Micromechanical strength formulae of unidirectional composites. Materials Letters 40(4), 164-169.

Huang, Z.M., Ramakrishna, S. and Tay, A.A.O. (1999b) A micromechanical approach to the tensile strength of a knitted fabric composite. J. Comp. Mater. 33(19), 1758-1791.

Huang, Z.M. (2000a) A unified micromechanical model for the mechanical properties of two constituent composite materials, Part I: Elastic behavior. J. Thermoplastic Comp. Mater. 13(4), 252-271.

Huang, Z.M. (2000b) A unified micromechanical model for the mechanical properties of two constituent composite materials. Part II: Plastic behavior. J. Thermoplastic Comp. Mater. 13(5), 344-362.

Huang, Z.M. (2000c) A unified micromechanical model for the mechanical properties of two constituent composite materials, Part V: Laminate strength. J. of Thermoplastic Comp. Mater. 13(3), 190-206.

Huang, Z.M. (2000d) Strength formulae of unidirectional composites including thermal residual stresses. Materials Letters, 43(1-2), 36-42.

Huang, Z.M. (2000e) Tensile strength of fibrous composites at elevated temperature. Mater. Sci. Tech. 16(1), 81-94.

Huang, Z.M. (2001a) Simulation of the mechanical properties of fibrous composites by the bridging micromechanics model. Composites Part A, 32(2), 143-172.

Huang, Z.M. (2001b) Micromechanical prediction of ultimate strength of transversely isotropic fibrous composites. International Journal of Solids & Structures, 38, 4147-4172.

Huang, Z.M. (2002) Micromechanical modeling of fatigue strength of unidirectional fibrous composites. International Journal of Fatigue, 24(6), 659-670.

Huang, Z.M. (2004) A bridging model prediction of the ultimate strength of composite laminates subjected to biaxial loads. Comp. Sci. Tech., 64, 395-448.

Labossiere, P. & Neal, K.W. (1987) Macroscopic failure criteria for fiber-reinforced composite materials. Solid Mech. Arch. 12, 439-450.

Levin, V.M. (1967) On the coefficients of thermal expansion of heterogeneous materials. Mekhanika Tverdovo Tela, 1, 88.

Matsuda, N. & Matsuura, K. (1997) High temperature deformation and fracture behavior of continuous Alumina Fiber Reinforced Aluminium Composites with different fiber orientation. Mater. Trans. JIM 38, 205-214.

Nahas, M.N. (1986) Survey of Failure and Post-Failure Theories of Laminated Fiber-Reinforced Composites. J. Comp. Tech. Res. 8(4), 138-153.

Pae, K.D. & Rhee, K.Y. (1995) Effects of hydrostatic pressure on the compressive behavior of thick laminated 45° and 90° unidirectional graphite-fiber/epoxy-matrix composites. Composites Science and Technology 53, 281-287.

Pipes, R. B. & Cole, B. W. (1973) On the off-axis strength test for anisotropic materials. *J. Comp. Mater.*,7: 246-256.

Pindera, M. J. & Herakovich, C. T. (1981) An endochronic theory for transversely isotropic fibrous composites. *VPI-E-81-27*, Virginia polytechnic institute and state university.

Rhee, K.Y., Chi, C. H., Park, S. J. (2003) Experimental investigation on the compressive characteristics of multi-directional graphite/epoxy composites under hydrostatic pressure environment. Materials Science and Engineering A360, 1-6.

Robertson, D. & Mall, S. (1997) Micromechanical analysis and modeling, in Titanium matrix composites—mechanical behavior (pp. 397-464), eds. by S. Mall & T. Nicholas. Lancaster, Basel:Technomic Publishing Co., Inc.

Rotem, A. & Hashin, Z. (1975) Failure modes of angle ply laminates. J. Comp. Mater. 9, 191-206.

Rotem, A. (1998) Prediction of laminate failure with the Rotem failure criterion. Comp. Sci. Tech. 58(7), 1083.

Rowlands, R.E. (1985) Strength (Failure) Theories and their experimental correlation, in G. C. Sih & A. M. Skudra (eds.), Failure Mechanics of Composites (pp. 71-125), North-Holland, Amsterdam.

Soden, P.D., Hinton, M.J., & Kaddour, A.S. (1998) A comparison of the predictive capabilities of current failure theories for composite laminates. Comp. Sci. Tech 58, 1225-1254.

Sun, C.T. & Tao, J.X. (1998) Prediction of failure envelopes and stress/strain behavior of composite laminates. Comp. Sci. Tech. 58(7), 1125.

Timoshenko, S.P. (1953) History of Strength of Materials. New York: McGraw Hill.

Tsai, S.W. & Wu, E.M. (1971) A General Theory of Strength for Anisotropic Materials. J. Comp. Mater. 5(1), 58-80.

Tsai, S.W. & Hahn, H.T. (1980) Introduction to Composite Materials. Lancaster, Basel:Technomic Publishing Co., Inc.

Zinoviev, P.A., Tsvetkov, S.V., Kulish, G.G., et al. (2001) The behavior of high-strength unidirectional composites under tension with superposed hydrostatic pressure. Comp. Sci. & Tech. 61(8), 1151-1161.

Zweben, C. (1990) Static strength and elastic properties, in Mechanical Behavior and Properties of Composite Materials, Delaware Composites Design Encyclopedia, eds. by C. Zweben, H. T. Hahn & T. W. Chou. Lancaster, Basel: Technomic Publishing Co., Inc. Vol. 1, pp. 49-70.

5

Strength of Multidirectional Laminates

5.1 Introduction

Most composite materials used in engineering applications are in the form of laminates, consisting of multiple laminae, or plies or layers, which are oriented in different directions and are bonded together in an integral structural unit. This is because UD composites are extremely weak in the direction transverse to the fibers, as a result of poor matrix properties. For any structural application, fiber materials should be placed in more than one direction. Otherwise, even a secondary load condition in the transverse direction can cause the composite to fail.

On the other hand, the mechanical characteristics of laminated composites are much more complicated than those of UD composites. The virtually limitless combinations of ply materials, ply orientations and ply stacking sequences offered by laminated construction considerably enhance the design flexibility inherent in composite structures and thus cause difficulty in laminate analysis, especially in the understanding of laminate load carrying capacity.

A fundamental reason for this difficulty is that while a UD lamina is generally statically determinate under e.g., a testing load condition, it becomes always statically indeterminate in the laminate. The load shared by the lamina cannot be determined only using an equilibrium condition. The lamina deformation should also be clearly understood. In general, a multiaxial stress state occurs in the lamina, even though the laminate itself is subjected to an overall uniaxial load. Experimental evidence is indicated in Fig. 5.1 (Zhang et al., 2000, 2001). Fig. 5.1(a) shows that when an isolated single layer of knitted fabric composite was subjected to a uniaxial tensile load, which was at an inclined angle of 45° in the fabric wale direction, the composite failed along the inclined plane of 45° in the loading direction. However, when the same composite layer was arranged into a laminate (Fig. 5.1(b)), which was subjected to the similar load condition as in the previous case, the composite layer failed along the plane perpendicular to the loading direction, as indicated in Fig. 5.1(c). Hence, the composite layer in the latter case, i.e. taken from the laminate, must have been subjected to an additional in-plane shear load (Zhang et al., 2001).

A further complication regarding laminate failure analysis comes from the fact that a different lamina in the laminate generally carries a different load share, as a result of a different stacking arrangement. Thus, some lamina must fail first before others. However, the failure of the laminated composite in any one ply does not imply failure of the others, nor does it mean the rupture of the whole laminate. A progressive failure process essentially exists in the laminate, and a stiffness discount must be applied to the failed lamina for the remaining laminate analysis.

In this chapter, the bridging model developed in Chapter 3 is combined with a laminate theory to simulate mechanical properties, especially strengths of multidirectional tape laminates subjected to various load conditions. The term "tape" used here refers to a laminate in which each layer is an angle-ply flat UD composite.

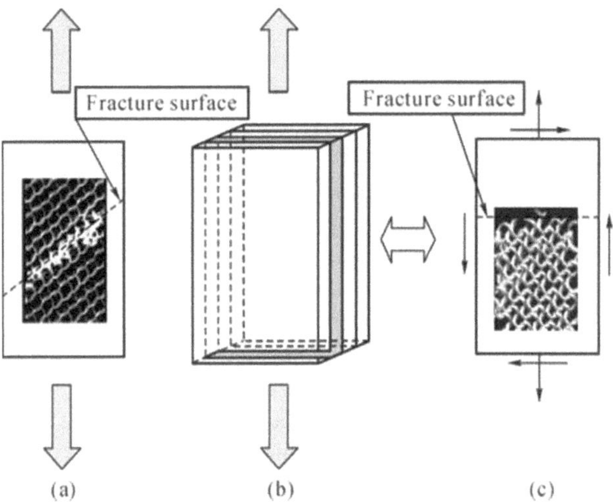

Fig. 5.1 A schematic diagram of (a) single layer knitted fabric composite under 45° off-axial tension; (b) [0/±45/0] laminate under uniaxial tension and (c) load sharing of 45° angle-ply lamina

5.2 Stacking Code and Global Coordinates

Given a laminate consisting of multi-layers of laminae (e.g., a three-layer laminate in Fig. 5.2(a)), a global coordinate system, (x, y, z), is assigned in such a way that x and y are on the middle surface of the laminate and z is in the laminate thickness direction, as shown in Fig. 5.2(b). There is no requirement that the global (x, y) should be located on the central plane of a lamina. In fact, each lamina layer can have a different thickness. It is only necessary that the (x, y) are on the middle-plane of the laminate so that the distances from the bottom and top surfaces to the plane of z=0 are equal.

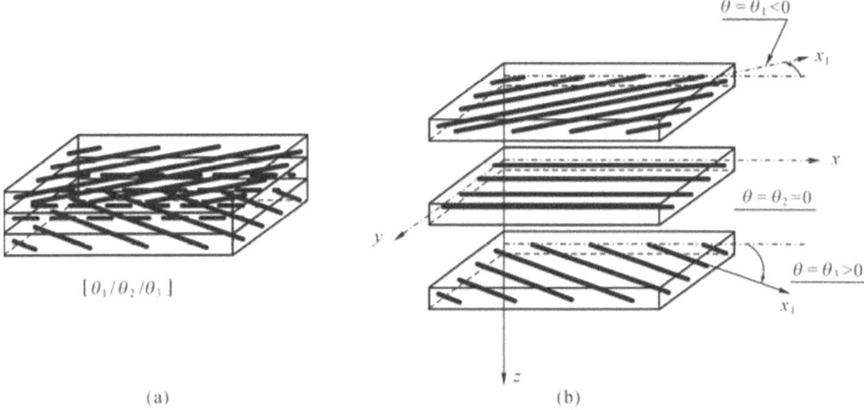

Fig. 5.2 Stacking code and global coordinates

Having chosen the global coordinate system, each ply in the laminate is identified by a ply angle θ, called a stacking angle and measured in degrees, between its fiber axial direction and the global x-axis. By convention, the ply angle is positive if traveling from the x-axis to the fiber direction is the same orientation as that from the x-axis to the y-axis. Otherwise the angle is negative, see Fig. 5.2(b) for a definition. The lamination sequence is written between a pair of square brackets, as $[\theta_1/\theta_2/\theta_3]$ in Fig. 5.2(a), starting with the first ply lay up. Here and throughout the book, the "first ply" as a geometrical definition is used to refer to the ply that has the smallest (the largest negative) z coordinate. Therefore, the lamination sequence can be specified only when a global coordinate system has been chosen. Although an upward or downward direction of the z-axis is immaterial, it is necessary that the coordinates (x, y, z) obey the right-hand screwing rule. The reader may have noticed the difference in the present definition for the lamination sequence compared to some others in the literature. However, the present definition has been incorporated into the computer codes for the laminate analysis attached to this book. Any correct use of those codes must follow the present definition. Thus, if the global z coordinate is chosen downward, the first ply refers to the top ply. However, if the z-axis is chosen upward, the first ply denotes the bottom ply. For example, the lamination sequence used in Fig. 5.2(a), $[\theta_1/\theta_2/\theta_3]$, indicates that the θ_1-ply has the largest negative z-coordinate whereas the θ_3-ply has the largest positive z-coordinate, with the z-coordinate of the θ_2-ply in between.

Adjacent parallel plies are denoted by a subscript which indicates the number of such plies. Adjacent non-parallel plies are separated using an oblique line. An overall subscript s outside the square bracket is used to designate a symmetric laminate, in which case only one-half of the lamination sequence, starting from the first-ply (i.e., the ply having the largest negative z coordinate) to the ply that has a zero z-coordinate, is listed. Adjacent plies of equal but opposite ply angle are denoted by $\pm\theta$ (or $\mp\theta$) with an understanding that the $+\theta$ ply (or the $-\theta$ ply) has a larger negative z coordinate than the $-\theta$ ply (or the $+\theta$ ply) does. The lamination sequences for which the fiber directions in adjacent plies are perpendicular to each other, for example $[0/90]_s$ and $[\pm45]_s$, are also called cross-plies. Non-perpendicular sequences are named as angle-plies. Some examples of lamination sequences are indicated in Fig. 5.3.

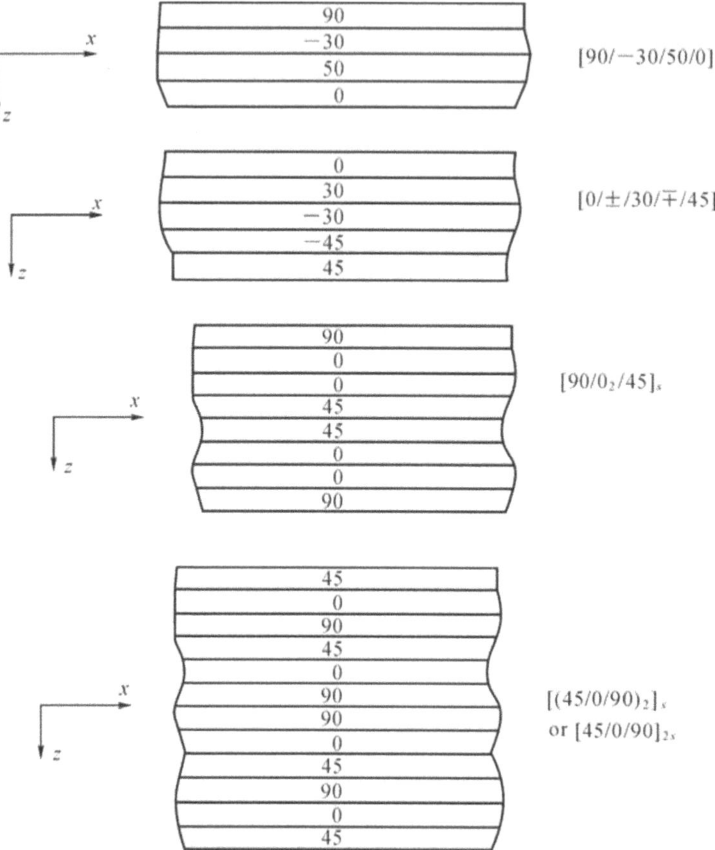

Fig. 5.3 Examples of lamination sequences

5.3 Classical Laminate Theory

In this section, the classical laminate theory is shown and is incorporated with the bridging model developed in Chapter 3 to analyze stiffness and, especially, strength of laminated composites. The section is subdivided into several sub-sections, dealing with respectively isothermal theory, convention for a positive shear stress, and thermal analysis, followed by a summary of all of the formulae for the thermal-mechanical analysis of a laminate.

5.3.1 Isothermal Theory

As the thickness of a laminate is generally small, relative to the laminate in-plane dimensions, the so-called classical lamination theory (Smith, 1953; Pister & Dong, 1959; Reissner & Stavsky, 1961; Stavsky, 1964; Lekhnitskii, 1968) is applicable. This theory has been developed essentially based on the two classical Kirchhoff hypotheses for the deformation of thin plate and shell structures (Timoshenko, 1940), which states that: (1) the strain components in the thickness direction are negligible and (2) a straight line normal to the laminate middle-surface remains straight and normal to the middle-surface of the laminate after deformation. Mathematically, the first assumption implies that there are only three in-plane strain components in the laminate whereas the second one designates that a displacement variation in the thickness direction, if any, is linear.

Suppose that the laminated composite, consisting of a number of unidirectional laminae and stacked at different ply-angles, together with the global coordinate system, is indicated in Fig. 5.2. According to the classical lamination theory (the first basic assumption stated above), only the in-plane stress and strain increments, i.e., $\{\mathrm{d}\sigma_i\}^G = \{\mathrm{d}\sigma_{xx}, \mathrm{d}\sigma_{yy}, \mathrm{d}\sigma_{xy}\}^{\mathrm{T}}$ and $\{\mathrm{d}\varepsilon_i\}^G = \{\mathrm{d}\varepsilon_{xx}, \mathrm{d}\varepsilon_{yy}, 2\mathrm{d}\varepsilon_{xy}\}^{\mathrm{T}}$, are retained, where "$G$" refers to the global coordinate system. All of the three out-of-plane strain increments, i.e., $\mathrm{d}\varepsilon_{xz}$, $\mathrm{d}\varepsilon_{yz}$ and $\mathrm{d}\varepsilon_{zz}$, are zero, whereas the out-of-plane stress increments can be obtained once the in-plane strain increments are determined using, e.g., an equation similar to Eq. (1.4.2). Let us now derive a relation correlating the in-plane stress and strain increments.

According to the basic assumptions, the incremental displacements at any material point, (x, y, z), can be expressed as

$$\begin{aligned} \mathrm{d}u &= \mathrm{d}u(x, y, z) = \mathrm{d}u^0(x, y) - z\mathrm{d}\varphi_x(x, y) \\ \mathrm{d}v &= \mathrm{d}v(x, y, z) = \mathrm{d}v^0(x, y) - z\mathrm{d}\varphi_y(x, y) \\ \mathrm{d}w &= \mathrm{d}w(x, y, z) \equiv \mathrm{d}w^0(x, y) \end{aligned} \tag{5.1}$$

where $\mathrm{d}u^0$, $\mathrm{d}v^0$, and $\mathrm{d}w^0$ are the displacement increments of the middle surface of the laminate along the x, y and z directions, respectively (total displacements are obtained just through simple updating). In Eq. (5.1), φ_x and φ_y are two in-plane functions, which are introduced to represent rotation angles of the normal lines of the middle surface.

Substituting Eq. (5.1) into the strain Eq. (1.1), we yield

$$2\mathrm{d}\varepsilon_{xz} = \frac{\partial(\mathrm{d}u)}{\partial z} + \frac{\partial(\mathrm{d}w)}{\partial x} = -\mathrm{d}\varphi_x(x, y) + \frac{\partial(\mathrm{d}w)}{\partial x},$$

$$2\mathrm{d}\varepsilon_{yz} = \frac{\partial(\mathrm{d}v)}{\partial z} + \frac{\partial(\mathrm{d}w)}{\partial y} = -\mathrm{d}\varphi_y(x, y) + \frac{\partial(\mathrm{d}w)}{\partial y}$$

Making use of the first assumption leads to

$$\mathrm{d}\varphi_x(x,y)=\frac{\partial(\mathrm{d}w)}{\partial x} \quad \text{and} \quad \mathrm{d}\varphi_y(x,y)=\frac{\partial(\mathrm{d}w)}{\partial y} \tag{5.2}$$

From Eq. (1.1) and by virtue of Eqs. (5.1) and (5.2), the global in-plane strain increments of the laminate at the material point (x, y, z) are expressed as

$$\mathrm{d}\varepsilon_{xx}=\mathrm{d}\varepsilon_{xx}^0+z\mathrm{d}\kappa_{xx}^0,\ \mathrm{d}\varepsilon_{yy}=\mathrm{d}\varepsilon_{yy}^0+z\mathrm{d}\kappa_{yy}^0,\ 2\mathrm{d}\varepsilon_{xy}=2\mathrm{d}\varepsilon_{xy}^0+2z\mathrm{d}\kappa_{xy}^0 \tag{5.3}$$

where $$\mathrm{d}\varepsilon_{xx}^0=\frac{\partial(\mathrm{d}u^0)}{\partial x},\ \mathrm{d}\varepsilon_{yy}^0=\frac{\partial(\mathrm{d}v^0)}{\partial y},\ \mathrm{d}\varepsilon_{xy}^0=\frac{1}{2}\left(\frac{\partial(\mathrm{d}u^0)}{\partial y}+\frac{\partial(\mathrm{d}v^0)}{\partial x}\right) \tag{5.4.1}$$

and $$\mathrm{d}\kappa_{xx}^0=-\frac{\partial^2(\mathrm{d}w^0)}{\partial x^2},\ \mathrm{d}\kappa_{yy}^0=-\frac{\partial^2(\mathrm{d}w^0)}{\partial y^2},\ \mathrm{d}\kappa_{xy}^0=-\frac{\partial^2(\mathrm{d}w^0)}{\partial x\partial y} \tag{5.4.2}$$

are the strain and the curvature increments of the middle surface, respectively.

Note that the above stress and strain components, $\{\mathrm{d}\sigma\}^G$ and $\{\mathrm{d}\varepsilon\}^G$, are expressed in the laminate global coordinate system. In order to perform further analysis, we need a constitutive relationship that connects the two quantities together. As each layer is a unidirectional lamina in its local coordinate system, the lamina constitutive relationship, Eq. (3.66.5), is applicable. However, Eq. (3.66.5) is expressed in a local coordinate system, whereas $\{\mathrm{d}\sigma_i\}^G$ and $\{\mathrm{d}\varepsilon_i\}^G$ are in the global one. Thus, a coordinate transformation is required. From Eqs. (1.76.2) and (1.80.1), we obtain

$$\{\mathrm{d}\sigma_i\}=([T_{ij}]_c)^{-1}\{\mathrm{d}\sigma_j^G\}=[T_{ij}]_s^{\mathrm{T}}\{\mathrm{d}\sigma_j\}^G \tag{5.5.1}$$

whereas from Eqs. (1.78) and (1.80.2), it follows that

$$\{\mathrm{d}\varepsilon_i\}=([T_{ij}]_s)^{-1}\{\mathrm{d}\varepsilon_j^G\}=[T_{ij}]_c^{\mathrm{T}}\{\mathrm{d}\varepsilon_j\}^G \tag{5.5.2}$$

Substituting Eqs. (5.5.1) and (5.5.2) into Eq. (3.66.5) and making necessary inversion yields

$$\{\mathrm{d}\sigma_i\}^G=([T_{in}]_c)_k([S_{nm}]_k)^{-1}([T_{mj}]_c^{\mathrm{T}})_k\{\mathrm{d}\varepsilon_j\}^G=[C_{ij}]_k^G\{\mathrm{d}\varepsilon_j\}^G=[(C_{ij}^G)_k]\{\mathrm{d}\varepsilon_j\}^G \tag{5.6}$$

where $$[(C_{ij}^G)_k]=([T_{in}]_c)_k([S_{nm}]_k)^{-1}([T_{mj}]_c^{\mathrm{T}})_k \tag{5.7}$$

is an instantaneous stiffness matrix of the kth lamina in the global coordinate system. For the current 2D problem, the coordinate transformation matrix, $[T_{ij}]_c$, is given by Eq. (1.86). It should be pointed out that the lamina compliance matrix, $[S_{nm}]_k$, used in Eq. (5.7) is a 3×3 matrix (i.e., n, m=1, 2, 3) for a 2D response (Section 3.9). Hence, the stiffness, $[(C_{ij}^G)_k]$, has the same dimension as that of the $[S_{nm}]_k$.

Suppose that the overall applied in-plane force increments per unit length (width) on the laminate are given by $\mathrm{d}N_{xx}$, $\mathrm{d}N_{yy}$ and $\mathrm{d}N_{xy}$ whereas the overall applied moment increments per unit length (width) are denoted by $\mathrm{d}M_{xx}$, $\mathrm{d}M_{yy}$ and $\mathrm{d}M_{xy}$. A schematic diagram of these forces and moments and their "positive direction" is shown in Fig. 5.4. These forces and moments must be balanced with the internal stress resultants. Thus, according to the equilibrium condition, we obtain

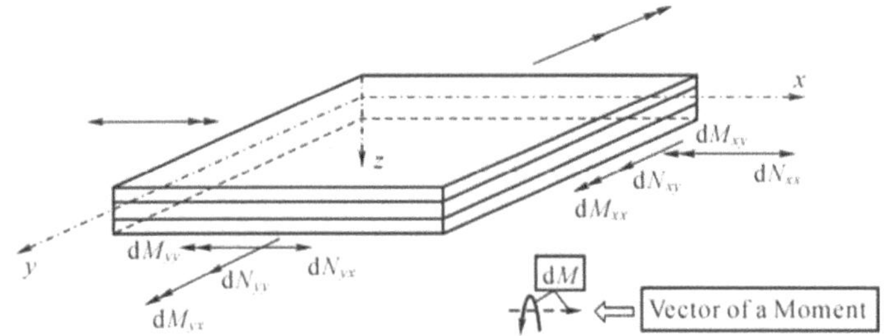

Fig. 5.4 Definition of stress resultants per unit length (width)

$$\begin{Bmatrix} \mathrm{d}N_{xx} \\ \mathrm{d}N_{yy} \\ \mathrm{d}N_{xy} \end{Bmatrix} = \int_{-h/2}^{h/2} \begin{Bmatrix} \mathrm{d}\sigma_{xx} \\ \mathrm{d}\sigma_{yy} \\ \mathrm{d}\sigma_{xy} \end{Bmatrix} \mathrm{d}z = \int_{-h/2}^{h/2} [(C_{ij}^{G})] \begin{Bmatrix} \mathrm{d}\varepsilon_{xx} \\ \mathrm{d}\varepsilon_{yy} \\ 2\mathrm{d}\varepsilon_{xy} \end{Bmatrix} \mathrm{d}z = \sum_{k=1}^{N} [(C_{ij}^{G})_k] \int_{z_{k-1}}^{z_k} \begin{Bmatrix} \mathrm{d}\varepsilon_{xx} \\ \mathrm{d}\varepsilon_{yy} \\ 2\mathrm{d}\varepsilon_{xy} \end{Bmatrix} \mathrm{d}z \tag{5.8.1}$$

$$\begin{Bmatrix} \mathrm{d}M_{xx} \\ \mathrm{d}M_{yy} \\ \mathrm{d}M_{xy} \end{Bmatrix} = \int_{-h/2}^{h/2} \begin{Bmatrix} \mathrm{d}\sigma_{xx} \\ \mathrm{d}\sigma_{yy} \\ \mathrm{d}\sigma_{xy} \end{Bmatrix} z\mathrm{d}z = \int_{-h/2}^{h/2} [(C_{ij}^{G})] \begin{Bmatrix} \mathrm{d}\varepsilon_{xx} \\ \mathrm{d}\varepsilon_{yy} \\ 2\mathrm{d}\varepsilon_{xy} \end{Bmatrix} z\mathrm{d}z = \sum_{k=1}^{N} [(C_{ij}^{G})_k] \int_{z_{k-1}}^{z_k} \begin{Bmatrix} \mathrm{d}\varepsilon_{xx} \\ \mathrm{d}\varepsilon_{yy} \\ 2\mathrm{d}\varepsilon_{xy} \end{Bmatrix} \mathrm{d}z \tag{5.8.2}$$

where $h = \sum_{k=1}^{N} (z_k - z_{k-1})$ (5.8.3)

is the thickness of the laminate. z_k and z_{k-1} are the global z coordinates of the top and bottom surfaces of the kth lamina, respectively, and N is the total number of lamina plies in the laminate. Writing out Eqs. (5.8.1) and (5.8.2) by using Eq. (5.3) explicitly gives

$$\begin{Bmatrix} \mathrm{d}N_{xx} \\ \mathrm{d}N_{yy} \\ \mathrm{d}N_{xy} \\ \mathrm{d}M_{xx} \\ \mathrm{d}M_{yy} \\ \mathrm{d}M_{xy} \end{Bmatrix} = \begin{bmatrix} Q_{11}^{I} & Q_{12}^{I} & Q_{13}^{I} & Q_{11}^{II} & Q_{12}^{II} & Q_{13}^{II} \\ Q_{12}^{I} & Q_{22}^{I} & Q_{23}^{I} & Q_{12}^{II} & Q_{22}^{II} & Q_{23}^{II} \\ Q_{13}^{I} & Q_{23}^{I} & Q_{33}^{I} & Q_{13}^{II} & Q_{23}^{II} & Q_{33}^{II} \\ Q_{11}^{II} & Q_{12}^{II} & Q_{13}^{II} & Q_{11}^{III} & Q_{12}^{III} & Q_{13}^{III} \\ Q_{12}^{II} & Q_{22}^{II} & Q_{23}^{II} & Q_{12}^{III} & Q_{22}^{III} & Q_{23}^{III} \\ Q_{13}^{II} & Q_{23}^{II} & Q_{33}^{II} & Q_{13}^{III} & Q_{23}^{III} & Q_{33}^{III} \end{bmatrix} \begin{Bmatrix} \mathrm{d}\varepsilon_{xx}^{0} \\ \mathrm{d}\varepsilon_{yy}^{0} \\ 2\mathrm{d}\varepsilon_{xy}^{0} \\ \mathrm{d}\kappa_{xx}^{0} \\ \mathrm{d}\kappa_{yy}^{0} \\ 2\mathrm{d}\kappa_{xy}^{0} \end{Bmatrix} \tag{5.9.1}$$

where

$$Q_{ij}^{I}=\sum_{k=1}^{N}(C_{ij}^{G})_{k}(z_{k}-z_{k-1}),Q_{ij}^{II}=\frac{1}{2}\sum_{k=1}^{N}(C_{ij}^{G})_{k}(z_{k}^{2}-z_{k-1}^{2}),Q_{ij}^{III}=\frac{1}{3}\sum_{k=1}^{N}(C_{ij}^{G})_{k}(z_{k}^{3}-z_{k-1}^{3}) \tag{5.9.2}$$

If the externally applied total in-plane stresses are $(\sigma_{xx}^{0},\sigma_{yy}^{0},\sigma_{xy}^{0})$, the quantities on the left hand side of Eq. (5.9.1) are defined as

$$\mathrm{d}N_{xx}=\int_{-h/2}^{h/2}(\mathrm{d}\sigma_{xx}^{0})\mathrm{d}z,\ \mathrm{d}N_{yy}=\int_{-h/2}^{h/2}(\mathrm{d}\sigma_{yy}^{0})\mathrm{d}z,\ \mathrm{d}N_{xy}=\int_{-h/2}^{h/2}(\mathrm{d}\sigma_{xy}^{0})\mathrm{d}z \tag{5.10.1}$$

$$\mathrm{d}M_{xx}=\int_{-h/2}^{h/2}(\mathrm{d}\sigma_{xx}^{0})z\mathrm{d}z,\ \mathrm{d}M_{yy}=\int_{-h/2}^{h/2}(\mathrm{d}\sigma_{yy}^{0})z\mathrm{d}z,\mathrm{d}M_{xy}=\int_{-h/2}^{h/2}(\mathrm{d}\sigma_{xy}^{0})z\mathrm{d}z \tag{5.10.2}$$

Having solved the middle-surface strain and curvature increments, $\mathrm{d}\varepsilon_{xx}^{0}$, $\mathrm{d}\varepsilon_{yy}^{0}$, $\mathrm{d}\varepsilon_{xy}^{0}$, $\mathrm{d}\kappa_{xx}^{0}$, $\mathrm{d}\kappa_{yy}^{0}$ and $\mathrm{d}\kappa_{xy}^{0}$, from Eq. (5.9.1), we can now evaluate the stress increments sustained by each lamina in the laminate, based on Eq. (5.6). As mentioned before, the composite theory is developed on a somewhat "averaged" basis. Taking the average for both sides of Eq. (5.6) in the thickness direction with respect to the z coordinate, the averaged stress increments in the kth lamina expressed in the global system are found to be

$$\{\mathrm{d}\sigma_{i}\}_{k}^{G}=([T_{in}]_{c})_{k}([S_{nm}]_{k})^{-1}([T_{mj}]_{c}^{\mathrm{T}})_{k}\{\mathrm{d}\varepsilon_{j}\}_{k}^{G} \tag{5.11.1}$$

where

$$\{\mathrm{d}\varepsilon_{i}\}_{k}^{G}=\{\mathrm{d}\varepsilon_{xx}^{0}+\frac{z_{k}+z_{k-1}}{2}\mathrm{d}\kappa_{xx}^{0},\mathrm{d}\varepsilon_{yy}^{0}+\frac{z_{k}+z_{k-1}}{2}\mathrm{d}\kappa_{yy}^{0},2\mathrm{d}\varepsilon_{xy}^{0}+(z_{k}+z_{k-1})\mathrm{d}\kappa_{xy}^{0}\}^{\mathrm{T}} \tag{5.12}$$

Substituting Eq. (5.11.1) into the right-hand side of the Eq. (5.5.1), the overall stress increments on the lamina in the local coordinate system are obtained as

$$\{d\sigma_{i}\}_{k}=([T_{ij}]_{s}^{\mathrm{T}})_{k}\{d\sigma_{j}\}_{k}^{G} \tag{5.13.1}$$

and the internal stress increments in the fiber and matrix materials of this lamina are further calculated from Eqs. (3.68.1) and (3.68.2). The total internal stress updating and failure status detection can be carried out accordingly, as per Eqs. (3.75.1) and (3.75.2) and those given in Chapter 4. We will go into these further in subsequent sections.

5.3.2 Convention for Positive Shear Stress

After the coordinate transformation, i.e., Eq. (5.13.1), has been made, a global shear stress applied on the lamina may become a normal stress exerted in the lamina local coordinate system. Because the load carrying capacities of a composite under tension and compression are essentially different, it is of great importance to distinguish correctly a positive value from a negative one for a shear stress. In this book, the first subscript attached to a shear stress, e.g., the x in σ_{xy}, indicates the plane where the stress is applied, whereas the second one, y in the present example, refers to the direction the shear stress is directed to. By convention, a shear stress applied on a plane with its outward normal in accordance with the positive (or negative) direction of the first axis is defined as positive if it is directed to the positive (or negative) direction of the second axis (Fig. 5.5). With this convention, the stress resultants indicated in Fig. 5.4 are all positive.

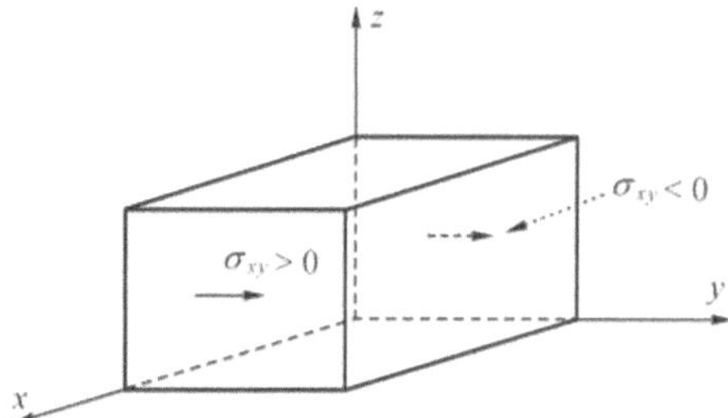

Fig. 5.5 Positive direction of a shear stress

As an example, let us consider transformation of a globally applied shear stress into stress components in a local coordinate system, as shown in Figs. 5.6(a) and 5.6(b) or Figs. 5.6(c) and 5.6(d).

By making use of Eqs. (1.76.2), (1.80.1) and (1.87), it is seen that the coordinate transformation formulae for the stress components from a global into the local coordinate systems are given by

$$\begin{Bmatrix} \sigma_{11} \\ \sigma_{22} \\ \sigma_{12} \end{Bmatrix} = \begin{bmatrix} l_1^2 & m_1^2 & 2l_1m_1 \\ l_2^2 & m_2^2 & 2l_2m_2 \\ l_1l_2 & m_1m_2 & l_1m_2 + l_2m_1 \end{bmatrix} \begin{Bmatrix} \sigma_{xx} \\ \sigma_{yy} \\ \sigma_{xy} \end{Bmatrix} \tag{5.13.2}$$

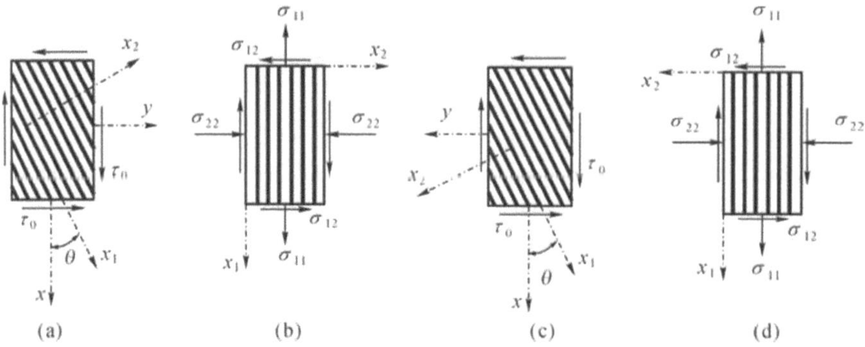

Fig. 5.6 Coordinate transformation for a globally applied shear stress

In the present case, we have $\sigma_{xx}=\sigma_{yy}=0$. If a global coordinate system as in Fig. 5.6(a) is chosen, we have $\sigma_{xy}=\tau_0$ according to the above convention for the positive shear stress. From Eq. (1.74), it is seen that $l_1=\cos(\theta)$, $l_2=-\sin(\theta)$, $m_1=\sin(\theta)$ and $m_2=\cos(\theta)$. Substituting them into Eq. (5.13.2) gives

$$\sigma_{11}=\tau_0\sin(2\theta),\ \sigma_{22}=-\tau_0\sin(2\theta) \text{ and } \sigma_{12}=\tau_0\cos(2\theta) \tag{5.14.1}$$

which are the stress components in the local coordinate system of Fig. 5.6(b). It is seen that the shear stress applied in Fig. 5.6(a) generates a transverse compression on the lamina in its local coordinate system.

On the other hand, if we choose the global coordinate system to be that shown in Fig. 5.6(c), we have $\sigma_{xy}=-\tau_0$ and $l_1=\cos(\theta)$, $l_2=\sin(\theta)$, $m_1=-\sin(\theta)$ and $m_2=\cos(\theta)$. The stress components in the local coordinate system of Fig. 5.6(d) are then found to be

$$\sigma_{11}=\tau_0\sin(2\theta),\ \sigma_{22}=-\tau_0\sin(2\theta) \text{ and } \sigma_{12}=-\tau_0\cos(2\theta) \tag{5.14.2}$$

Comparing Eq. (5.14.2) with Eq. (5.14.1), we can see that the normal stresses are the same whereas the shear stresses have a different sign in a different local coordinate system. It is important that the two (global and local) coordinate systems obey the right-hand screw rule and have their third coordinates in the same orientation.

5.3.3 Thermal Analysis

When a unidirectional lamina is subjected to only a temperature variation, there will be no overall stress (i.e., no thermal load) on the lamina (Eq. (3.85)), although its constituents generally sustain thermal stresses. In such a case, the constituent thermal stresses are evaluated using Eqs. (3.84.1) and (3.84.2), together with Eq. (3.88.1) or Eq. (3.88.2). However, if the laminate is subjected to a temperature variation, each lamina in the laminate will be subjected to overall thermal stresses (called thermal loads) in general. In this latter case, the thermal stresses in the constituents of the lamina will consist of two parts: one part is still given by Eqs. (3.84.1) and (3.84.2), and another part should be evaluated based on Eqs. (3.68.1) and (3.68.2). The second part is due to the composite lamination or stacking constraint. Namely, each lamina has different global coefficients of thermal expansion, although the local ones may be the same. Hence, it is only necessary to calculate the thermal loads shared by the lamina in the laminate analysis.

For the kth lamina, the thermal stress and strain increments in the global coordinate system satisfy (refer to Eq. (3.83.3))

$$\{\mathrm{d}\sigma_i\}_k^{G,(T)} = [C_{ij}]_k^G \{\mathrm{d}\varepsilon_j\}_k^{G,(T)} - \{\beta_i\}_k^G \mathrm{d}T \tag{5.15.1}$$

where $$\{\beta_i\}_k^G = \{(\beta_1)_k^G, (\beta_2)_k^G, (\beta_3)_k^G\}^{\mathrm{T}} = ([T_{in}]_c)_k ([S_{nj}]_k)^{-1} \{\alpha_j\}_k \tag{5.15.2}$$

$\{\alpha_i\}_k$ are the thermal expansion coefficients of the lamina calculated from Eq. (3.87). On the other hand, substituting Eq. (5.15.1) onto the right hand side of Eq. (5.13.1) yields the thermal stress increments applied on the kth lamina in its local coordinate system. These stress increments must be further substituted onto the right hand sides of Eqs. (3.90.1) and (3.90.2), rather than Eqs. (3.68.1) and (3.68.2), to determine the thermal stress increments in the constituent fiber and matrix materials. Namely, the two parts of the thermal stresses are evaluated simultaneously. In this way, the thermal residual stresses in each lamina in the laminate can be obtained.

It is noted that the thermal strain increments, $\{\mathrm{d}\varepsilon_i\}_k^{G,(T)}$, in Eq. (5.15.1) are defined as (refer to Eq. (5.12))

$$\begin{aligned}\{\mathrm{d}\varepsilon_i\}_k^{G,(T)} = \{&\mathrm{d}\varepsilon_{xx}^{0,(T)} + \frac{z_k + z_{k-1}}{2}\mathrm{d}\kappa_{xx}^{0,(T)}, \mathrm{d}\varepsilon_{yy}^{0,(T)} + \frac{z_k + z_{k-1}}{2}\mathrm{d}\kappa_{yy}^{0,(T)}, \\ &2\mathrm{d}\varepsilon_{xy}^{0,(T)} + (z_k + z_{k-1})\mathrm{d}\kappa_{xy}^{0,(T)}\}^T\end{aligned} \tag{5.16}$$

By using the condition that the stress resultants from Eq. (5.15.1) are all zero because of no external load applied on the laminate, the middle surface strain and curvature increments due to the temperature variation are obtained from

$$\begin{Bmatrix} \mathrm{d}\Omega_1^I \\ \mathrm{d}\Omega_2^I \\ \mathrm{d}\Omega_3^I \\ \mathrm{d}\Omega_1^{II} \\ \mathrm{d}\Omega_2^{II} \\ \mathrm{d}\Omega_3^{II} \end{Bmatrix} = \begin{bmatrix} Q_{11}^I & Q_{12}^I & Q_{13}^I & Q_{11}^{II} & Q_{12}^{II} & Q_{13}^{II} \\ Q_{12}^I & Q_{22}^I & Q_{23}^I & Q_{12}^{II} & Q_{22}^{II} & Q_{23}^{II} \\ Q_{13}^I & Q_{23}^I & Q_{33}^I & Q_{13}^{II} & Q_{23}^{II} & Q_{33}^{II} \\ Q_{11}^{II} & Q_{12}^{II} & Q_{13}^{II} & Q_{11}^{III} & Q_{12}^{III} & Q_{13}^{III} \\ Q_{12}^{II} & Q_{22}^{II} & Q_{23}^{II} & Q_{12}^{III} & Q_{22}^{III} & Q_{23}^{III} \\ Q_{13}^{II} & Q_{23}^{II} & Q_{33}^{II} & Q_{13}^{III} & Q_{23}^{III} & Q_{33}^{III} \end{bmatrix} \begin{Bmatrix} \mathrm{d}\varepsilon_{xx}^{0,(T)} \\ \mathrm{d}\varepsilon_{yy}^{0,(T)} \\ 2\mathrm{d}\varepsilon_{xy}^{0,(T)} \\ \mathrm{d}\kappa_{xx}^{0,(T)} \\ \mathrm{d}\kappa_{yy}^{0,(T)} \\ 2\mathrm{d}\kappa_{xy}^{0,(T)} \end{Bmatrix} \tag{5.17}$$

where $\mathrm{d}\Omega_i^{\mathrm{I}} = \sum_{k=1}^{N} (\beta_i)_k^G (z_k - z_{k-1}) \mathrm{d}T$ and $\mathrm{d}\Omega_i^{\mathrm{II}} = \frac{1}{2}\sum_{k=1}^{N} (\beta_i)_k^G (z_k^2 - z_{k-1}^2) \mathrm{d}T$ (5.18)

can be regarded as equivalent thermal loads. The overall stiffness elements in Eq. (5.17) are exactly the same as those in Eq. (5.9.1), i.e., given by Eq. (5.9.2).

5.3.4 Coupled Thermal-Mechanical Analysis

When the laminate is subjected to both mechanical loads and a temperature variation, the resulting stresses in a lamina, as well as in its constituent fiber and matrix materials, can be obtained by the superimposition of contributions from both the equivalent thermal and mechanical loads. For the convenience of the reader, the relevant equations for analyzing a coupled thermal-mechanical problem are summarized below.

(1) The stress increments shared by the kth lamina in its local coordinate system are given by

$$\{\mathrm{d}\sigma_i\}_k = ([T_{ij}]_s^{\mathrm{T}})_k \{\mathrm{d}\sigma_j\}_k^G$$

(2) The stress increments shared by the k-th lamina in the global coordinate system read

$$\{\mathrm{d}\sigma_i\}_k^G = [C_{ij}]_k^G \{\mathrm{d}\varepsilon_j\}_k^G - \{\beta_i\}_k^G \mathrm{d}T$$

where $\{\beta_i\}_k^G = \{(\beta_1)_k^G, (\beta_2)_k^G, (\beta_3)_k^G\}^{\mathrm{T}} = ([T_{in}]_c)_k ([S_{nj}]_k)^{-1} \{\alpha_j\}_k$

and $\{\mathrm{d}\varepsilon_i\}_k^G = \{\mathrm{d}\varepsilon_{xx}^0 + \frac{z_k + z_{k-1}}{2}\mathrm{d}\kappa_{xx}^0, \mathrm{d}\varepsilon_{yy}^0 + \frac{z_k + z_{k-1}}{2}\mathrm{d}\kappa_{yy}^0, 2\mathrm{d}\varepsilon_{xy}^0 + (z_k + z_{k-1})\mathrm{d}\kappa_{xy}^0\}^{\mathrm{T}}$

(3) The strain and curvature increments on the middle-surface of the laminate are solved from

$$\begin{Bmatrix} \mathrm{d}\Omega_1^I + \mathrm{d}N_{xx} \\ \mathrm{d}\Omega_2^I + \mathrm{d}N_{yy} \\ \mathrm{d}\Omega_3^I + \mathrm{d}N_{xy} \\ \mathrm{d}\Omega_1^{II} + \mathrm{d}M_{xx} \\ \mathrm{d}\Omega_2^{II} + \mathrm{d}M_{yy} \\ \mathrm{d}\Omega_3^{II} + \mathrm{d}M_{xy} \end{Bmatrix} = \begin{bmatrix} Q_{11}^I & Q_{12}^I & Q_{13}^I & Q_{11}^{II} & Q_{12}^{II} & Q_{13}^{II} \\ Q_{12}^I & Q_{22}^I & Q_{23}^I & Q_{12}^{II} & Q_{22}^{II} & Q_{23}^{II} \\ Q_{13}^I & Q_{23}^I & Q_{33}^I & Q_{13}^{II} & Q_{23}^{II} & Q_{33}^{II} \\ Q_{11}^{II} & Q_{12}^{II} & Q_{13}^{II} & Q_{11}^{III} & Q_{12}^{III} & Q_{13}^{III} \\ Q_{12}^{II} & Q_{22}^{II} & Q_{23}^{II} & Q_{12}^{III} & Q_{22}^{III} & Q_{23}^{III} \\ Q_{13}^{II} & Q_{23}^{II} & Q_{33}^{II} & Q_{13}^{III} & Q_{23}^{III} & Q_{33}^{III} \end{bmatrix} \begin{Bmatrix} \mathrm{d}\varepsilon_{xx}^0 \\ \mathrm{d}\varepsilon_{yy}^0 \\ 2\mathrm{d}\varepsilon_{xy}^0 \\ \mathrm{d}\kappa_{xx}^0 \\ \mathrm{d}\kappa_{yy}^0 \\ 2\mathrm{d}\kappa_{xy}^0 \end{Bmatrix}$$

(4) The stress increments in the fiber and matrix materials of the k-th lamina are obtained as

$$\{\mathrm{d}\sigma_i^f\}_k = (V_f[I]+V_m[A_{ij}]_k)^{-1}\{\mathrm{d}\sigma_j\}_k + \{b_i^f\}_k \mathrm{d}T$$

$$\{\mathrm{d}\sigma_i^m\}_k = [A_{ij}]_k(V_f[I]+V_m[A_{ij}]_k)^{-1}\{\mathrm{d}\sigma_j\}_k + \{b_i^m\}_k \mathrm{d}T$$

where the thermal stress concentration factors, $\{b_i^m\}_k$ and $\{b_i^f\}_k$, are determined through Eqs. (3.88.1) and (3.88.2).

5.4 Fatal or Nonfatal Failure

Once the stress increments in the fiber and matrix materials of each lamina in the laminate have been obtained, they are substituted into the following stress-updating formulae to evaluate the total stresses at the current load level

$$\{\sigma_i^f\}_k = \{\sigma_i^f\}_k + \{\mathrm{d}\sigma_i^f\}_k, \quad k=1, \ldots, N \tag{5.19.1}$$

$$\{\sigma_i^m\}_k = \{\sigma_i^m\}_k + \{\mathrm{d}\sigma_i^m\}_k, \quad k=1, \ldots, N \tag{5.19.2}$$

From them, the maximum and minimum stresses can be easily evaluated, as per

$$\sigma_{\max}^f = \frac{\sigma_{11}^f + \sigma_{22}^f}{2} + \frac{1}{2}\sqrt{(\sigma_{11}^f - \sigma_{22}^f)^2 + 4(\sigma_{12}^f)^2} \tag{5.20.1}$$

$$\sigma_{\min}^f = \frac{\sigma_{11}^f + \sigma_{22}^f}{2} - \frac{1}{2}\sqrt{(\sigma_{11}^f - \sigma_{22}^f)^2 + 4(\sigma_{12}^f)^2} \tag{5.20.2}$$

$$\sigma_{\max}^m = \frac{\sigma_{11}^m + \sigma_{22}^m}{2} + \frac{1}{2}\sqrt{(\sigma_{11}^m - \sigma_{22}^m)^2 + 4(\sigma_{12}^m)^2} \tag{5.21.1}$$

$$\sigma_{\min}^m = \frac{\sigma_{11}^m + \sigma_{22}^m}{2} - \frac{1}{2}\sqrt{(\sigma_{11}^m - \sigma_{22}^m)^2 + 4(\sigma_{12}^m)^2} \tag{5.21.2}$$

whereas the principal stresses of the fiber and the matrix are arranged according to algebraic values of the corresponding maximum, minimum and zero stresses, respectively. These principal stresses can be substituted into Eqs. (4.5) and (4.6) to check if the fiber or the matrix attains a failure stress state. If either the fiber or the matrix fails, the corresponding lamina is said to have failed, and the stress state sustained by the laminate is called a failure strength. By definition, the first ply failure strength is referred to as a stress state applied step-by-step on the laminate under which one of the laminae has failed and no previous failure has occurred. It should be noticed that the term "first ply" used here in a failure characterization means the first failed ply. It has no relation to the "first ply" in the previous geometrical definition. The first failed ply may or may not be the geometrical first-ply of the laminate. Moreover, it is very possible that the laminate can still sustain additional loads after the first ply failure. At some other, generally higher, load level, a second ply failure occurs and the laminate is said to attain the second ply failure strength. In this way, a progressive failure process is developed for the laminate.

According to the micromechanical failure criteria adopted in this book, i.e., Eqs. (4.5) and (4.6), there are four types of failures pertaining to a lamina failure. They are a fiber tensile failure, a fiber compressive failure, a matrix tensile failure and a matrix compressive failure. Apparently, different failures have different effects on the load carrying capacity of the laminate. Some failures may cause the laminate to completely lose its load sustaining ability. In such a case, the laminate is said to have attained a fatal failure. No additional load is applicable to the laminate when a fatal failure occurs, and a simulation procedure must be terminated. On the other hand, some other failures may only partially affect the load carrying capacity of the laminate and those failures are called nonfatal failures.

Much experimental evidence has shown that when a fiber failure occurs the corresponding composite generally cannot sustain any further load. In this book, any fiber failure, regardless of tensile or compressive failure, is considered as a fatal failure for the laminate. A fatal failure is also referred to as an ultimate failure, whereas the corresponding failure strength is taken as an ultimate strength of the laminate.

In additional to fiber failures, researchers have also found that a transverse compressive failure of a lamina generally results in an ultimate failure (Zinoviev et al., 1998; Liu & Tsai, 1998). As such a failure is generally caused by a matrix compressive failure, it is considered in this book that any matrix compressive failure in the laminate corresponds to a fatal failure as well (Zhou & Huang, 2008). We thus have three kinds of fatal failures: fiber tensile failure, fiber compressive failure and matrix compressive failure. Whichever occurs, the laminate is considered to have attained an ultimate failure.

On the other hand, a tensile failure of a matrix involves many more loading conditions and happens most often. Many experimental results support the view that a matrix tensile failure, such as a matrix crack or fiber-matrix interface debonding in the composite, generally does not lead to an ultimate failure of the laminate. For this reason, this book considers a matrix tensile failure in the laminate to be a nonfatal failure. A stiffness discount should be applied when a nonfatal failure occurs.

5.5 Stiffness Degradation

As mentioned before, a lamina failure due to a matrix tensile failure generally does not cause the laminate to lose its load sustaining ability completely. Instead, a partial effect takes place, as though the laminate was somewhat softened. In other words, the laminate stiffness should be degraded once a nonfatal failure occurs.

There are different ways to degrade a laminate stiffness. The simplest one is to disregard the failed lamina for ever. Namely, no stiffness contribution from the failed lamina will be considered henceforth. This kind of stiffness degradation is called a total discount scheme. By this scheme, $N-1$ steps, at most, in stiffness degradation are applicable where N is the number of laminae in the laminate. As such, a total stiffness discount method should not be applicable with the present nonfatal failure. Otherwise, the N-th ply failure would automatically be a fatal failure and the simulation would have to be terminated after such a ply failure as no stiffness could afford further simulation.

Recognizing that the hardening modulus of a matrix material on a tensile stress-strain curve will be much smaller than its elastic counterpart when the material is under a failure stress state, a simple partial stiffness discount scheme is incorporated in this book to represent a tensile failure of the matrix. This scheme is carried out by multiplying a small constant factor, called a degradation factor, with the current modulus of the matrix to deteriorate the stiffness of the failed lamina. Namely,

$$E^m = \Delta E_0^m \tag{5.22}$$

where Δ is the degradation factor and E_0^m is the matrix modulus just before the failure takes place. All of the other constituent properties including elastic property parameters of the fiber and Poisson's ratio of the matrix are kept unchanged. Then, the stiffness matrix of the failed lamina can be determined using the degraded matrix modulus together with other properties of the fiber and matrix materials.

An important issue would be the determination of a proper degradation factor. It must be realized that no rigorous proof is available for showing which degradation factor is the best. Numerical experiments have been carried out for a problem taken from the worldwide failure exercise (WWFE-I) problems (Soden et al., 1998a). The problem under consideration is a $[\pm 55°]_s$ laminate made of silenka E-glass and MY750/HY917 /DY063 epoxy subjected to a combined σ_{xx} and σ_{yy} load with σ_{yy}/σ_{xx}=2. Mechanical properties of the constituent materials are given in Table 5.1. Stress-strain curves along x and y directions predicted by using different degradation factors are plotted in Fig. 5.7, in which the experimental data taken from Soden et al. (2002) are also shown for comparison. From the figure, it is seen that the predicted ultimate failure strengths of the laminate are almost the same, regardless of which degradation factor from 0.005 to 0.05 is used. However, the predicted stress-strain curves of the laminate do depend on the used factors heavily.

As expected, the initial segments of all of the predicted curves are coincident with each other, and agree well with the experiments. Afterwards, a significant difference has been observed. It is seen that the predicted stress-strain curve in the x direction based on the degradation factor of 0.05 differs obviously from the experiments. Even more, there exists a larger discrepancy between the shapes of the predicted and measured stress-strain curves. The measured strains in the y direction were larger than those in the x direction under small loads. When the load exceeded 400 MPa, the measured x-directional strains became obviously larger than the y-directional ones. However, the predicted y-directional strains with the degradation factor of 0.05 are always larger than those in the x-direction. Thus, the factor 0.05 seems not very suitable for this problem. The degradation factor 0.005 results in the best fit for the x-directional curve, but the predicted y-directional curve is much different from that in the experiment. Although there is still a moderate discrepancy between the predicted and the measured strains by using the degradation factor 0.01, the overall fit seems better compared with the overall fits based on the factors 0.005 and 0.05. Similar numerical experiments have been carried out using other examples taken from the WWFE-I. The following conclusions can be made from these experiments.

(1) As long as a degradation factor is less than 0.1, the predicted ultimate failure strength of a laminate will not be affected significantly. Thus, if only an ultimate strength of the laminate is concerned, an arbitrary value less than 0.1 can be used as the degradation factor.

(2) The predicted stress-strain curves vary remarkably when different degradation factors are used. It is found that a value of around 0.01 can give a reasonably good predictive accuracy for most laminates. In the following examples in this book, 0.01 is chosen as the degradation factor if no other comment is made.

Table 5.1 Constituent properties of an E-glass/MY750/HY917/DY063 lamina (V_f=0.60)

Properties of silenka E-glass fiber (Huang, 2004a)

E_{11} (GPa)	E_{22} (GPa)	ν_{12}	G_{12} (GPa)	ν_{23}	σ_u (MPa)	$\sigma_{u,c}$ (MPa)
74	74	0.2	30.8	0.2	2092.8	1311.8

Elastic-plastic parameters of MY750/HY917/DY063 epoxy matrix (Huang, 2004a)

(ν^m=0.35, σ_u^m=60.9 MPa and $\sigma_{u,c}^m$=74.8 MPa)

i=	1*	2	3	4	5	6	7	8
$(\sigma_Y^m)_i$ (MPa)	32.6	39.9	46.8	52.0	55.6	58.0	60.1	62.0
$(E_T^m)_i$ (GPa)	3.35	1.698	1.387	0.918	0.542	0.317	0.244	0.186

* Note: $(E_T^m)_1 \equiv E^m$ and $(\sigma_Y^m)_1 \equiv \sigma_Y^m$.

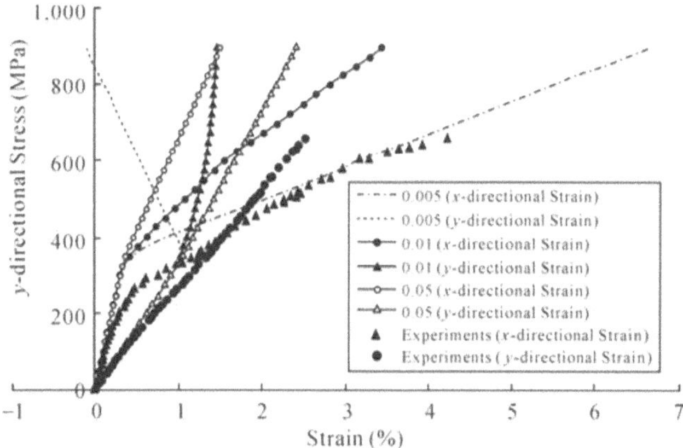

Fig. 5.7 Influence of different degradation factors on predicted stress-strain curves of a laminate (from Zhou & Huang, 2008)

It is further noticed that the matrix modulus, not only in obtaining an instantaneous compliance matrix $[S_{ij}^m]$ but also in defining bridging matrix elements, must be modified when a stiffness discount is applied. As the transverse and shear moduli of a lamina are predominantly influenced by the matrix modulus, the use of the degradation factor 0.01 will result in a similar amount of reduction in these moduli. This in turn may lead to an excessively large deformation or strain when the laminate is subjected to some loading condition. As can be understood, too large a deformation or strain is not acceptable for the safety of a structure. Therefore, an additional ultimate failure condition should be also adopted in the simulation of a progressive failure process in a laminate. This condition is a maximum deformation or strain constraint, which will be dealt with in more detail in a subsequent section.

5.6 Inter-layer in between Adjacent Laminae

It has been recognized that during the fabrication of a multidirectional laminate there is a rich matrix region in between two adjacent laminae. In fact, the two laminae can be well regarded as being bonded together by the matrix, and an interface layer in between them consisting of the pure matrix does exist from a micromechanical viewpoint, although the thickness of such a layer is small. The mechanical behaviour of the laminate is likely to be influenced by the status of the interfaces. For this reason, an inter-layer made of the pure matrix is introduced to represent the rich matrix region at the interface between two adjacent laminae. With such an introduction, the number of layers in the laminate is changed as though it consisted of $2N-1$ layers rather than N layers. All of the equations derived in Sections 5.3 and 5.4 are applicable except that the number N should be replaced by $2N-1$.

The introduction of the inter-layers is reasonable in terms of the experimental evidence but the thickness of such a layer is difficult or even impossible to measure exactly. It has been shown that when the thickness of an inter-layer is chosen to be 1% to 10% that of the primary layers, where a primary layer refers to an original lamina constituting the laminate, the predicted elastic properties and failure strengths of the laminate do not vary significantly (Zhou & Huang, 2008). As such, 5% of the primary layer thickness can be considered as the thickness of the inter-layer, as shown in Fig. 5.8. Supposing that the total thickness and the fiber volume fraction of the laminate are kept unchanged, the original thickness and the fiber volume fraction of each primary layer should be adjusted accordingly. For a laminate made of equally thick primary layers with an equal fiber volume fraction, the adjustment is done as follows. All of the primary layers are divided into two categories, mid-layers and surface layers respectively (Fig. 5.8). The thickness of the mid-layers is changed from h_0 to $0.95h_0$ whereas the fiber volume fraction is increased from V_f^0 to $\frac{100}{95}V_f^0$. On the other hand, the thickness of the surface layers varies from h_0 to $0.975h_0$, with the fiber volume fraction of the surface layers increasing from V_f^0 to $\frac{100}{97.5}V_f^0$.

Mechanical properties of the inter-layers are assumed to be exactly the same as those of the matrix material used for the laminate. If the laminate is made of hybrid materials, for example, if the matrix materials used in the adjacent layers are different, the inter-layer is defined to be a third isotropic material whose mechanical properties are determined by averaging those of the matrix materials in the adjacent laminae. In simulation, the inter-layers can be considered as either isotropic matrix material or an equivalent lamina with a zero fiber volume fraction.

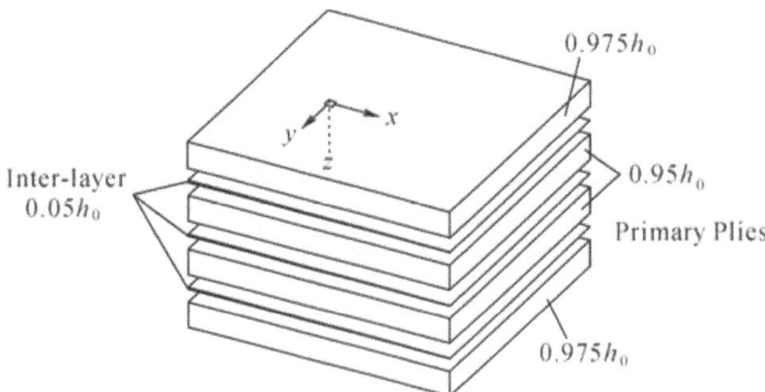

Fig. 5.8 Schematic of a laminate containing pure matrix inter-layers (from Zhou & Huang, 2008)

The influence of the introduced pure matrix inter-layers on the mechanical behavior of a laminate can be three-fold. Firstly, the stresses shared by each lamina will be changed by the introduction of the inter-layers. Secondly, the failure modes and the progressive failure process may be more complicated. Finally, due to the difference in the thicknesses and fiber volume fractions between the mid-layers and the surface layers, the ultimate strength of the laminate may be affected by the total number of the layers, although this effect is likely to be insignificant.

An example consisting of the same lamina properties (Table 5.1) and with the same lamination angle, i.e., ±45° as considered before, is re-examined to show the afore-mentioned third kind of effect. The ultimate strength of the laminate with a given number of primary layers is evaluated (ultimate failure criteria for laminates will be addressed in the following section) and a relationship between the ultimate strengths and the number of different primary layers is plotted in Fig. 5.9. From the figure, it is seen that while the number of the primary layers is increased from 4 to 20, the ultimate strength of the laminate is decreased from 674.2 MPa to 668.9 MPa. The maximum variation is only 0.8% and, especially when the number of the primary layers is larger than 10, the predicted ultimate strength is essentially the same.

It must be pointed out that essentially no fatal failure should be applicable to a pure matrix inter-layer, even though it may fail due to an excessive compression. This is because the inter-layers are artificially introduced only to represent rich matrix regions in between the adjacent primary layers, and the failure of such a layer is generally not an ultimate failure for an actual structure. Thus, no matter that an inter-layer failure is caused by tensile or compressive stress, its stiffness deteriorates by multiplying the same factor of 0.01 with its current modulus, as similarly done for a tensile failure of a primary layer. Continuity in such degradation for an inter-layer stiffness may eventually result in strains that are too high, which would then cause the laminate strains or deformations to be unreasonably large. A question arises as to whether a strain constraint is necessary. From Eq. (5.15), one can recognize that when the strain of an inter-layer becomes very large, so does the primary layer strain. However, an ultimate strain constraint should have already been applied to the primary layers. Thus, no additional strain constraint is needed for the pure matrix inter-layers when they are incorporated into a laminate analysis under the framework of the classical laminate theory. Some key points for the introduction of pure matrix inter-layers are summarized in Table 5.2.

Nevertheless, it deserves mentioning that when a three-dimensional laminate theory is employed, as illustrated in Section 5.8, some additional constraint may be required with respect to the strain calculation of the inter-layers. More details will be shown later.

Table 5.2 Key points for pure matrix inter-layers

Point	Description
1	Thickness of an inter-layer is 5% that of a primary layer
2	An inter-layer assumes the matrix properties, but no failure corresponds to an ultimate failure
3	Either a tensile or compressive failure of an inter-layer is accompanied by a stiffness discount

Remark 5.1

A pure matrix inter-layer that is introduced only accounts for the effect of the interface between two adjacent laminae on the mechanical behaviour of a laminate. No interface between fiber and matrix materials of a lamina is concerned explicitly in this book. However, the effect of the latter interface on the mechanical properties of the lamina has been somewhat implicitly taken into account by properly choosing the bridging parameters α and β.

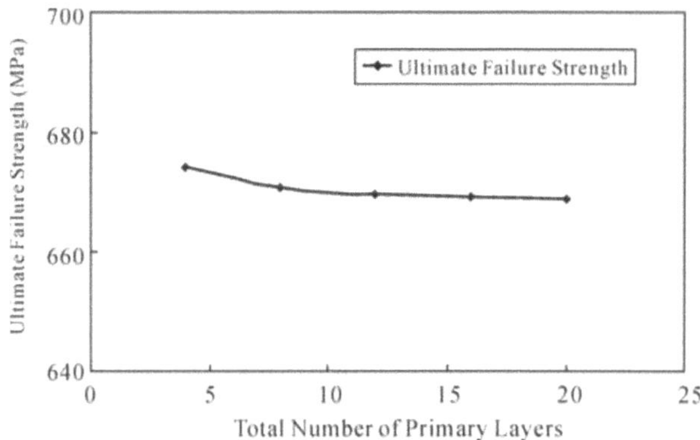

Fig. 5.9 Final failure strengths versus number of primary plies (from Zhou & Huang, 2008)

5.7 Ultimate Failure Criteria

From what has been discussed in the previous sections, it can be understood that an ultimate failure of a laminate can be caused by either an excessive stress or an excessive strain. The stress-based ultimate failures have been clearly addressed in Section 5.4. On the other hand, a strain-based ultimate failure has not been quantitatively characterized yet. This must be done before a complete set of ultimate failure criteria can be established.

There are two kinds of layers in a laminate under consideration. One kind consists of primary layers, which are the original lamina layers, and another consists of pure matrix inter-layers, which are introduced artificially for simulation purposes. A strain constraint, i.e., a strain-based ultimate failure criterion, is applicable only to the primary layers. For the primary layers, an excessive strain may occur no matter whether a stiffness discount has been applied to a failed lamina or not. However, an actual structure cannot pertain to a strain value greater than some limitation. Zinoviev et al. (1998) pointed out that modern composite laminates of ±45° structures usually showed an ultimate strain under uniaxial tension of around 10% – 15%. For illustration, let us consider a $[\pm 55°]_s$ E-glass/MY750/HY917/DY063 laminate taken from the WWFE-I (Soden et al., 1998a) subjected to combined σ_{yy} and σ_{xx} loads. The constituents of the laminate have the same mechanical properties as those given in Table 5.1. For this problem, a large strain occurs before a stress-based ultimate failure can be assumed for the laminate. Different critical strain values varying from 10% to 20% have been used as an additional constraint. If any strain in absolute value in a lamina of the laminate is greater than this constraint the laminate is considered to have attained an ultimate failure. The thus predicted failure strength envelopes are plotted in Fig. 5.10, which are compared with the experimental data also shown in the figure (Soden et al., 2002). It is seen that only a negligibly small amount of discrepancy exists. This is because a large deformation or strain usually results from tensile failures of the matrix in more than one layer. When such a case occurs, the strain of the laminate may enlarge so rapidly that a gap of 10% strain will not significantly affect the ultimate strength of the laminate. In the present example, this means that any value in between 10% and 20% seems reasonable for detecting an ultimate failure. By further numerical experiments on problems of the WWFE-I, a value of 12% is found to be most suitable as the critical strain value for controlling an ultimate failure.

In summary, there are four cases which may cause a laminate to attain an ultimate failure. Any one case can be considered as an ultimate failure criterion. They are summarized in Table 5.3 for easy reference.

Table 5.3 Ultimate failure criteria for a laminate

Case	Description
1	A fiber tensile failure occurs in any primary layer
2	A fiber compressive failure occurs in any primary layer
3	A matrix compressive failure occurs in any primary layer
4	A strain of any primary layer in absolute value is equal to or greater than 12%

It can be understood from Table 5.3 that an ultimate failure of the laminate may or may not correspond to the first ply failure. The stress or the load on the laminate under which an ultimate failure is assumed is named as an ultimate strength.

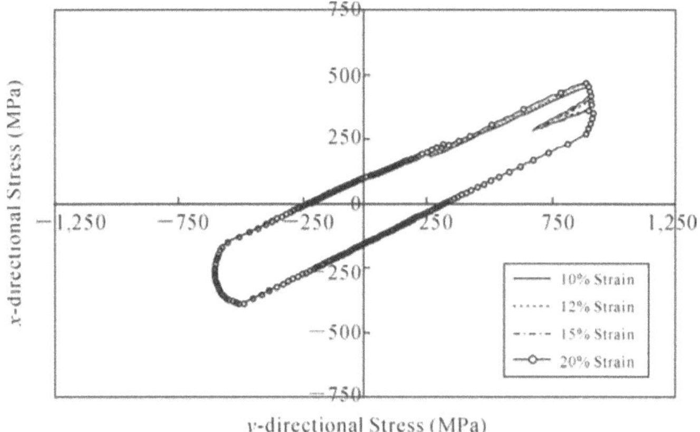

Fig. 5.10 Comparison between the usages of different ultimate strains (from Zhou & Huang, 2008)

Remark 5.2

One should be careful when the 12% strain constraint specified in Table 5.3 is used for detecting an ultimate failure of a laminate. In some cases, an ultimate strain other than 12% may be more pertinent in controlling the ultimate failure of the laminate. For instance, if a rubber or a ceramic material is used as a matrix, the 12% strain limitation must be too small or too great to achieve an accurate prediction for an ultimate failure of the resulting composite.

5.8 Pseudo 3D Laminate Theory

In most cases, a laminate can be well regarded as being subjected to a planar stress state for which the classical laminate theory (2D theory) described in Section 5.3 is applicable and is accurate enough. However, if an external load applied in the third direction, i.e., in the laminate thickness direction is not a small quantity, a three-dimensional (3D) stress state within the laminate must be identified. For this purpose, a pseudo 3D laminate theory is developed.

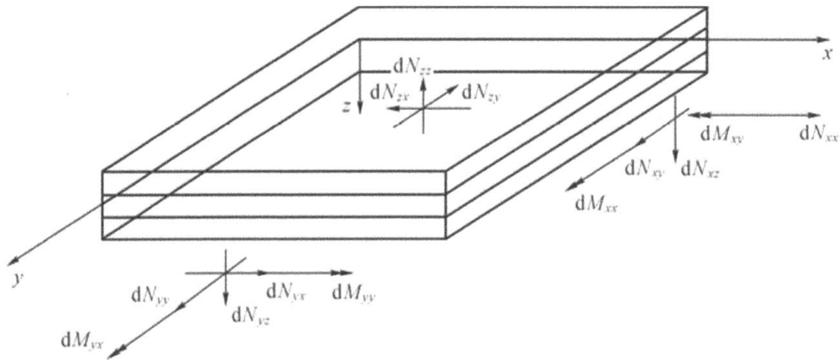

Fig. 5.11 A laminate global coordinate system with positive applied load increments (3D form)

In developing the pseudo 3D theory, a basic assumption similar to the second one used in the classical laminate theory is adopted, which states that a straight line normal to the laminate middle-surface remains straight after deformation. Mathematically, this implies that the incremental displacements at any point (x, y, z) can be expressed as

$$\mathrm{d}u(x, y, z) = \mathrm{d}u^0(x, y) - z\mathrm{d}\varphi_x(x, y) \tag{5.23.1}$$

$$\mathrm{d}v(x, y, z) = \mathrm{d}v^0(x, y) - z\mathrm{d}\varphi_y(x, y) \tag{5.23.2}$$

$$\mathrm{d}w(x, y, z) = \mathrm{d}w^0(x, y, z) \tag{5.23.3}$$

Compared with the classical laminate theory, the first assumption adopted there is no longer applicable. Namely, the strain components in the thickness direction are not negligible and Eq. (5.2) is invalid.

Substituting Eqs. (5.23.1) – (5.23.3) into Eq. (1.1) gives the following expressions

$$\mathrm{d}\varepsilon_{xx} = \frac{\partial(\mathrm{d}u_0)}{\partial x} - z\frac{\partial(\mathrm{d}\varphi_x)}{\partial x} = \mathrm{d}\varepsilon_{xx}^0 - z\frac{\partial(\mathrm{d}\varphi_x)}{\partial x} = \mathrm{d}\varepsilon_{xx}^0 + z\mathrm{d}\kappa_{xx}^0 \tag{5.24.1}$$

$$\mathrm{d}\varepsilon_{yy} = \frac{\partial(\mathrm{d}v_0)}{\partial y} - z\frac{\partial(\mathrm{d}\varphi_y)}{\partial y} = \mathrm{d}\varepsilon_{yy}^0 - z\frac{\partial(\mathrm{d}\varphi_y)}{\partial y} = \mathrm{d}\varepsilon_{yy}^0 + z\mathrm{d}\kappa_{yy}^0 \tag{5.24.2}$$

$$2\mathrm{d}\varepsilon_{xy} = \frac{\partial(\mathrm{d}u_0)}{\partial y} + \frac{\partial(\mathrm{d}v_0)}{\partial x} - z\left(\frac{\partial(\mathrm{d}\varphi_x)}{\partial y} + \frac{\partial(\mathrm{d}\varphi_y)}{\partial x}\right) = 2\mathrm{d}\varepsilon_{xy}^0 + 2z\mathrm{d}\kappa_{xy}^0 \tag{5.24.3}$$

$$2\mathrm{d}\varepsilon_{xz} = \frac{\partial \mathrm{d}w}{\partial x} - \mathrm{d}\varphi_x \tag{5.24.4}$$

$$2\mathrm{d}\varepsilon_{yz} = \frac{\partial \mathrm{d}w}{\partial y} - \mathrm{d}\varphi_y \tag{5.24.5}$$

$$\mathrm{d}\varepsilon_{zz} = \frac{\partial(\mathrm{d}w)}{\partial z} \tag{5.24.6}$$

where $\mathrm{d}\varepsilon_{xx}^0$, $\mathrm{d}\varepsilon_{yy}^0$ and $2\mathrm{d}\varepsilon_{xy}^0$ are the in-plane strain increments and $\mathrm{d}\kappa_{xx}^0 = -\frac{\partial(\mathrm{d}\varphi_x)}{\partial x}$, $\mathrm{d}\kappa_{yy}^0 = -\frac{\partial(\mathrm{d}\varphi_y)}{\partial y}$ and $2\mathrm{d}\kappa_{xy}^0 = -(\frac{\partial(\mathrm{d}\varphi_x)}{\partial y} + \frac{\partial(\mathrm{d}\varphi_y)}{\partial x})$ are the curvature increments on the middle surface, respectively. The constitutive relationship that interrelates the stress and strain increments, Eqs. (5.6) and (5.7), remains unchanged, i.e.,

$$\{\mathrm{d}\sigma_i\}^G = ([T_{in}]_c)_k([S_{nm}]_k)^{-1}([T_{mj}]_c^{\mathrm{T}})_k\{\mathrm{d}\varepsilon_j\}^G = [(C_{ij}^G)_k]\{\mathrm{d}\varepsilon_j\}^G, \tag{5.11.2}$$

where $[(C_{ij}^G)_k] = ([T_{in}]_c)_k([S_{nm}]_k)^{-1}([T_{mj}]_c^{\mathrm{T}})_k$.

However, the stress and strain increments in Eq. (5.11.2) are three-dimensional, i.e., $\{\mathrm{d}\sigma_i\}^G=\{\mathrm{d}\sigma_{xx}, \mathrm{d}\sigma_{yy}, \mathrm{d}\sigma_{zz}, \mathrm{d}\sigma_{yz}, \mathrm{d}\sigma_{xz}, \mathrm{d}\sigma_{xy}\}^{\mathrm{T}}$ and $\{\mathrm{d}\varepsilon_j\}^G=\{\mathrm{d}\varepsilon_{xx}, \mathrm{d}\varepsilon_{yy}, \mathrm{d}\varepsilon_{zz}, 2\mathrm{d}\varepsilon_{yz}, 2\mathrm{d}\varepsilon_{xz}, 2\mathrm{d}\varepsilon_{xy}\}^{\mathrm{T}}$ and an instantaneous compliance matrix, $[S_{nm}]$, must be defined using a 3D bridging matrix, Eq. (3.70). For the convenience of further analysis, let us re-arrange the components of $\{\mathrm{d}\sigma_i\}^G$ and $\{\mathrm{d}\varepsilon_j\}^G$ as per

$$\begin{Bmatrix} \mathrm{d}\sigma_{xx} \\ \mathrm{d}\sigma_{yy} \\ \mathrm{d}\sigma_{xy} \\ \mathrm{d}\sigma_{yz} \\ \mathrm{d}\sigma_{xz} \\ \mathrm{d}\sigma_{zz} \end{Bmatrix} = \left[\left(C_{ij}^{G}\right)_{k}^{*}\right] \begin{Bmatrix} \mathrm{d}\varepsilon_{xx} \\ \mathrm{d}\varepsilon_{yy} \\ 2\mathrm{d}\varepsilon_{xy} \\ 2\mathrm{d}\varepsilon_{yz} \\ 2\mathrm{d}\varepsilon_{xz} \\ \mathrm{d}\varepsilon_{zz} \end{Bmatrix} = \begin{bmatrix} \left(C_{ij}^{G}\right)_{k}^{1} & \left(C_{ij}^{G}\right)_{k}^{2} \\ \left(C_{ij}^{G}\right)_{k}^{3} & \left(C_{ij}^{G}\right)_{k}^{4} \end{bmatrix} \begin{Bmatrix} \mathrm{d}\varepsilon_{xx} \\ \mathrm{d}\varepsilon_{yy} \\ 2\mathrm{d}\varepsilon_{xy} \\ 2\mathrm{d}\varepsilon_{yz} \\ 2\mathrm{d}\varepsilon_{xz} \\ \mathrm{d}\varepsilon_{zz} \end{Bmatrix} \tag{5.25}$$

where $\left[\left(C_{ij}^{G}\right)_{k}^{*}\right]$ is re-arranged from the matrix $[(C_{ij}^{G})_{k}] = ([T_{in}]_{c})_{k} \ \ ([S_{nm}]_{k})^{-1}([T_{mj}]_{c}^{\mathrm{T}})_{k}$ and $\left(C_{ij}^{G}\right)_{k}^{1}$, $\left(C_{ij}^{G}\right)_{k}^{2}$, $\left(C_{ij}^{G}\right)_{k}^{3}$ and $\left(C_{ij}^{G}\right)_{k}^{4}$ are 3×3 sub-matrices of $\left[\left(C_{ij}^{G}\right)_{k}^{*}\right]$.

Suppose that the stress resultant increments in unit length applied on the laminate are denoted by dN_{xx}, dN_{yy}, dN_{zz}, dN_{xy}, dN_{xz}, dN_{yz}, dM_{xx}, dM_{yy} and dM_{xy}, as shown in Fig. 5.11. The balances of in-plane forces result in the following equations

$$\begin{Bmatrix} \mathrm{d}N_{xx} \\ \mathrm{d}N_{yy} \\ \mathrm{d}N_{xy} \end{Bmatrix} = \int_{-h/2}^{h/2} \begin{Bmatrix} \mathrm{d}\sigma_{xx} \\ \mathrm{d}\sigma_{yy} \\ \mathrm{d}\sigma_{xy} \end{Bmatrix} \mathrm{d}z$$

$$= \int_{-h/2}^{h/2}[(C_{ij}^{G})^{1}] \begin{Bmatrix} \mathrm{d}\varepsilon_{xx}^{0} \\ \mathrm{d}\varepsilon_{yy}^{0} \\ 2\mathrm{d}\varepsilon_{xy}^{0} \end{Bmatrix} \mathrm{d}z + \int_{-h/2}^{h/2}[(C_{ij}^{G})^{1}] \begin{Bmatrix} \mathrm{d}\kappa_{xx}^{0} \\ \mathrm{d}\kappa_{yy}^{0} \\ 2\mathrm{d}\kappa_{xy}^{0} \end{Bmatrix} z\mathrm{d}z + \int_{-h/2}^{h/2}[(C_{ij}^{G})^{2}] \begin{Bmatrix} 2\mathrm{d}\varepsilon_{yz} \\ 2\mathrm{d}\varepsilon_{xz} \\ \mathrm{d}\varepsilon_{zz} \end{Bmatrix} \mathrm{d}z$$

$$= \sum_{k=1}^{n}[(C_{ij}^{G})_{k}^{1}] \begin{Bmatrix} \mathrm{d}\varepsilon_{xx}^{0} \\ \mathrm{d}\varepsilon_{yy}^{0} \\ 2\mathrm{d}\varepsilon_{xy}^{0} \end{Bmatrix} (z_{k} - z_{k-1}) + \frac{1}{2}\sum_{k=1}^{n}[(C_{ij}^{G})_{k}^{1}] \begin{Bmatrix} \mathrm{d}\kappa_{xx}^{0} \\ \mathrm{d}\kappa_{yy}^{0} \\ 2\mathrm{d}\kappa_{xy}^{0} \end{Bmatrix} (z_{k}^{2} - z_{k-1}^{2})$$

$$+ \sum_{k=1}^{n}[(C_{ij}^{G})_{k}^{2}] \begin{Bmatrix} 2\mathrm{d}\varepsilon_{yz}^{(k)} \\ 2\mathrm{d}\varepsilon_{xz}^{(k)} \\ \mathrm{d}\varepsilon_{zz}^{(k)} \end{Bmatrix} (z_{k} - z_{k-1})$$

$$= [Q_{ij}^{I}] \begin{Bmatrix} \mathrm{d}\varepsilon_{xx}^{0} \\ \mathrm{d}\varepsilon_{yy}^{0} \\ 2\mathrm{d}\varepsilon_{xy}^{0} \end{Bmatrix} + [Q_{ij}^{II}] \begin{Bmatrix} \mathrm{d}\kappa_{xx}^{0} \\ \mathrm{d}\kappa_{yy}^{0} \\ 2\mathrm{d}\kappa_{xy}^{0} \end{Bmatrix} + \sum_{k=1}^{n}([Q_{ij}^{IV}])_{(k)} \begin{Bmatrix} 2\mathrm{d}\varepsilon_{yz}^{(k)} \\ 2\mathrm{d}\varepsilon_{xz}^{(k)} \\ \mathrm{d}\varepsilon_{zz}^{(k)} \end{Bmatrix} \tag{5.26}$$

where $([Q_{ij}^{IV}])_{(k)} = [(C_{ij}^{G})_{k}^{2}](z_{k} - z_{k-1})$ and n=2N−1 after incorporation of the inter-layers.

The in-plane moment equilibrium conditions give

$$
\begin{Bmatrix} \mathrm{d}M_{xx} \\ \mathrm{d}M_{yy} \\ \mathrm{d}M_{xy} \end{Bmatrix} = \int_{-h/2}^{h/2} \begin{Bmatrix} \mathrm{d}\sigma_{xx} \\ \mathrm{d}\sigma_{yy} \\ \mathrm{d}\sigma_{xy} \end{Bmatrix} z\mathrm{d}z
$$

$$
= \int_{-h/2}^{h/2} [(C_{ij}^{G})^{1}] \begin{Bmatrix} \mathrm{d}\varepsilon_{xx}^{0} \\ \mathrm{d}\varepsilon_{yy}^{0} \\ 2\mathrm{d}\varepsilon_{xy}^{0} \end{Bmatrix} z\mathrm{d}z + \int_{-h/2}^{h/2} [(C_{ij}^{G})^{1}] \begin{Bmatrix} \mathrm{d}\kappa_{xx}^{0} \\ \mathrm{d}\kappa_{yy}^{0} \\ 2\mathrm{d}\kappa_{xy}^{0} \end{Bmatrix} z^{2}\mathrm{d}z + \int_{-h/2}^{h/2} [(C_{ij}^{G})^{2}] \begin{Bmatrix} 2\mathrm{d}\varepsilon_{yz} \\ 2\mathrm{d}\varepsilon_{xz} \\ \mathrm{d}\varepsilon_{zz} \end{Bmatrix} z\mathrm{d}z
$$

$$
= \frac{1}{2}\sum_{k=1}^{n} [(C_{ij}^{G})_{k}^{1}] \begin{Bmatrix} \mathrm{d}\varepsilon_{xx}^{0} \\ \mathrm{d}\varepsilon_{yy}^{0} \\ 2\mathrm{d}\varepsilon_{xy}^{0} \end{Bmatrix} (z_{k}^{2} - z_{k-1}^{2}) + \frac{1}{3}\sum_{k=1}^{n} [(C_{ij}^{G})_{k}^{1}] \begin{Bmatrix} \mathrm{d}\kappa_{xx}^{0} \\ \mathrm{d}\kappa_{yy}^{0} \\ 2\mathrm{d}\kappa_{xy}^{0} \end{Bmatrix} (z_{k}^{3} - z_{k-1}^{3})
$$

$$
+ \frac{1}{2}\sum_{k=1}^{n} [(C_{ij}^{G})_{k}^{2}] \begin{Bmatrix} 2\mathrm{d}\varepsilon_{yz}^{(k)} \\ 2\mathrm{d}\varepsilon_{xz}^{(k)} \\ \mathrm{d}\varepsilon_{zz}^{(k)} \end{Bmatrix} (z_{k}^{2} - z_{k-1}^{2})
$$

$$
= [Q_{ij}^{\mathrm{II}}] \begin{Bmatrix} \mathrm{d}\varepsilon_{xx}^{0} \\ \mathrm{d}\varepsilon_{yy}^{0} \\ 2\mathrm{d}\varepsilon_{xy}^{0} \end{Bmatrix} + [Q_{ij}^{\mathrm{III}}] \begin{Bmatrix} \mathrm{d}\kappa_{xx}^{0} \\ \mathrm{d}\kappa_{yy}^{0} \\ 2\mathrm{d}\kappa_{xy}^{0} \end{Bmatrix} + \sum_{k=1}^{n} ([Q_{ij}^{\mathrm{V}}])_{(k)} \begin{Bmatrix} 2\mathrm{d}\varepsilon_{yz}^{(k)} \\ 2\mathrm{d}\varepsilon_{xz}^{(k)} \\ \mathrm{d}\varepsilon_{zz}^{(k)} \end{Bmatrix} \tag{5.27}
$$

where $([Q_{ij}^{V}])_{(k)} = \frac{1}{2}[(C_{ij}^{G})_{k}^{2}](z_{k}^{2} - z_{k-1}^{2})$.

In the above equations, it is noted that the out-of-plane strain increments given by Eqs. (5.24.4) – (5.24.6), $2\mathrm{d}\varepsilon_{xz} = \frac{\partial(\mathrm{d}w)}{\partial x} - \mathrm{d}\varphi_{x}$, $2\mathrm{d}\varepsilon_{yz} = \frac{\partial(\mathrm{d}w)}{\partial y} - \mathrm{d}\varphi_{y}$ and $\mathrm{d}\varepsilon_{zz} = \frac{\partial(\mathrm{d}w)}{\partial z}$ are varied through the thickness direction because the displacement w is dependent on the out-of-plane variable z. However, the thickness of each layer of the composite laminate is thin in general, and the variation in the out-of-plane strains is confined within the layer. Thus, instead of using the variable quantities, the out-of-plane strain increments at the middle surface of each layer, $2\mathrm{d}\varepsilon_{yz}^{(k)}$, $2\mathrm{d}\varepsilon_{xz}^{(k)}$ and $\mathrm{d}\varepsilon_{zz}^{(k)}$, are chosen as representative of that layer. These strains are constant throughout the thickness of each layer, but may vary from layer to layer.

Suppose that the applied out-of-plane stress increments are denoted by $\mathrm{d}\sigma_{yz}$, $\mathrm{d}\sigma_{xz}$ and $\mathrm{d}\sigma_{zz}$. The out-of-plane equilibrium equations are found to be

$$\begin{aligned} d\sigma_{yz}^{(k)} &= d\sigma_{yz} \\ d\sigma_{xz}^{(k)} &= d\sigma_{xz} \qquad k=1, 2, \ldots, n \\ d\sigma_{zz}^{(k)} &= d\sigma_{zz} \end{aligned} \tag{5.28}$$

Substituting Eq. (5.24) into Eq. (5.28) leads to additional $3n$ equations

$$\begin{Bmatrix} d\sigma_{yz} \\ d\sigma_{xz} \\ d\sigma_{zz} \end{Bmatrix} = [(G_{ij}^G)_k^3] \begin{Bmatrix} d\varepsilon_{xx}^{(k)} \\ d\varepsilon_{yy}^{(k)} \\ 2d\varepsilon_{xy}^{(k)} \end{Bmatrix} + [(G_{ij}^G)_k^4] \begin{Bmatrix} 2d\varepsilon_{yz}^{(k)} \\ 2d\varepsilon_{xz}^{(k)} \\ d\varepsilon_{zz}^{(k)} \end{Bmatrix}$$

$$= [(C_{ij}^G)_k^3] \begin{Bmatrix} d\varepsilon_{xx}^0 \\ d\varepsilon_{yy}^0 \\ 2d\varepsilon_{xy}^0 \end{Bmatrix} + [(C_{ij}^G)_k^3] \begin{Bmatrix} d\kappa_{xx}^0 \\ d\kappa_{yy}^0 \\ 2d\kappa_{xy}^0 \end{Bmatrix} \frac{z_{k-1} + z_k}{2} + [(C_{ij}^G)_k^4] \begin{Bmatrix} 2d\varepsilon_{yz}^{(k)} \\ 2d\varepsilon_{xz}^{(k)} \\ d\varepsilon_{zz}^{(k)} \end{Bmatrix},$$

$$k=1, 2, \ldots, n, \tag{5.29}$$

From Eqs. (5.26), (5.27) and (5.29), it is seen that there are altogether $3n+6$ equations to determine the same number of unknown quantities, i.e., $d\varepsilon_{xx}^0$, $d\varepsilon_{yy}^0$, $2d\varepsilon_{xy}^0$, $d\kappa_{xx}^0$, $d\kappa_{yy}^0$ and $2d\kappa_{xy}^0$ and $2d\varepsilon_{yz}^{(k)}$, $2d\varepsilon_{xz}^{(k)}$ and $d\varepsilon_{zz}^{(k)}$ with k=1, 2, …, n. Thus, the $3n+6$ unknown strain increments can be obtained by solving the equations simultaneously.

After determination of the strain quantities, the averaged strain increments in each lamina are given by

$$\{d\varepsilon\}_k^G = \begin{Bmatrix} d\varepsilon_{xx}^0 + \frac{z_k + z_{k-1}}{2} d\kappa_{xx}^0, d\varepsilon_{yy}^0 + \frac{z_k + z_{k-1}}{2} d\kappa_{yy}^0, \\ d\varepsilon_{zz}^{(k)}, d\varepsilon_{yz}^{(k)}, d\varepsilon_{xz}^{(k)}, 2d\varepsilon_{xy}^0 + (z_k + z_{k-1}) d\kappa_{xy}^0 \end{Bmatrix}, \quad k=1, 2, \ldots, n \tag{5.30}$$

Substituting Eq. (5.30) into Eq. (5.11.2) gives the averaged 3D stress increments in each lamina in the global system. Further substituting the resulting equation into a 3D equation similar to Eq. (5.13.2) (refer to Eq. (1.76.2)), the stress increments shared by the lamina in its local coordinate system are obtained. The internal stresses in the constituent materials are then derived by using the bridging model as described in Chapter 3, whereas a failure of the lamina is detected using the failure criteria developed in Chapter 4. The ultimate failure criteria given in Table 5.3 are used to determine an ultimate strength of the laminate, or the partial stiffness discount scheme represented by Eq. (5.22) is applied to a failed layer including an inter-layer if no ultimate failure is assumed.

In the present case, however, an inter-layer may possibly attain an unlimited large out-of plane strain even though any strain component of a primary layer has been constrained to a limitation value of e.g., 12%. Physically, when the modulus of an inter-layer is very small, an externally applied load in the thickness direction may cause the layer to deform very greatly along the same direction. In order to avoid the laminate assuming an unreasonably large strain or deformation in the thickness direction, any strain of an inter-layer in the thickness direction greater than 100% will not be included in the following equation to evaluate the laminate out-of plane strain:

$$\mathrm{d}\varepsilon_{zz} = \frac{\mathrm{d}w}{h} = \frac{\sum_{k=1}^{n}(\mathrm{d}w_k)}{\sum_{k=1}^{n}(z_k - z_{k-1})} = \frac{\sum_{k=1}^{n}\mathrm{d}\varepsilon_{zz}^{k}(z_k - z_{k-1})}{\sum_{k=1}^{n}(z_k - z_{k-1})} \tag{5.31}$$

Remark 5.3

When all of the out-of plane stress increments, $\mathrm{d}\sigma_{yz}$, $\mathrm{d}\sigma_{xz}$, and $\mathrm{d}\sigma_{zz}$, are zero, the pseudo 3D laminate theory will deteriorate to the classical laminate theory.

5.9 Constituent Properties

It can be realized that a key issue in applying the bridging model to a composite analysis is to define input parameters correctly. Two classes of input data are required. One is the constituent fiber and matrix properties and another is the laminate geometric parameters. The second class of input data can be defined or taken as design variables according to the composite *in situ* fabrication condition. These data are generally specified or provided beforehand and hence are easily obtainable. The first class data, i.e., the fiber and matrix properties, however, need to be measured through experiments. The measurement of monolithic matrix properties is rather easy: the stress-strain curves of uniaxial tension, uniaxial compression and four- or three-point bending tests will be sufficient for most static problems. If a thermal load (i.e., a temperature variation) is involved, the matrix stress-strain curves at every temperature in the variation range should be measured. On the other hand, the measurement of monolithic fiber properties is comparatively difficult and, if done, large deviations in the experimental data can be recognized. This is because individual fibers generally have very small diameters. For instance, the diameter of a carbon, Kevlar, or glass fiber is in a range of 0.003 mm to 0.02 mm (Watts, 1984). Even a relatively thick boron fiber only has a diameter of 0.1 mm to 0.2 mm (Watts, 1984). Such thin fibers are difficult to test to obtain properties other than longitudinal tensile ones. Instead of direct measurements, the fiber properties are generally back-calculated from the overall responses of a composite (usually a unidirectional lamina) using some micromechanics theory. In reality, relatively large deviations in the measurement for composite properties, especially for ultimate strengths, can be observed. Even with the same constituent material system and the same fiber volume fraction, different manufacturers can produce composites with different mechanical responses. Thus, it is highly possible that different material data may be found in the literature for the same fiber material. This means that the *in situ* constituent properties should be used whenever possible. The *in situ* constituent properties can be measured (for the matrix material) or calibrated against some overall responses of the composites. However, as a general rule of thumb, when a constituent (especially fiber) material has been widely used in the composite industry, its representative properties can be obtained from the literature.

Let us consider two such materials. One is a Union Carbide T300 fiber, which has been used in the fabrication of composites for various applications (NASA, 1975; Rotem & Nelson, 1981). In this book, extensive comparisons are made between predictions and experiments for the composites made from T300 fibers and different epoxy matrices. The elastic properties of the T300 carbon fibers given by Soden et al. (1998a) and summarized in Table 5.4 are used throughout this book. The carbon fibers can be well regarded as linearly elastic until rupture. Soden et al. (1998a) also provided longitudinal tensile and compressive strengths of a UD composite made from T300 fibers and an epoxy (BSL 914C) matrix having a fiber volume fraction of 60%. The BSL 914C matrix properties reported are (Soden et al., 1998a): E^m=4.0 GPa, ν^m=0.35, σ_u^m=75 MPa and ε_u^m=4%, where ε_u^m is the ultimate tensile strain of the matrix. Supposing that the matrix exhibits a bilinear stress-strain curve and assumes a typical yield strength of 50MPa, its hardening modulus is found to be E_T^m=0.91 GPa. Substituting these parameters into Eqs. (4.11) – (4.15) and using the composite longitudinal tensile strength, 1,500 MPa, the fiber tensile strength is found to be 2,467.7 MPa. Similarly, based on the composite longitudinal compressive strength, 900 MPa, the retrieved fiber compressive strength is 1,470.4 MPa. Both of them are listed in Table 5.4.

Table 5.4 Properties of T300 carbon fibers (Soden et al., 1998a)

E_{11} (Gpa)	E_{22} (GPa)	ν_{12}	G_{12} (GPa)	ν_{23}	σ_u (MPa)	$\sigma_{u,c}$ (MPa)
230	15	0.2	15	0.07	2467.7	1470.4

Another material considered is a Narmco 5208 epoxy used as matrix (NASA, 1975; Rotem & Nelson, 1981). Sendeckyj et al. (1975) carried out extensive experiments on the laminates made of the T300/5208 material system with various lamination lay-ups. The overall in-plane shear stress-shear strain curve measured from a UD composite (Sendeckyj et al., 1975) is used to retrieve the tensile stress-strain curve of the 5208 matrix. It is noted that among three candidate stress-strain curves measured from a UD composite under, respectively, a longitudinal, a transverse and an in-plane shear load, which may be used to retrieve the matrix stress-strain data, the in-plane shear curve should be taken as the first choice. The reasons are apparent. Whenever possible, the overall longitudinal curve should not be used to back-calculate the matrix stress-strain data up to failure, since the longitudinal failure of a UD composite is generally caused by a fiber failure and the matrix may have not been loaded to its ultimate level. Furthermore, an in-plane shear stress-strain curve usually displays more nonlinear behavior than a transverse one and hence the retrieved matrix stress-strain curve thus made can exhibit more distinctly nonlinear characteristics. The retrieved matrix stress-strain curve based on the in-plane shear response of the UD composite is expressed in 10 piecewise linear segments (Huang, 2000b), with tangential moduli and critical stress data being summarized in Table 5.5. Thus, under any load condition, the corresponding matrix hardening (tangential) modulus is defined as

$$E_T^m = \left(E_T^m\right)_{i+1} \quad \text{when} \quad \left(\sigma_Y^m\right)_i \le \sigma_e^m \le \left(\sigma_Y^m\right)_{i+1}, \; i=0,1,\ldots,9 \quad \left(\sigma_Y^m\right)_0=0 \tag{5.32.1}$$

and

$$E_T^m = \left(E_T^m\right)_{10}, \text{ when } \sigma_e^m \ge \left(\sigma_Y^m\right)_{10} \tag{5.32.2}$$

with $(E_T^m)_1 \equiv E^m$ and $(\sigma_Y^m)_1 \equiv \sigma_Y^m$. These plastic parameters are assumed to be applicable to both tension and compression in this book if no additional mention is made.

Sendeckyj et al. (1975) also reported the longitudinal tension, transverse tension and in-plane shear strengths of the UD composite, being 1,619 MPa, 49 MPa and 76 MPa respectively. The composite had a volume fraction of V_f=0.664. Using 49 MPa and Eqs. (4.17) – (4.18), the retrieved matrix tensile strength is 36 MPa, whereas using 76 MPa and according to Eqs. (4.19) – (4.20), the retrieved matrix strength is 48 MPa. Taking a simple average, the tensile strength of 42 MPa is used for the 5,208 epoxy matrix in the following calculation. Furthermore, the matrix compressive strength, 108.3 MPa, is back-calculated from the ultimate tensile strength of a 30 degree angle ply laminate, $[\pm 30°]_{2s}$, also given by Sendeckyj et al. (1975). This is because under the longitudinal tensile load (applied in 0° direction) the matrix material in the $[\pm 30°]_s$ laminate is subjected to an essential compression (Remark 2.3).

Table 5.5 Elastic-plastic parameters of 5208 epoxy matrix (Huang, 2000b)
(v^m=0.35, σ_u^m=42 MPa and $\sigma_{u,c}^m$=108.3 MPa)

i=	1*	2	3	4	5	6	7	8	9	10
$(\sigma_Y^m)_i$ (MPa)	28.0	34.8	42.2	49.4	56.4	63.0	69.1	74.8	80.3	83.0
$(E_T^m)_i$ (GPa)	4.50	3.30	3.07	2.63	2.22	1.81	1.45	1.20	0.99	0.42

* Note: $(E_T^m)_1 \equiv E^m$ and $(\sigma_Y^m)_1 \equiv \sigma_Y^m$.

5.10 Inelastic Response

With the constituent properties given in Tables 5.4 and 5.5, stress-strain responses of different laminates, $[0]_8$, $[\pm 30°]_{2S}$, $[\pm 45°]_{2S}$, $[\pm 60°]_{2S}$, $[\pm 90°]_{2S}$, $[0°/\pm 30°/0°]_S$, $[0°/\pm 45°/0°]_S$, $[0°/\pm 60°/0°]_S$ and $[90°/\pm 45°/90°]_S$, made from T300 fibers and 5208 matrix materials and subjected to uniaxial tension up to failure are evaluated. All of the laminae involved assume approximately the same fiber volume fraction, V_f=0.664. As all of the laminates are symmetrically arranged and are only subjected to in-plane load, the classical laminate theory is applicable and no bending curvature will occur. According to Eq. (5.3), the overall strains of each layer in the respective laminate are the same, equal to the middle-surface strain values, ε_{xx}^{0}, ε_{yy}^{0} and ε_{xy}^{0}. The loading direction is longitudinal, i.e., along the 0°-direction of the laminates. Under such load condition, the constituents are not likely to be subjected to bi-axial tension or compression. Thus, the classical maximum normal stress criteria, designated by inequalities Eqs. (4.3) and (4.6), are applied to detect each ply failure. Each lamina in the laminate is assumed to have the same thickness. No thermal residual stress is incorporated in the present analysis, since no related parameters (e.g., thermal expansion coefficients of the constituents and the stress-free temperature) were given in Sendeckyj et al. (1975) and since the retrieved constituent properties, given in Tables 5.4 and 5.5, have been obtained without considering any temperature influence. During the calculation, a stress-strain curve is terminated as long as an ultimate failure of the laminate is attained. The calculation can also be done by using the attached computer code listed in Chapter 6, which is developed based on the 3D theories. Sendeckyj et al. (1975) measured the responses of the laminates experimentally, whose results will be used as a benchmark to check the accuracy of the calculations.

First, the unidirectional laminate, $[0]_8$, subjected to longitudinal tension, transverse tension (equivalent to longitudinal tension on the $[\pm 90]_{2S}$ laminate) and in-plane shear are evaluated. The theoretical curves are plotted in Figs. 5.12 – 5.14, in which the measured stress-strain data (Sendeckyj et al., 1975) are also shown. It is seen that the predicted stress-strain curves when loaded in the longitudinal direction agree perfectly with the experiments. On the other hand, the predicted curve when loaded in the transverse direction is higher than the measured one, whereas the predicted in-plane shear stress-strain curve is lower than the experimental data. This is because the failure of the UD lamina when loaded in transverse tension or in-plane shear is due to the fracture of the matrix material. The used matrix strength, 42 MPa, is higher than the matrix strength retrieved from the transverse tensile strength and lower than that back-calculated from the in-plane shear strength. The longitudinal load on the $[0]_8$ laminate causes its fibers to fail first, whereas under transverse tension or in-plane shear load condition the primary layers of the laminate fail first, due to a tensile failure of the matrix followed by tensile failures of the inter-layers. In the latter two cases, excessively large deformations occur after the failures of all of the layers, yet a small additional load can be further sustained as a result of the partial stiffness discount applied to a failed lamina. It can be seen from Figs. 5.13 and 5.14 that horizontal curves occur after the failures of all of the layers, which have been caused by the matrix tensile failure. The ultimate failure is detected when a laminate strain attains the constraint. However, as the constraint strain has been specified as 12%, much larger than the strain data recorded, the later parts of the horizontal curves are not plotted in the figures.

Ideally, the performance of a multilayer, unidirectional laminate, e.g., $[0]_8$, under an in-plane load condition should be equivalent to that of a single layer, unidirectional lamina [0]. The failure of a UD lamina under a transverse or in-plane shear load is generally caused by that of matrix material, as shown in Chapter 4, and the predicted transverse and in-plane shear stress-strain curves will not exhibit a horizontal segment as no pure matrix interface layer will be introduced into a single layer. On the other hand, the present simulation has incorporated pure matrix interface layers into the multilayer laminate, $[0]_8$, resulting in a difference in the predicted transverse and in-plane shear stress-strain curves. A summary of the predicted failure features of the T300/5208 $[0]_8$ laminate under different load conditions is given in Table 5.6.

Table 5.6 Predicted failure features of the T300/5208 $[0]_8$ laminate

Load	Failure Description
Longitudinal Stress	Almost linearly elastic up to an ultimate failure, caused by the tensile failure of the fiber
Transverse Stress	Before an initial failure occurs, the stress-strain curve is almost linear. The initial failure is caused by a matrix tensile failure in the middle layers at 56.2 MPa, followed by a matrix tensile failure in the surface layers at 56.4 MPa. Then the inter-layers attain their tensile failure at 57.3 MPa. After all of the primary layers have failed, the predicted strains increase rapidly, resulting in an almost horizontal segment of curve. The ultimate failure of the laminate is detected by the critical strain constraint.
In-plane shear Stress	An initial failure occurs in the middle layers due to matrix tensile failure at 65 MPa, before which the laminate nonlinear behavior is caused by the plastic deformation of the matrix. The surface layers fail as a result of matrix tensile failure at 65.4 MPa, and then rapidly increasing strains are recognized. A failure of the inter-layers occurs at 66.6 MPa, due to the critical strain condition being attained. Similarly,as loaded transversely, a nearly horizontal line on the stress-strain curve is seen after the failures of the surface layers.

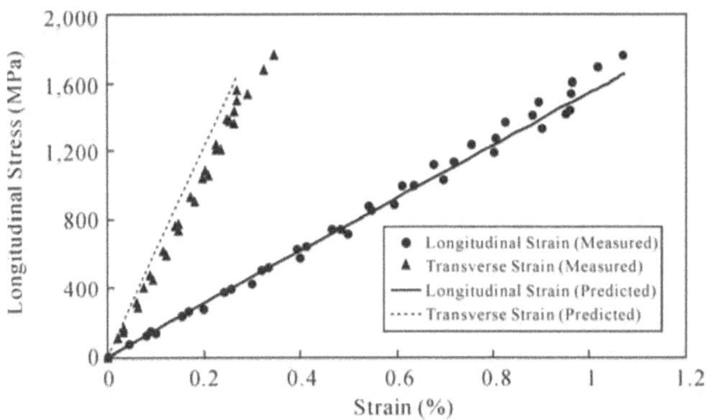

Fig. 5.12 Longitudinal stress versus longitudinal and transverse strains for T300/5208 $[0]_8$ laminate

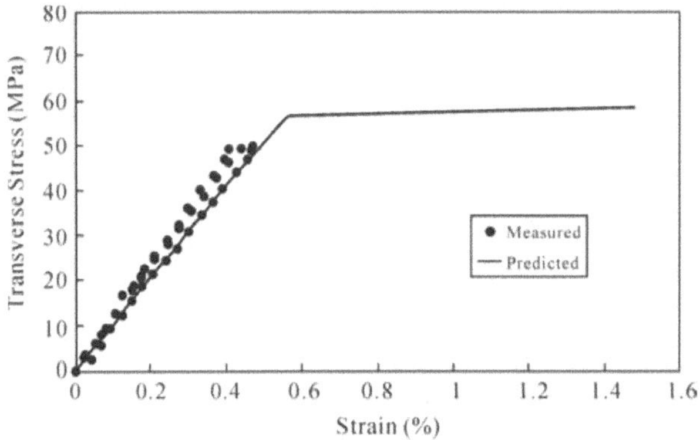

Fig. 5.13 Transverse stress versus transverse strain for T300/5208 $[0]_8$ laminate

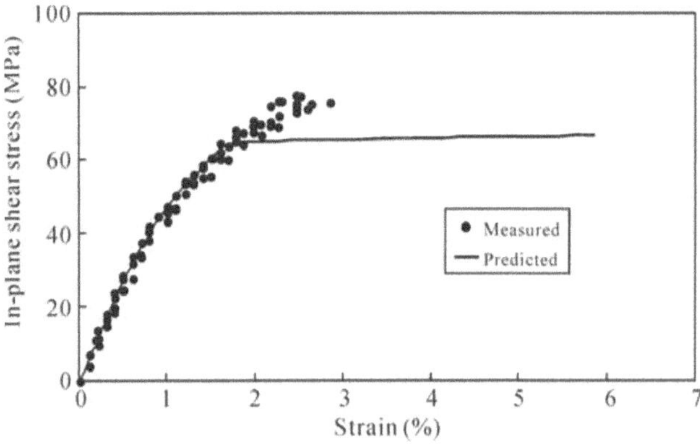

Fig. 5.14 In-plane shear stress versus in-plane shear strain for T300/5208 $[0]_8$ laminate

For all of the other laminates, both the longitudinal (x-directional) and transverse (y-directional) strains, i.e., ε_{xx}^{0} and ε_{yy}^{0}, are evaluated versus the longitudinally applied stress. The theoretical stress-strain curves are graphed in Fig. 5.15 through Fig. 5.21, respectively, which are compared with the measured counterparts taken from Sendeckyj et al. (1975). It must be pointed out that under a longitudinal tensile load all of the laminates undergo a transverse contraction. Thus, the predicted and the measured transverse strains, ε_{yy}^{0}, are actually negative. However, for the convenience of presentation, the transverse strains have been plotted in the positive region of the corresponding figures. The comparison clearly indicates that most of the predicted stress-strain curves agree well with the experimental data along not only the longitudinal but also the transverse directions. Distinct discrepancies have been found for only two laminates, i.e., the $[\pm45°]_{2S}$ and the $[90°/\pm45°/90°]_{S}$ laminates, as shown in Figs. 5.16 and 5.21, respectively.

For the $[\pm45°]_{2S}$ laminate, the predicted failure of the laminate is initiated from a tensile failure of the matrix in the primary layers. Following this, a tensile failure of the introduced pure matrix inter-layers occurs. After all of the layers have failed, an excessive strain is attained, as shown in Fig. 5.16. Similarly, as in Figs. 5.13 and 5.14, only a part of the horizontal curve is graphed in the figure. The predicted ultimate strength is lower than that given by the experiment. This may be attributed to the lower matrix strength retrieved from the transverse and in-plane shear strengths of the lamina, because the transverse strength of a lamina is usually lower than that used in the laminate, as illustrated in Chapter 3.

For the $[90°/\pm45°/90°]_{2S}$ laminate, the first-ply failure occurs to the (+45°) and (–45°) laminae at 149.2 MPa whereas the second-ply failure takes place in the (90°) laminae at 153.8 MPa, both of which are caused by a matrix tensile failure. Then the failure of the inter-layers occurs at 216.5 MPa. After all these failures, the stress-strain curves of the laminate still exhibit linearity. This is because the fibers in the (±45°) layers sustain most of the applied load during a subsequent loading process. An ultimate failure of the laminate is caused by the tensile failure of the fibers in the (±45°) layers at 469 MPa. The predicted ultimate strength is significantly higher than the measured one. The measured strength is close to the stress level at which all of the layers fail due to the matrix tensile failure. This example is intended to show that although part of the predicted stress-strain curves agrees well with the majority of the measured data for all of the laminates, there is a possibility of a discrepancy between the predicted and measured ultimate strengths. The bridging model developed in Chapter 3 is a unified elastic-plastic constitutive theory for fiber reinforced composites. A failure and strength prediction for the composites also depends on a lot of other issues such as the failure criteria and stiffness discount schemes used. In an elementary mechanics of materials textbook, four ultimate failure criteria for isotropic materials are generally introduced based on which we have the first, second, third and fourth strength theories, respectively. It has been well known that no strength theory is perfectly applicable to every structure made of isotropic materials. Each theory has its merits as well as drawbacks. Similarly, the failure criteria and strength theory developed in this book for fiber reinforced composite materials and structures may not be sufficiently accurate in every case.

In the analysis, the progressive failure process of a laminate can be captured clearly. Taking the $[0°/\pm60°/0°]_S$ laminate (Fig. 5.20) as an example, its progressive failures are recorded as follows. When the laminate is subjected to a longitudinal load, the initial failure occurs at the (±60°) layers at a stress level of σ_{xx}=467 MPa due to a matrix tensile failure. Then the pure matrix inter-layers fail at σ_{xx}=826 MPa. An ultimate failure is caused by the tensile failure of the fibers in the [0°] layers at a stress level of 860 MPa, which is the ultimate strength of the laminate. The progressive failure features for this and other laminates considered in this section are summarized in Table 5.7.

Table 5.7 Predicted progressive failure features of the T300/5208 laminates

Laminate Type	Failure Description (Under a Longitudinal Tensile Load)
$[\pm30°]_{2S}$	No progressive failure is detected during the whole loading process. The ultimate failure is caused by a compressive failure of the matrix in the primary layers. Nonlinear stress-strain behavior is due to plastic deformations of the matrix material
$[\pm45°]_{2S}$	An initial failure occurs at 116.1 MPa caused by a matrix tensile failure in the middle (±45°) layers. Failures of the surface (±45°) layers and the inter-layers occur at 116.8 MPa and 119.4 MPa, respectively, both due to the matrix tensile failure. After all of the primary layers fail, predicted strains increase rapidly and an ultimate failure is detected by the critical strain constraint. The nonlinearity of the stress-strain curve before the initial failure results from the plastic deformation of the matrix material
$[\pm60°]_{2S}$	Initial failure occurs in the middle (±60°) layers at 77.1 MPa, and is immediately followed by the surface (±60°) layer failures at 77.3 MPa, both due to the matrix tensile failure. Then, a rapid increase in predicted strains is observed and the ultimate failure is detected by the critical strain constraint. It is noted that even up to the ultimate failure, the inter-layers have not failed and hence the stress-strain curves after the primary layer failures are not horizontal but at an inclined angle
$[0°/\pm30°/0°]_S$	The ultimate failure is caused by a tensile failure of the fibers in the (0°) layers and no other failure mode is detected. It is observed that a transverse strain is even larger than a corresponding longitudinal one
$[0°/\pm45°/0°]_S$	The predicted stress-strain curves are almost linear until an ultimate failure occurs at 904 MPa. The initial failure is caused by a matrix tensile failure in the (±45°) layers at 622 MPa, whereas the ultimate failure of the laminate is due to a fiber tensile failure in the (0°) layers
$[0°/\pm60°/0°]_S$	An initial failure occurs in the (±60°) layers caused by a matrix tensile failure at 467 MPa, followed by tensile failure of the inter-layers at 826 MPa. The ultimate failure in the (0°) layers takes place at 860 MPa due to a fiber tensile failure. It is noted that although the progressive failures are detected the stress-strain curves are almost linear because the fibers in the (0°) layers sustain most of the applied load
$[90°/\pm45°/90°]_S$	The first-ply failure occurs in the (±45°) layers at 149.2 MPa caused by a matrix tensile failure, followed by the second-ply failure in the (90°) layers at 153.8 MPa, also due to a matrix tensile failure. Then, the inter-layers attain their tensile failures at 216.5 MPa. Afterwards, the laminate can still sustain an additional load until a fiber tensile failure occurs in the (±45°) layers at 469 MPa

An interesting feature can be seen from the results shown in Figs. 5.16 and 5.18. We all know that an isotropic material generally cannot have a Poisson's ratio greater than 0.5. However, this is not true for an anisotropic composite material. Figs. 5.16 and 5.18 clearly show that a composite can have a Poisson's ratio, ν_{xy} (which is defined as the negative transverse strain over the longitudinal strain, both corresponding to the same but initial stress level), greater than 0.5. Moreover, a negative Poisson's ratio can also occur in some composites (Lakes et al., 2001; Webber et al., 2000; Zhang et al., 1999).

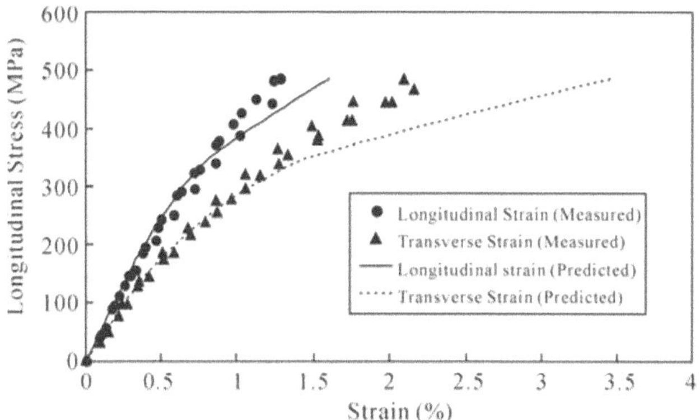

Fig. 5.15 Longitudinal stress versus longitudinal and transverse strains for T300/5208 $[\pm 30°]_{2S}$ laminate

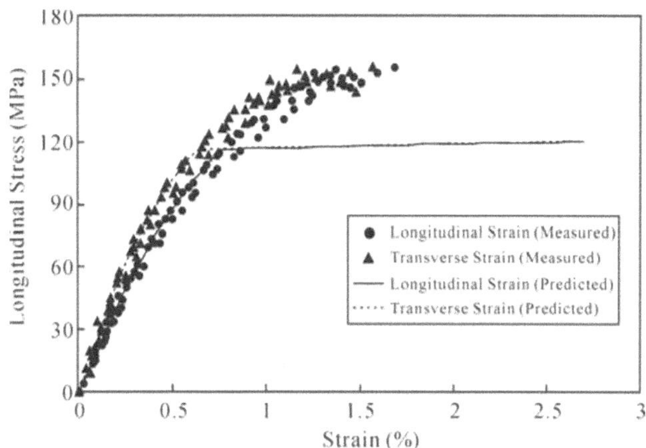

Fig. 5.16 Longitudinal stress versus longitudinal and transverse strains for T300/5208 $[\pm 45°]_{2S}$ laminate

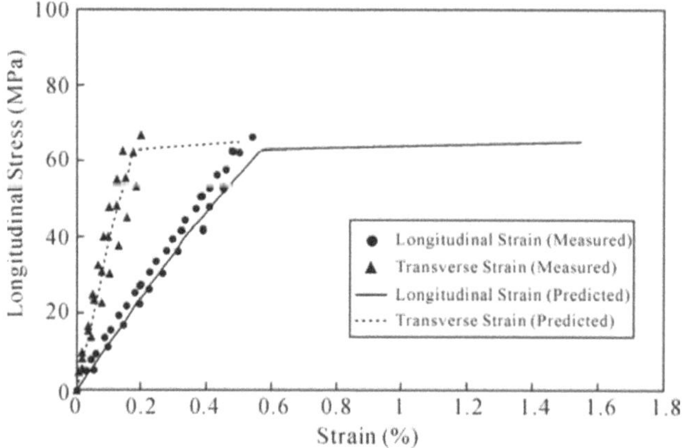

Fig. 5.17 Longitudinal stress versus longitudinal and transverse strains for T300/5208 $[\pm 60°]_{2S}$ laminate

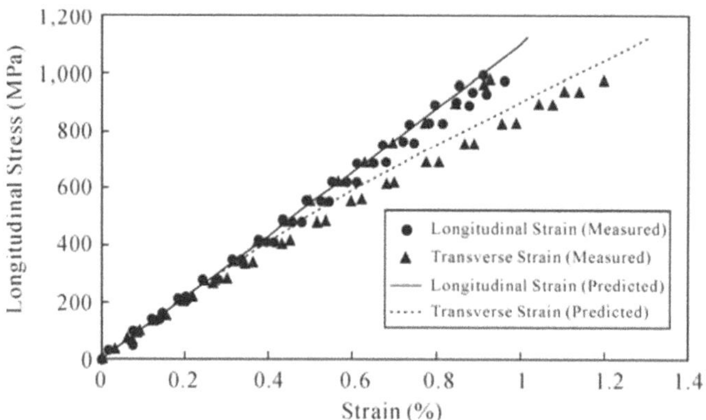

Fig. 5.18 Longitudinal stress versus longitudinal and transverse strains for T300/5208 $[0°/\pm 30°/ 0°]_S$ laminate

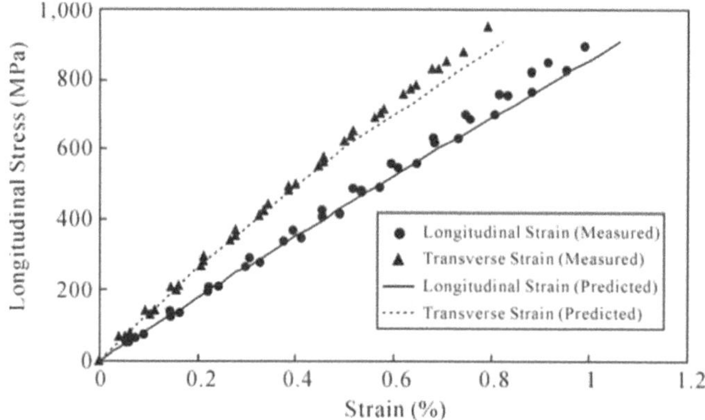

Fig. 5.19 Longitudinal stress versus longitudinal and transverse strains for T300/5208 $[0°/\pm 45°/0°]_S$ laminate

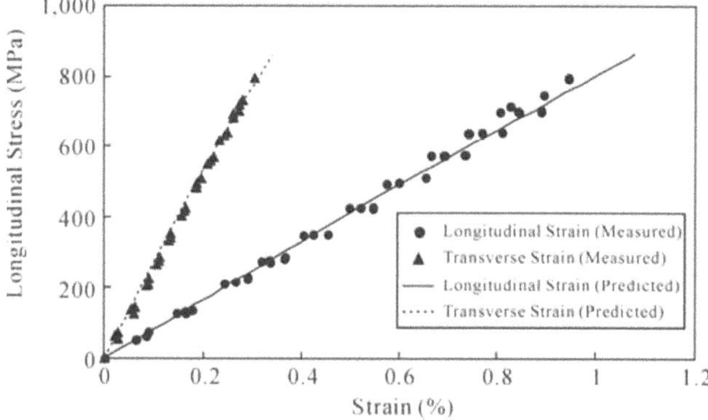

Fig. 5.20 Longitudinal stress versus longitudinal and transverse strains for T300/5208 [0°/±60°/0°]$_S$ laminate

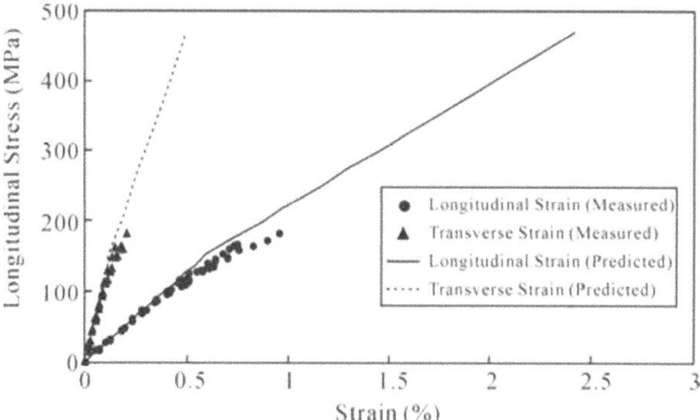

Fig. 5.21 Longitudinal stress versus longitudinal and transverse strains for T300/5208 [90°/±45°/90°]$_S$ laminate

5.11 Biaxial Strength Envelope

In most cases, a composite laminate in engineering applications is subjected to a combined load condition, e.g., combined uniaxial tension and compression in two or three orthogonal directions, or combined uniaxial tension/compression and shear loads. To understand the load carrying capacity of the laminate subjected to possibly various load combinations, failure envelopes are determined. A point on such an envelope represents a stress state at which a failure of the laminate occurs. In other words, if a stress combination is located inside the envelope, no corresponding failure of the laminate will occur. On the other hand, if a stress combination is outside the envelope, the laminate has attained a failure status. It is noted that each kind of failure can correspond to a failure envelope, e.g., the first ply failure envelope or the ultimate failure envelope. Such a failure envelope is also called a strength envelope.

With the bridging model, a failure stress state of a composite laminate under any multiaxial stress combination can be easily estimated. In this section, only in-plane biaxial loads are concerned. Namely, there is no bending moment or out-off plane load applied to the laminate. Similarly, as done in Section 4.5, a failure stress state for the composite laminate can be determined by setting the ratio of the applied two directional stresses to a given value. Varying this ratio, the predicted failure stress states constitute an envelope in the plane of the two stresses. Taking $\sigma_{yy}-\sigma_{xy}$ failure envelope as an example, the applied stresses can be assigned by

$$\{\sigma_i\}=\{\sigma_{xx}, \sigma_{yy}, \sigma_{xy}\}^T=\sigma\{0, \cos(\theta), \sin(\theta)\}^T, 0°\leq\theta\leq90°.$$

Two angle plied laminates, $[\pm45°]_S$ and $[\pm55°]_S$, made of an E-glass/MY750/HY917/DY063 system, are considered for illustration. The constituent properties are the same as those given in Table 5.1. However, the fiber volume fractions of these two laminates are different, V_f=0.504 for the $[\pm45°]_S$ and V_f=0.602 for the $[\pm55°]_S$ laminates respectively. Both laminates are subjected to combined loads of σ_{xx} and σ_{yy}, and the predicted strength envelopes are plotted in Figs. 5.22 and 5.23 respectively. Experimental data reported by Soden et al. (1993) are also shown in the figures for comparison.

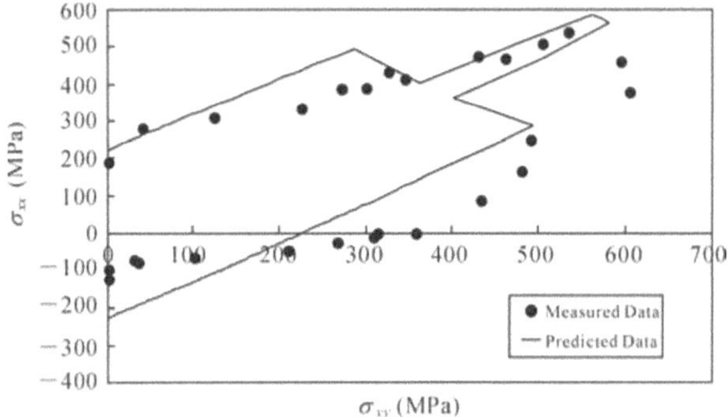

Fig. 5.22 Predicted and measured failure envelopes of a $[\pm45°]_S$ glass/epoxy laminate

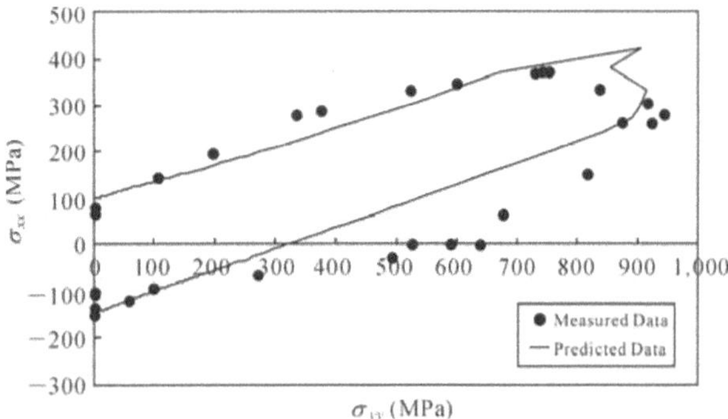

Fig. 5.23 Predicted and measured failure envelopes of a $[\pm55°]_S$ glass/epoxy laminate

In these two examples, bi-axial tension stresses can occur in the constituents and hence the generalized maximum normal stress theory is applied to detect the failures of the laminates with a power-index q=3. As a whole, the predicted envelopes agree well with the experiments. For the $[\pm45°]_S$ laminate, an ultimate failure caused by a fiber failure occurs only when it is subjected to a load combination with a load ratio of σ_{xx}:σ_{yy} in between 0.966 and 1.035. The ultimate failures of the laminate subjected to all of the other load combinations are caused by excessive strains. For the $[\pm55°]_S$ laminate, when a load ratio of σ_{xx} over σ_{yy} is less than 0.287, its ultimate failure is caused by a compressive failure of the matrix in the primary layers. When the ratio is in between 0.287 and 0.466, an ultimate failure is due to the failure of the fibers. Finally, if the ratio is greater than 0.466, an excessively large deformation occurs after a tensile failure of the matrix in all of the layers and the ultimate failure is detected by the additional critical strain constraint.

It deserves mentioning that the ultimate failure strength is sensitive to a load combination, especially when the failure mode is changed from one to another. Thus, some parts of the predicted failure envelopes are not smooth, as shown in Figs. 5.22 and 5.23.

5.12 Strength Under Thermo-Mechanical Load

In quite a number of cases when a thermoset polymer (represented by an epoxy) matrix composite laminate is analyzed, the influence of thermal residual stresses on the mechanical responses of the laminate can be neglected. There are two reasons for doing so. The first reason is that an epoxy matrix based composite is generally fabricated at or near to room temperature (RT). The variation between the stress-free and working (taken as room temperature in the previous analyses) temperatures would not be very large. The resulting thermal residual stresses, if any, are negligibly small. The second reason is that some or even all of the constituent properties of the laminate have been determined through retrieval of the laminate responses. The retrieved constituent properties have already compensated to some extent the influence of the thermal residual stresses on the mechanical behavior of the laminate. However, when the composite is made of a metal or some thermoplastic polymer matrix material, the thermal residual stresses in the composite can be high enough so that a neglect of them may cause a large prediction error. This is because the metal matrix or the thermoplastic polymer matrix composite has been fabricated at a relatively high temperature and a large difference exists between the composite processing (stress-free) and working temperatures. In some cases, the composite, especially the metal matrix composite, can be subjected to severe mechanical loads together with dramatic variations in working temperatures during its service duration. Knowledge of the mechanical performance of such a laminate under an extreme thermo-mechanical load condition is necessary.

Titanium alloy based metal matrix composite laminates offer excellent potential for modern aerospace applications (Newaz & Majumdar, 1994; Mall & Nicholas, 1997), where high-temperature working conditions are generally expected. Let us consider one such example, in which four laminates of different lay-ups, i.e., $[0]_8$, $[0/90]_{2s}$, $[0_2/\pm45]_s$ and $[0/\pm45/90]_s$, are made from ceramic silicon-carbide (SCS-6) fibers and Ti-15-3 matrix. Measured uniaxial (in the x-direction) tensile strengths of the laminates at two or three different temperatures have been reported in the literature (Robertson & Mall, 1996), and are summarized in Table 5.8 for illustration. All of the laminae in the laminates have the same fiber volume fraction, V_f=0.34 (Robertson & Mall, 1996), and the same thickness. According to Robertson & Mall (1996), the laminates assume a stress-free processing temperature at 815 °C. Therefore, thermal residual stresses are first generated in the fiber and matrix materials when the laminates are cooled down from 815 °C to room temperature (25 °C) before subsequent thermal and mechanical loads are applied. These thermal residual stresses for the laminates with different lay-ups at RT can be calculated by using the bridging model and temperature-dependent properties of the constituent materials. The calculated thermal residual stresses are listed in Table 5.9.

Table 5.8 Measured (Robertson & Mall, 1996) and predicted tensile strengths of SCS-6/Ti-15-3 composite laminates under uniaxial loads (V_f=0.34 and each layer being of the same thickness)

Lay-up	Temperature	Measured (MPa)	Predicted (MPa)
$[0]_8$	RT*	1,336 – 1,517	1,392
	427 °C	1,365 – 1,387	1,249
	650 °C	948	949
$[0/90]_{2s}$	RT	945 – 1,060	1,060
	650 °C	548	600
$[0_2/\pm45]_s$	RT	1,069	1,082
	650 °C	554	648
$[0/\pm45/90]_s$	RT	752	940
	650 °C	421	564

*RT= room temperature, taken as 25 °C.

Table 5.9 Thermal residual stresses at RT in fiber and matrix materials of the laminates with different lay-ups

Lay-up	Layer	σ_{11}^f (MPa)	σ_{22}^f (MPa)	σ_{12}^f (MPa)	σ_{11}^m (MPa)	σ_{22}^m (MPa)	σ_{12}^m (MPa)
$[0]_8$*	Surface [0]	−529.66	−228.73	0	271.87	115.69	0
	Middle [0]	−531.03	−235.04	0	270.63	111.95	0
	Inter-layer	–	–	–	307.74	223.81	0
$[0/90]_{2s}$*	Surface [0]	−763.88	−99.82	0	250.83	191.65	0
	Middle [0]	−764.98	−104.93	0	249.71	188.78	0
	90	−765.82	−104.64	0	249.82	188.61	0
	Inter-layer	–	–	–	279.96	279.84	0
$[0_2/\pm45]_s$*	Surface [0]	−612.38	−164.16	0	267.05	153.72	0
	Middle [0]	−613.62	−169.89	0	265.92	150.32	0
	+45	−784.01	−113.67	−54.80	245.35	183.50	−31.94
	-45	−784.01	−113.67	54.80	245.35	183.50	31.94
	Inter-layer	–	–	–	299.64	252.11	0
$[0/\pm45/90]_s$	0	−763.81	−99.82	0	250.84	191.64	0
	+45	−765.40	−104.79	−0.16	249.77	188.69	−0.10
	-45	−765.40	−104.79	0.16	249.77	188.69	0.10
	90	−765.90	−104.62	0	249.70	188.79	0
	Inter-layer	–	–	–	279.97	279.83	0

* as indicated in Section 5.6, thermal residual stresses in the primary surface and middle layers are different as a result of the introduced inter-layers, although the differences are insignificant.

The SCS-6 fiber can be considered as isotropic and linearly elastic until rupture (Robertson & Mall, 1996, 1998; Bigelow, 1993), whereas the Ti-15-3 matrix is taken as isotropic and bi-linearly elastic-plastic in the present analysis. The temperature-dependent constituent properties of the SCS-6 fiber and the Ti-15-3 matrix, except for their ultimate strengths, are taken from Robertson & Mall (1998). These parameters are listed in Table 5.10 for the fiber, and Table 5.11 for the matrix. The fiber tensile strengths at RT and 650 °C are recovered using the measured strengths of the $[0]_8$ lay-up laminate at RT and 650 °C, which were 2,584 MPa and 2,380 MPa, respectively. An inherent assumption made in the recovery is that the failure of the unidirectional composite both at RT and at 650 °C resulted from the fiber fracture. Thus, using the other constituent parameters given in Tables 5.10 and 5.11, the fiber tensile strengths are inversely determined. The fiber tensile strengths at other temperatures are defined simply through linear interpolation or extrapolation. Results are shown in Table 5.10. In the literature (Bigelow, 1993), the ultimate tensile strengths of the Ti-15-3 matrix at room temperature (RT) and at 538 °C are measured, being 948 MPa and 500 MPa, respectively. However, as the ultimate strength values of the constituent materials are the most important parameters for the strength predictions, the matrix strength at RT has been slightly amended to 783 MPa, which has been calibrated using the RT strength of the $[0/90]_{2s}$ laminate. This is because the predicted ultimate failure strength of the $[0/90]_{2s}$ laminate at RT is significantly affected by the matrix failure if the constituent parameters given in Tables 5.10 and 5.11 are employed, although the ultimate failure of the laminate is caused by the fiber failure (the progressive failures will be illustrated in the following paragraphs). Supposing that the predicted RT strength of the $[0/90]_{2s}$ laminate, without varying any other parameter in Tables 5.10 and 5.11, is equal to the measured value, the matrix tensile strength of 783 MPa is obtained. The tensile strength of the matrix at any other temperature, $\sigma_u^m(T)$, is assumed to be

$$\sigma_u^m(T) = \alpha(T)\sigma_Y^m(T), \quad \alpha(T) = \alpha_1 + \frac{T-25}{538-25}(\alpha_2 - \alpha_1) \tag{5.33}$$

where α_1=783/763=1.0262 and α_2=500/447=1.1186. Eq. (5.33) implies that the ratio between ultimate tensile and yield strengths of the matrix at any temperature can be defined through a linear interpolation or extrapolation by using the corresponding strengths at room temperature and at 538 °C. These strength data are summarised in Table 5.11. Furthermore, the ultimate compressive strengths of the fiber and the matrix at any temperature are considered to be equal to their corresponding tensile strengths, respectively. The generalized maximum normal stress theory is employed in the analysis with a power-index of q=3.

Table 5.10 Thermoelastic properties of SCS-6 fibers (Robertson & Mall, 1998)

T (°C)	E^f (GPa)	ν^f	σ_u^f (MPa)	α_f (10^{-6}/°C)
25	393	0.25	2,584 [a]	3.564
93	390	0.25	2,562 [b]	3.660
204	386	0.25	2,526 [b]	3.618
316	382	0.25	2,489 [b]	3.638
427	378	0.25	2,453 [b]	3.687
538	374	0.25	2,417 [b]	3.752
650	370	0.25	2,380 [a]	3.826
760	365	0.25	2,344 [b]	3.903
871	361	0.25	2,308 [b]	3.980
1,093	354	0.25	2,235 [b]	4.103

(a) Retrieved using the tensile strength of $[0]_8$ laminate;
(b) Interpolation/extrapolation value.

Table 5.11 Thermoelastic properties of Ti-15-3 matrix (Robertson & Mall, 1998)

T (°C)	E^m (GPa)	σ_Y^m (MPa)	E_T^m (GPa)	ν^m	σ_u^m (MPa)	α_m (10^{-6}/°C)
25	83.6	763	3.32	0.36	783 [a]	8.48
315	80.4	645 [b]	3.54 [b]	0.36	696 [c]	9.16
482	72.2	577	3.67	0.36	640 [c]	9.71
538	67.8	447	2.69	0.36	500	9.89
566	64.4	287	2.39	0.36	322 [c]	9.98
650	53.0	198	1.12	0.36	225 [c]	10.26
900	25.0	20 [b]	0.8 [b]	0.36	23 [c]	10.50

(a) Retrieved using the ultimate tensile strength of $[0/90]_{2s}$ laminate;
(b) Interpolation/extrapolation value;
(c) Determined according to Eq. (5.33).

Using the constituent parameters given in Tables 5.10 and 5.11, the ultimate tensile strengths of the laminates subjected to different thermo-mechanical loads are estimated. Incorporated with the influence of the thermal residual stresses, the predicted uniaxial tensile strengths of the four laminates at different temperatures are given in Table 5.9. On the whole, the predictions agree reasonably well with the experiments. Some discrepancies can be attributed to the inaccurate constituent properties used, especially the ultimate strength parameters involved. Preferably, the ultimate tensile strengths of the fiber and the matrix at each temperature should be calibrated using two overall tensile strengths of a lamina/laminate, such as the longitudinal and transverse strengths of a unidirectional lamina, at the same temperature. This is especially true if the composites under consideration are not perfectly fabricated. The present bridging model is developed based on a perfect bonding assumption for the fiber/matrix interface up to failure. However, it has been recognised that Silicon-carbide fiber-Titanium matrix composites generally do not fulfil the perfect bonding assumption (Newaz & Majumdar, 1994) as their fabrication is commonly made by hot-pressing the titanium sheets in between which the silicon-carbide fibers are arranged. To compensate for any defect involved, one simple way is to redefine the constituent properties using calibrated data, such as using 783 MPa to replace the RT strength of the monolithic matrix, 948 MPa, as shown in the above. Since not enough experimental data for the unidirectional laminae of the considered materials are available, no attempt to further improve the prediction accuracy is made in this book.

Table 5.12 Predicted progressive failure process of $[0/90]_{2s}$ laminate at RT and 650 °C when subjected to uniaxial loading

Lay-up	Temperature	Failure Order	Failure Strength (MPa)	Failed Ply	Failure Status
$[0/90]_{2s}$	RT	First-ply Failure	857	[90]	Matrix, Tension
		Second-ply Failure	900	Inter-layer	Matrix, Tension
		Third-ply Failure	944	[0]	Matrix, Tension
		Ultimate Failure	1,098	[0]	Fiber, Tension
	650 °C	First-ply Failure	324	[90]	Matrix, Tension
		Second-ply Failure	382	Inter-layer	Matrix, Tension
		Third-ply Failure	385	[0]	Matrix, Tension
		Ultimate Failure	609	[0]	Fiber, Tension

Predicted progressive failure process of $[0/\pm45/90]_s$ laminate at RT and 650 °C when subjected to uniaxial loading

Lay-up	Temperature	Failure Order	Failure Strength (MPa)	Failed Ply	Failure Status
$[0/\pm45/90]_s$	RT	First-ply Failure	814	[90]	Matrix Tension
		Second-ply Failure	815	[45]	Matrix Tension
		Third-ply Failure	854	Inter-layer	Matrix Tension
		Fourth-ply Failure	895	[0]	Matrix Tension
		Ultimate Failure	989	[0]	Fiber Tension
	650 °C	First-ply Failure	310	[90]	Matrix Tension
		Ultimate Failure	553	[0]	Fiber Tension

All of the intermediate failure strengths of the laminates and the corresponding failure modes, i.e., the sources which cause the failures, can be identified using the bridging model. Taking the $[0/90]_{2s}$ and $[0/\pm45/90]_s$ laminates as examples, detailed progressive failure results at RT and at 650 °C are summarised in Table 5.12. For the $[0/90]_{2s}$ laminate, the progressive failure modes at 650 °C are the same as those at RT. The initial failure is caused by a non-fatal failure of the 90-laminae due to a tensile failure of the matrix, followed by tensile failure of the inter-layers. Then the 0-laminae also fail due to the matrix tensile failure. After the tensile failure of the matrix in all of the layers, the laminate attains its ultimate failure, which is caused by a tensile failure of the fibers in the 0-laminae. For the $[0/\pm45/90]_s$ laminate, an interesting feature to recognize is that the failure modes of a composite can be different at different temperatures. For instance, all of the layers of the $[0/\pm45/90]_s$ laminate at RT have failed before an ultimate failure of the laminate, which is caused by fiber failure in the 0-laminae, is attained.

However, the $[0/\pm45/90]_s$ laminate at 650 °C attains its ultimate failure only after the 90-layers have failed.

5.13 Fatigue Life Prediction

The bridging model combined with the classical or the pseudo 3D lamination theory can be used to predict an S-N relationship of a multidirectional composite laminate under a fatigue load condition, where S refers to a stress level and N denotes a cycling number. As done in Section 4.7, a fatigue life prediction for the laminate is carried out based on its constituent properties corresponding to different cycling numbers with the same load conditions specified.

Rotem and Hashin (1976) experimentally measured S-N data of a series of angle plied glass/epoxy laminates, $[\pm\theta]_{2S}$, with θ=30°, 35°, 41°, 45°, 49°, 55° and 66°. Each lamina in the laminate has the same thickness and the same fiber volume fraction of V_f=0.65. The laminates are subjected to uniaxial fatigue loads along the x-direction, with a stress ratio of R=0.1 and cyclic frequency of ω=19. Both stress and strain controls are used during their measurements.

Elastic properties of the glass fiber and epoxy matrix are chosen from Table 4.4, i.e., E^f=73 GPa, ν^f=0.22, E^m=3.45 GPa and ν^m=0.35. These elastic properties are assumed to be unchanged during the whole cyclic loading. The fiber tensile fatigue strengths are taken from Table 4.13, which are retrieved from the fatigue properties (S-N data) of the longitudinal (0°) fatigue performances of a lamina made of the same constituent material system. These properties are relisted in Table 5.13. As aforementioned, only tensile fatigue properties are retrieved because the composites are subjected to tensile fatigue and the resulting compressive stresses in the fibers are so small that they could hardly cause any fiber to fail.

On the other hand, a significant compression occurs in the matrix, especially when the ply-angle θ is around 30°. As shown in Section 5.10, the angle plied laminate $[\pm30°]_{2s}$ fails due to a matrix compressive failure when the laminate is subjected to a longitudinal (0°-directional) tension. Therefore, not only the tensile but also the compressive fatigue properties of the matrix are required. In the present analysis, the S-N data of the $[\pm30°]_{2s}$ laminate are used to determine the matrix compressive strengths whereas those of the $[\pm60°]_{2s}$ laminate are employed to retrieve the matrix tensile strengths at different cycling numbers. Due to scatter in the measured fatigue data (Rotem and Hashin, 1976), a linear interpolation approximation based on Excel trend-line is used to represent an S-N curve of the laminate, as shown in Fig. 5.24. The fatigue strengths of the laminate at several chosen cycle numbers obtained from the interpolations are listed in Table 5.14.

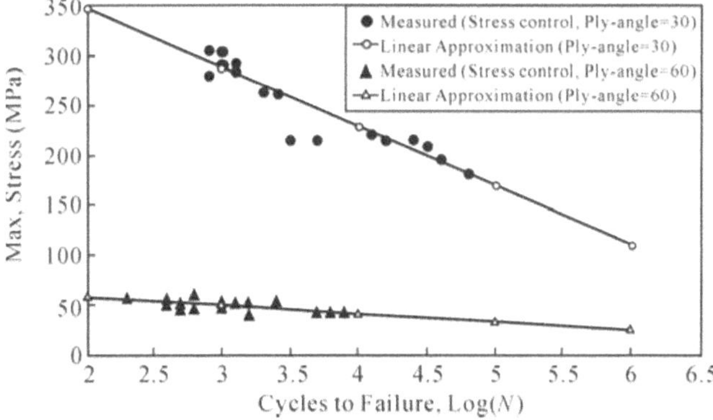

Fig. 5.24 Measured (Rotem and Hashin, 1976) and linear approximation S-N curves of the $[\pm 30°]_{2S}$ and $[\pm 60°]_{2S}$ laminates

Table 5.13 Retrieved constituent fatigue properties of glass/epoxy laminates used for life prediction

	Cycles to Failure, N					
	0	10^2	10^3	10^4	10^5	10^6
σ_u^f (MPa)	2,055	1,460	1,235	1,013	790	570
σ_u^m (MPa)	35	35	31.9	26.5	21.0	15.7
$\sigma_{u,c}^m$ (MPa)	70.9	70.9	59.4	47.0	34.3	21.4
σ_Y^m (MPa)	16	16	16	16	15.5	11.5
(GPa)	0.86	0.86	0.86	0.86	0.86	0.86

Table 5.14 Measured failure strength (Rotem and Hashin, 1976) of angle plied laminates (MPa)

Lay-up	Cycles to Failure, N					
	0	10^2	10^3	10^4	10^5	10^6
$[\pm 30°]_{2S}$	343.5	343.5 [a]	288.6	229.2	169.9	110.5
$[\pm 60°]_{2S}$	53.6	53.6 [b]	48.9	40.6	32.3	24

(a) Linear extrapolation=347.9;
(b) Linear extrapolation=57.3.

As measured stress-strain curves of the composites under static tension displayed nonlinear behavior (Rotem and Hashin, 1976), the matrix used could not be considered as linearly elastic until rupture. A bilinear elastic-plastic behavior is assumed in the retrieval. Mechanical parameters of the $[\pm 60°]_{2s}$ laminate are used to back-calculate the tensile properties of the matrix. When a uniaxial load is applied to the $[\pm 60°]_{2s}$ laminate, its initial failure is caused by a tensile failure of the matrix in the (±60°) layers. Then, an excessive deformation occurs after the tensile failure of the matrix in all of the lamina layers, resulting in an ultimate failure in the laminate. However, the measured ultimate strain of the $[\pm 60°]_{2s}$ laminate is 0.4022% (Rotem and Hashin, 1976), which is quite small. This is similar to the situations shown in Figs. 5.13, 5.14, and 5.16. Thus, for angle plied laminates, a horizontal curve in a stress-strain plot should be terminated earlier. The matrix tensile strength at a given cycle number is set to the maximum normal stress of the laminate when it is subjected to a corresponding tensile load. Due to limited information, the determination of the elastic-plastic parameters of the matrix (i.e., σ_Y^m and E_T^m) is somewhat arbitrary. They are determined in such a way that the predicted failure strain of the [±60°]$_{2s}$ laminate should be as close to the measured one as possible and that the predicted unidirectional tensile strength of the lamina based on the so-defined matrix parameters and the other given constituent properties must be equal to the measured value (Huang, 2001). For example, with the constituent properties of σ_u^f =2,055 MPa, σ_u^m =35 MPa, σ_Y^m = 16 MPa and E_T^m =0.86 GPa and a fiber volume fraction of V_f=0.60, the predicted longitudinal strength of the unidirectional lamina is 1247 MPa, which is equal to the measured value at N=0 (Hashin & Rotem, 1976), whereas the predicted ultimate strain of the [±60°]$_{2S}$ laminate is 0.8688%, higher than the measured strain, 0.4022% (Rotem & Hashin, 1976). However, if we take E_T^m =2.46 GPa, the predicted longitudinal strength of the lamina is 599 MPa which is incorrect, although the predicted ultimate strain of the laminate has been improved to 0.4668%. Thus, the matrix plastic and tensile strength parameters corresponding to N=0 are chosen as σ_Y^m =16 MPa, E_T^m =0.86 GPa and σ_u^m =35 MPa, respectively. These plastic parameters are kept unchanged for all of the cycle numbers unless the corresponding longitudinal load is unable to be applied to the unidirectional lamina. In the latter cases (N=10^5 and N=10^6), the matrix yield strength should be adjusted accordingly. For example, with σ_Y^m =16 MPa, E_T^m =0.86 GPa, σ_u^m =21 MPa and σ_u^f =790 MPa, the predicted longitudinal strength of the unidirectional lamina is lower than 480 MPa, which is the longitudinal tensile strength of the lamina at N=10^5 (Hashin & Rotem, 1976). Thus, the yield strength of the matrix is adjusted to 15.5 MPa, at which the predicted longitudinal strength of the lamina is 480 MPa. The retrieved matrix tensile strengths and plastic parameters at cycle numbers of N=0 to N=10^6 are listed in Table 5.13.

For simplicity and mainly because no other information is available, the matrix plasticity at compression is assumed to be the same as that at tension. The compressive strengths of the matrix at the chosen cycle numbers are back-calculated from those of the [±30°]$_{2s}$ laminate. By applying uniaxial loads at respective cycle numbers to the [±30°]$_{2s}$ laminate, the negative maximum compressive stress ($-\sigma^3$) in the matrix is taken as the matrix compressive strength at the corresponding cycle numbers.

Using the constituent properties given in Table 5.13, the predicted S-N curves of the angle plied laminates [±θ]$_{2s}$ with θ =35°, 41°, 45°, 49° and 55° are plotted in Figs. 5.25 – 5.29, respectively. The measured data given by Rotem and Hashin (1976) are also shown in the figures. It is seen that correlation between all of the predictions and the experiments is satisfactorily high.

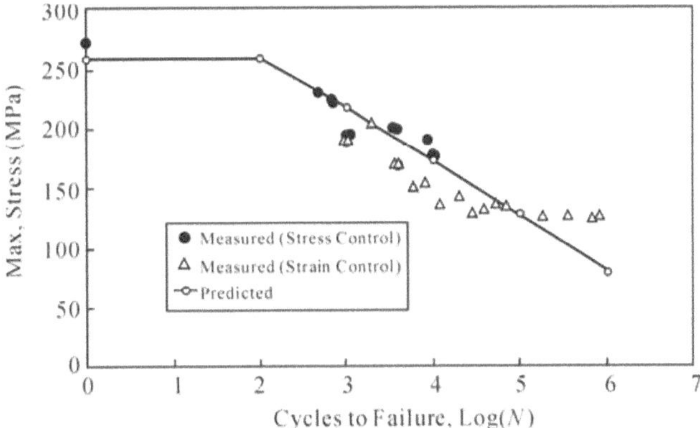

Fig. 5.25 Predicted and measured (Rotem & Hashin, 1976) S-N curves of a glass/epoxy $[\pm 35°]_{2s}$ laminate (from Huang, 2001)

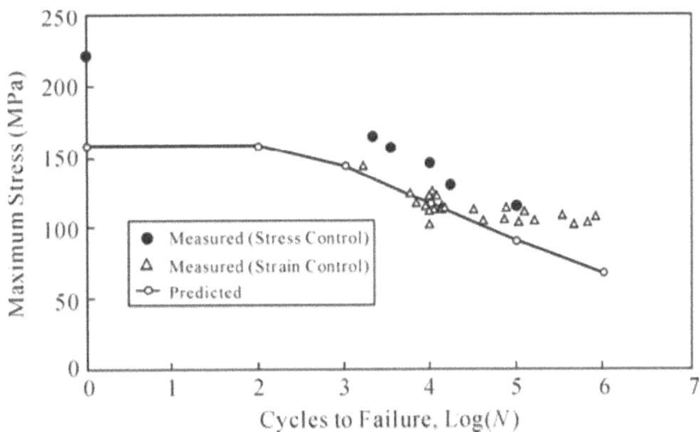

Fig. 5.26 Predicted and measured (Rotem & Hashin, 1976) S-N curves of a glass/epoxy $[\pm 41°]_{2s}$ laminate (from Huang, 2001)

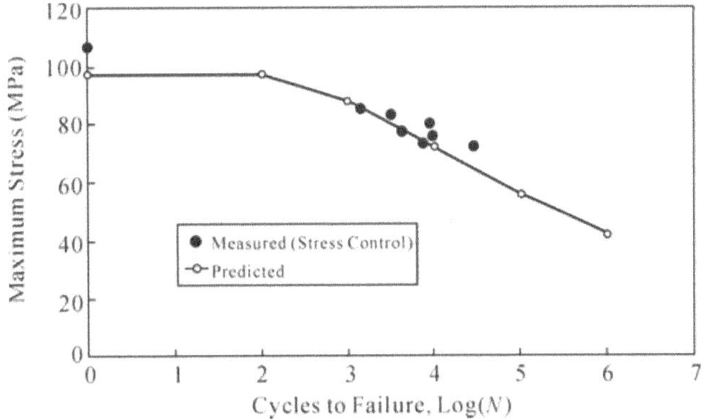

Fig. 5.27 Predicted and measured (Rotem & Hashin, 1976) S-N curves of a glass/epoxy $[\pm 45°]_{2s}$ laminate (from Huang, 2001)

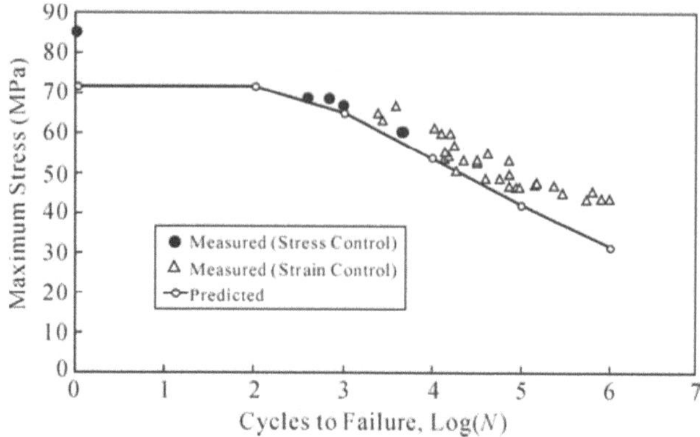

Fig. 5.28 Predicted and measured (Rotem & Hashin, 1976) S-N curves of a glass/epoxy $[\pm 49°]_{2s}$ laminate (from Huang, 2001)

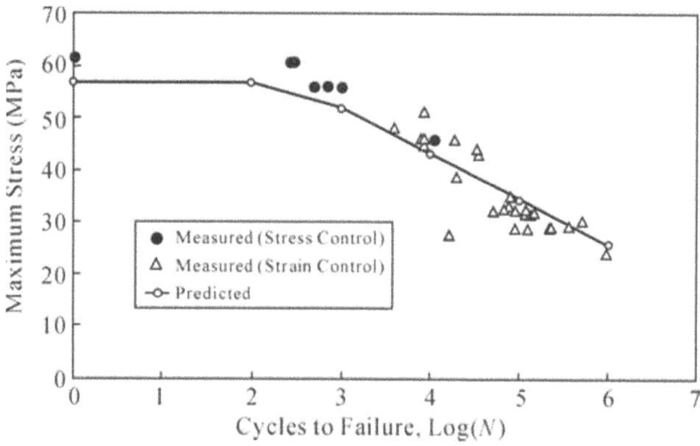

Fig. 5.29 Predicted and measured (Rotem & Hashin, 1976) S-N curves of a glass/epoxy $[\pm 55°]_{2s}$ laminate (from Huang, 2001)

5.14 Prediction for WWFE-I Problems

The first world-wide failure exercise (WWFE-I) organized by Hinton, Soden and Kadour (Hinton, et al., 1998, 2002, 2004; Soden, et al., 1998a, 1998b) was accomplished in order to assess predictive capabilities of the current failure theories for composite laminates. A total number of 14 problems are set forth, which are classified into 5 categories with 125 points marked. All of the laminates involved are subjected to uniaxial or bi-axial in-plane loads up to failure.

There are four types of constituent material systems of which the laminates are composed. They are E-glass/LY556/HT907/DY063, E-glass/MY750/HY917/DY063, AS4/3501-6 epoxy and T300/BSL914C epoxy. Determination of the constituent properties for the bridging model simulation is similar to that illustrated in Section 5.9, and is summarized in Tables 5.15 – 5.19, which is taken from Huang (2004a).

The elastic properties of the fiber and matrix materials in Table 5.15 are exactly the same as provided (Soden et al., 1998a). The elastic-plastic parameters for the matrix materials are back-calculated from the inelastic responses of the corresponding composite laminae. Eight linear segments are used to represent the retrieved stress-strain curves of the matrices, with their hardening moduli and yield strengths given in Table 5.16.

Tensile and compressive strengths of the fibers, as shown in Table 5.17, are retrieved from the longitudinal tensile and compressive strengths of the unidirectional laminae, respectively. Compressive strengths of the matrix materials are back-calculated from the transverse compressive strengths of the corresponding composite laminae. However, determination of the tensile strengths of the matrices is a little more complicated. Let the ultimate tensile stresses of a matrix back-calculated from the longitudinal tensile, transverse tensile and in-plane shear strengths of the corresponding lamina be denoted by $\sigma^1_{m,T}$, $\sigma^2_{m,T}$ and $\sigma^3_{m,T}$, respectively. If $\sigma^1_{m,T} \geq \dfrac{\sigma^2_{m,T}+\sigma^3_{m,T}}{2}$, $\sigma^1_{m,T}$ is chosen as the ultimate tensile strength of the matrix; otherwise, $(\sigma^2_{m,T}+\sigma^3_{m,T})/2$ is defined as the matrix tensile strength. For the material systems under consideration, $\sigma^1_{m,T}$ is always larger than $(\sigma^2_{m,T}+\sigma^3_{m,T})/2$. This means that the tensile strengths of both the fiber and the matrix materials are retrieved from the longitudinal tensile strengths of the corresponding laminae. Namely, when the laminae attain longitudinal tensile strengths both the fiber and the matrix materials assume their ultimate tensile strengths. The strength parameters of the matrix materials thus obtained are listed in Table 5.17.

The effect of thermal residual stresses on the failure and strength behavior of the laminates has been incorporated in the analysis. The thermal expansion coefficients of the constituent materials given by Soden et al. (1998a), together with the estimated constituent thermal residual stresses in a lamina are summarized in Table 5.18. The bridging parameters β and α for each lamina are calibrated using its transverse and in-plane shear moduli provided (Soden et al., 1998a), and are given in Table 5.19. The problem descriptions, the lamination lay ups, the materials used and the loading conditions for all of the exercise problems have been summarized in Table 5.20 (Soden et al., 1998a).

Table 5.15 Elastic Properties of the fiber and matrix materials used in the WWFE-I (Soden et al., 1998a)

Properties	E-glass/LY556 /HT907/DY063		E-glass/MY750 /HY917/DY063		AS4/3501-6		T300/BSL914C	
	Fiber	Matrix	Fiber	Matrix	Fiber	Matrix	Fiber	Matrix
E_{11} (GPa)	80	3.35	74	3.35	225	4.2	230	4.0
E_{22} (GPa)	80	3.35	74	3.35	15	4.2	15	4.0
G_{12} (GPa)	33.33	1.24	30.8	1.24	15	1.567	15	1.481
ν_{12}	0.2	0.35	0.2	0.35	0.2	0.34	0.2	0.35
ν_{23}	0.2	0.35	0.2	0.35	0.07	0.34	0.07	0.35

Table 5.16 Elastic-plastic parameters of the matrix materials used in the WWFE-I

	LY556/HT907 /DY063	MY750/HY917 /DY063	3501-6	BSL914C
$(\sigma_Y)_1$ (MPa)	31.9	32.6	38.1	41.6
$(\sigma_Y)_2$ (MPa)	38.4	39.9	41.8	49.6
$(\sigma_Y)_3$ (MPa)	44.7	46.8	46.1	55.8
$(\sigma_Y)_4$ (MPa)	49.9	52.0	50.1	59.9
$(\sigma_Y)_5$ (MPa)	53.6	55.6	54.0	63.1
$(\sigma_Y)_6$ (MPa)	56.1	58.0	57.6	66.3
$(\sigma_Y)_7$ (MPa)	58.1	60.1	61.2	68.9
$(\sigma_Y)_8$ (MPa)	60.0	62.0	64.6	71.4
$(E_T)_0$ (MPa)	3350	3350	4200	4000
$(E_T)_1$ (MPa)	1566	1698	2507	2015
$(E_T)_2$ (MPa)	1337	1387	2530	1384
$(E_T)_3$ (MPa)	944	918	2072	769
$(E_T)_4$ (MPa)	584	542	1721	548
$(E_T)_5$ (MPa)	338	317	1409	457
$(E_T)_6$ (MPa)	245	244	1202	324
$(E_T)_7$ (MPa)	197	186	991	275

* $E_T^m = E_T^{m,j}$ if $\sigma_Y^{m,j} \le \sigma_e^m \le \sigma_Y^{m,j+1}$, j=0, 1, …,7, with $E_T^{m,0} = E^m$, $\sigma_Y^{m,0} = 0$ and $E_T^m = E_T^{m,7}$ if $\sigma_e^m \ge \sigma_Y^{m,8}$.

Table 5.17 Strength parameters of the fibers and matrices used in the WWFE-I

Properties	E-glass/LY556 /HT907/DY063		E-glass/MY750 /HY917/DY063		AS4/3501-6		T300/BSL914C	
	Fiber	Matrix	Fiber	Matrix	Fiber	Matrix	Fiber	Matrix
σ_u (MPa)	1804.1	56.5	2092.8	60.9	3206.4	65.6	2462.5	56.4
$\sigma_{u,c}$ (MPa)	908.9	55.7	1311.8	74.8	2458.6	116.4	1499.4	116.8

Table 5.18 Thermal properties of the fiber and matrix materials used in the WWFE-I

Properties	E-glass/LY556 /HT907/DY063		E-glass/MY750 /HY917/DY063		AS4/3501-6		T300/BSL914C	
	Fiber	Matrix	Fiber	Matrix	Fiber	Matrix	Fiber	Matrix
α_1 ($\times 10^{-6}$/°C)	4.9	58	4.9	58	-0.5	45	-0.7	55
α_2 ($\times 10^{-6}$/°C)	4.9	58	4.9	58	15	45	12	55
CT [a] (°C)	120		120		177		120	
WT [b] (°C)	25		25		25		25	
$\sigma_{11}^{(T)c}$(MPa)	−12.55	20.47	−13.55	20.33	−22.23	33.35	−16.59	24.89
$\sigma_{22}^{(T)}$ (Mpa)	−6.86	11.2	−7.32	10.98	−9.06	13.6	−7.49	11.24
$\sigma_{12}^{(T)}$ (Mpa)	0	0	0	0	0	0	0	0

(a) CT= curing temperature (also called stress free temperature);
(b) WT= working temperature;
(c) (T) refers to the thermal residual stresses.

Table 5.19 Bridging parameters and fiber volume fractions of the composite systems used in the WWFE-I

	E-glass/LY556 /HT907/DY063	E-glass/MY750 /HY917/DY063	AS4/3501-6	T300/BSL914C
A	0.45	0.45	0.45	0.45
B	0.35	0.35	0.30	0.35
V_f	0.62	0.60	0.60	0.60

Table 5.20 Summary of laminate types, ply thicknesses, material types, loading and plots required

No.	Lamination	Ply Thickness	Materials Used	Loading	Plot Required
1			Gevetex E-glass /LY556 epoxy	σ_{yy} & σ_{xy}	Strength envelope*
2	Unidirectional lamina $[0°]_n$	any but a specified positive magnitude	T300/914C epoxy	σ_{xx} & σ_{xy}	Strength envelope
3			Silenka/ MY750 Epoxy	σ_{xx} & σ_{yy}	Strength envelope
4		90° plies: each 0.172 mm	Gevetex E-glass	σ_{xx} & σ_{yy}	Strength envelope
5	$[90°/30°/-30°]_s$	30° plies: each 0.414 mm	/LY556 epoxy	σ_{xx} & σ_{xy}	Strength envelope
6				σ_{xx} & σ_{yy}	Strength envelope
7	$[90°/\pm45°/0°]_s$	Each ply: 0.1375 mm	AS4/3501-6 epoxy	σ_{yy} only	σ_{yy}-ε_{yy} & σ_{yy}-ε_{xx} curves
8				σ_{yy}:σ_{xx}=2:1	σ_{yy}-ε_{yy} & σ_{yy}-ε_{xx} curves
9			Silenka/ MY750	σ_{xx} & σ_{yy}	Strength envelope
10	$[55°/-55°]_s$	Each ply: 0.25 mm	Epoxy	σ_{yy} only	σ_{yy}-ε_{yy} & σ_{yy}-ε_{xx} curves
11				σ_{yy}:σ_{xx}=2:1	σ_{yy}-ε_{yy} & σ_{yy}-ε_{xx} curves
12	[0°/90°/0°]	0° plies: each 0.26 mm 90° ply: 0.52 mm	Silenka/ MY750 Epoxy	σ_{xx} only	σ_{xx}-ε_{yy} & σ_{xx}-ε_{xx} curves
13				σ_{yy}:σ_{xx}=1:1	σ_{yy}-ε_{yy} curve
14	$[45°/-45°]_s$	Each ply: 0.25 mm	Silenka/ MY750 Epoxy	σ_{yy}:σ_{xx}=1:-1	σ_{yy}-ε_{yy} & σ_{yy}-ε_{xx} curves

With the information provided in Tables 5.15 – 5.20, the predicted results using the bridging model are plotted in Figs. 5.30 – 5.43 (Zhou & Huang, 2008). The measured data provided by the exercise organizers (Soden et al, 2002) are also shown in the figures. In general, satisfactorily good predictions for most of the exercise problems have been achieved. For some problems, e.g., Problem 7, the stress-strain curves of the $[90°/\pm45°/0°]_s$ laminate subjected to the σ_{yy} load (Fig. 5.36) and Problem 12, the stress-strain curves of the [0°/90°/0°] laminate subjected to the σ_{xx} load (Fig. 5.41), the predicted results are very close to the measured ones.

However, there are still some weak points in the predictions. For the UD laminae under some load combinations, the predicted strengths are excessively larger than those measured from the experiments, especially under in-plane shear and transverse tensile loads, see Figs. 5.30 and 5.31. The errors may be caused by the inaccurate strength parameters used for the constituent, mainly matrix, materials. The constituent strengths are obtained through back-calculation. As seen in the above, the matrix strengths retrieved from the longitudinal, transverse and in-plane shear strengths of a UD lamina are essentially different. Thus, the determination of some part of the strength envelope of the UD lamina having a load combination of transverse or in-plane shear stresses would be in error by nature. For some laminates, e.g., Problem 10, the $[55°/{-55°}]_s$ laminate subjected to the σ_{yy} load up to failure, the prediction of the bridging model shows a relatively large discrepancy with the experiment (Fig. 5.39). However, it should be pointed out that the experiments for Problem 10 have been carried out using tubular specimens with a longitudinal load combined with a liquid pressure (Soden et al., 2002). When the inner surface of a specimen was not covered with a rubber line, the specimen failed due to liquid weeping at a stress level close to the bridging model prediction. On the other hand, once covered with a rubber line, the specimen was loaded to failure at a much larger stress level with a failure mode of fiber fracture. However, the rubber line influence on the load carrying capacity of the specimen is not our concern in the present book, because the solution to that involves a structural analysis of the interaction between the rubber line and the composite tube and can be achieved through finite element analysis in general. For Problem 11, the predicted stress-strain curves of the $[55°/{-55°}]_s$ laminate subjected to the load combination of σ_{yy}:σ_{xx}=2:1 also do not fit well with the measured ones (Fig. 5. 40), and a similar recognition exists for this problem as for Problem 10.

In spite of the existing weak points, the bridging model predictions provide the best correlations with the experiments. According to the rules of the WWEF-I made by the organizers, the correlation between a theoretical prediction and an experiment has been quantitatively designated as Grade A, Grade B, Grade C or Grade NA, where Grade A represents a prediction which lies within ±10% of the corresponding experimental value, Grade B denotes a prediction in agreement between ±10% and ±50% of the experimental value, Grade C stands for a prediction outside ±50% of the experimental value and Grade NA means that no solution is offered by the theory (Hinton et al., 2002). The total number of Grade A plus Grade B scores achieved by the theory is employed to rank its position among all of the strength theories that were assessed in the exercise. The theory developed by Zinoviev gave good predictions for most of the exercise problems and was ranked as the champion (the most accurate) among the 19 best known theories included in the exercise, with a total score of Grade A + Grade B being 96 (Hinton et al., 2004). However, the score of Grade A + Grade B achieved by the present predictions (shown in Figs. 5.30 – 5.43) is 102 (Zhou & Huang, 2008), even higher than the score attained by Zinoviev's theory, which was a phenomenological strength theory for composites (Zinoviev et al., 1998).

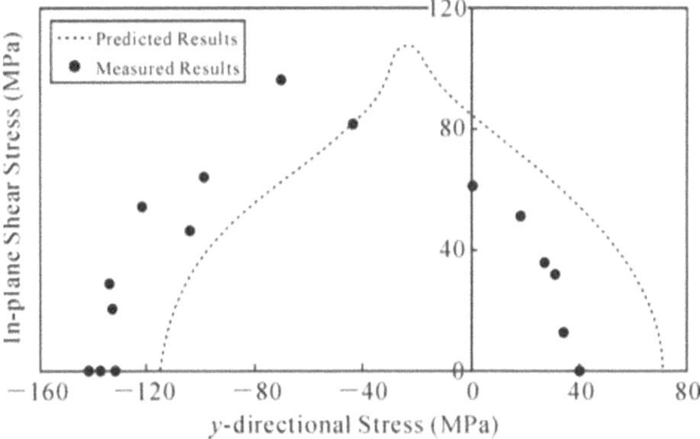

Fig. 5.30 Comparison between predicted and measured failure envelopes for the exercise Problem 1 (A UD lamina subjected to σ_{yy} *vs* σ_{xy} loads) (from Huang, 2004c)

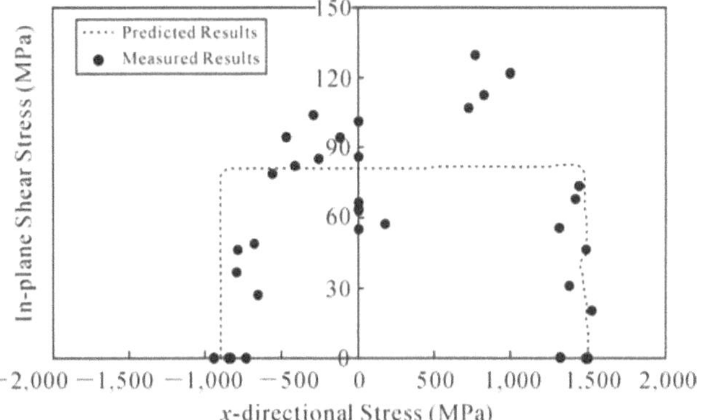

Fig. 5.31 Comparison between predicted and measured failure envelopes for the exercise Problem 2 (A UD lamina subjected to σ_{xx} *vs* σ_{xy} loads) (from Huang, 2004c)

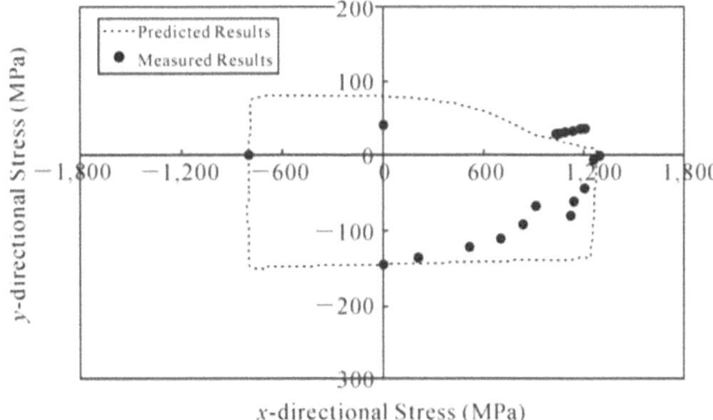

Fig. 5.32 Comparison between predicted and measured failure envelopes for the exercise Problem 3 (A UD lamina subjected to σ_{xx} *vs* σ_{yy} loads) (from Huang, 2004c)

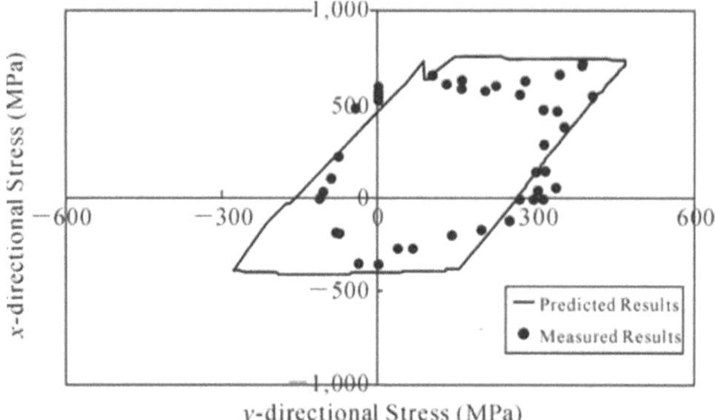

Fig. 5.33 Comparison between predicted and measured failure envelopes for the exercise Problem 4 ([90°/30°/−30°]$_s$ laminate subjected to σ_{xx} *vs* σ_{yy} loads) (from Zhou & Huang, 2008)

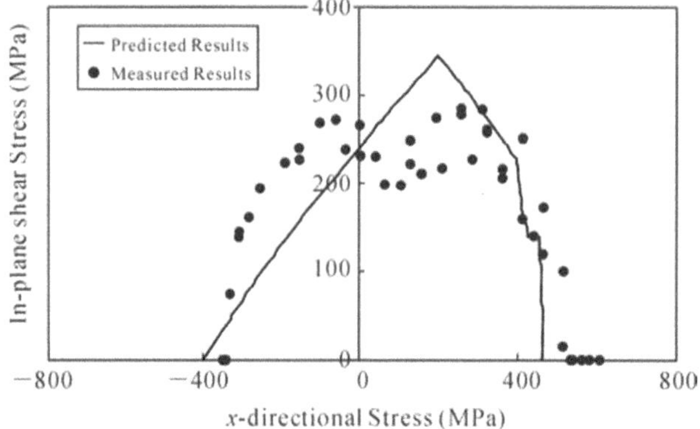

Fig. 5.34 Comparison between predicted and measured failure envelopes for the exercise Problem 5 ([90°/30°/−30°]$_s$ laminate subjected to σ_{xx} *vs* σ_{xy} loads) (from Zhou & Huang, 2008)

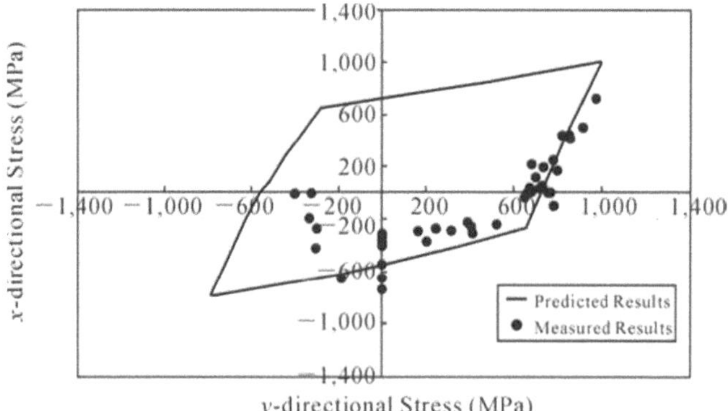

Fig. 5.35 Comparison between predicted and measured failure envelopes for the exercise Problem 6 ([90°/±45°/0°]$_s$ laminate subjected to σ_{xx} *vs* σ_{yy} loads) (from Zhou & Huang, 2008)

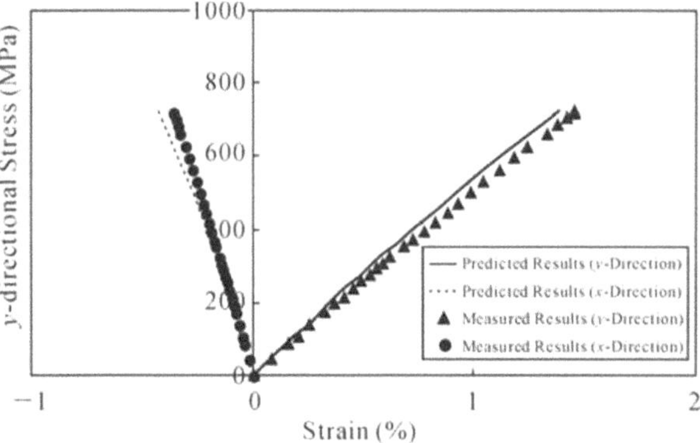

Fig. 5.36 Comparison between predicted and measured stress-strain curves for the exercise Problem 7 ([90°/±45°/0°]$_s$ laminate subjected to σ_{yy} loads) (from Zhou & Huang, 2008)

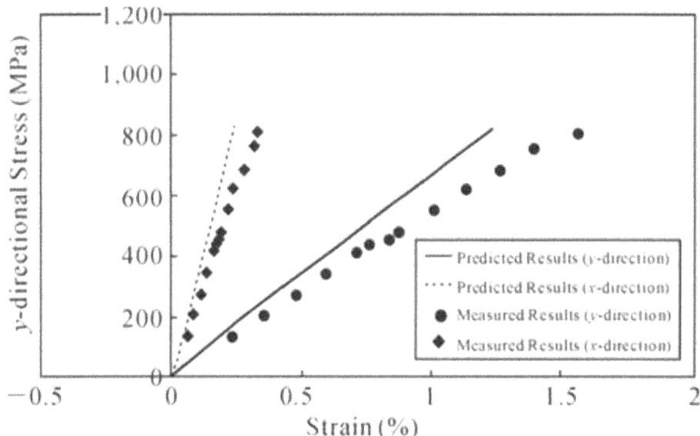

Fig. 5.37 Comparison between predicted and measured stress-strain curves for the exercise Problem 8 ([90°/±45°/0°]$_s$ laminate subjected to σ_{yy} and σ_{xx} loads: σ_{yy} :σ_{xx} =2:1) (from Zhou & Huang, 2008)

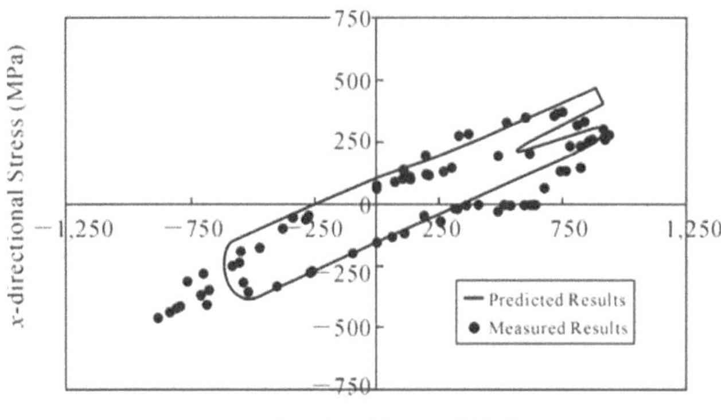

Fig. 5.38 Comparison between predicted and measured failure envelopes for the exercise Problem 9 ([55°/−55°]$_s$ laminate subjected to σ_{xx} *vs* σ_{yy} loads) (from Zhou & Huang, 2008)

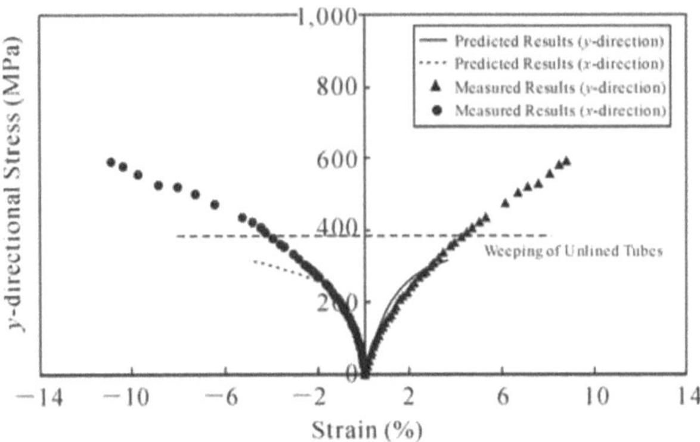

Fig. 5.39 Comparison between predicted and measured stress-strain curves for the exercise Problem 10 ([55°/−55°]$_s$ laminate subjected to σ_{yy} load) (from Zhou & Huang, 2008)

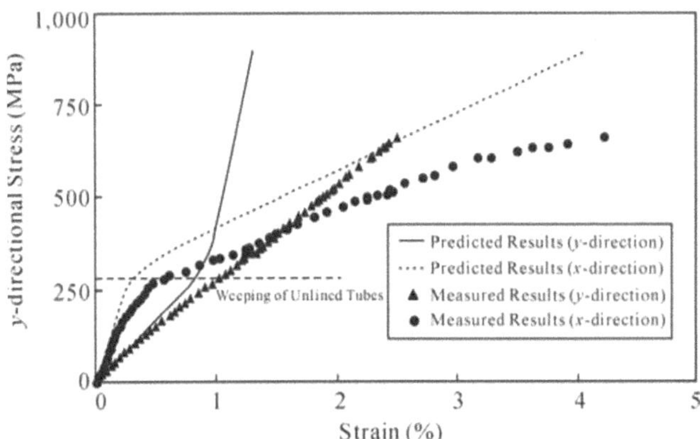

Fig. 5.40 Comparison between predicted and measured stress-strain curves for the exercise Problem 11 ([55°/−55°]$_s$ laminate subjected to σ_{yy} and σ_{xx} loads: σ_{yy} :σ_{xx} =2:1) (from Zhou & Huang, 2008)

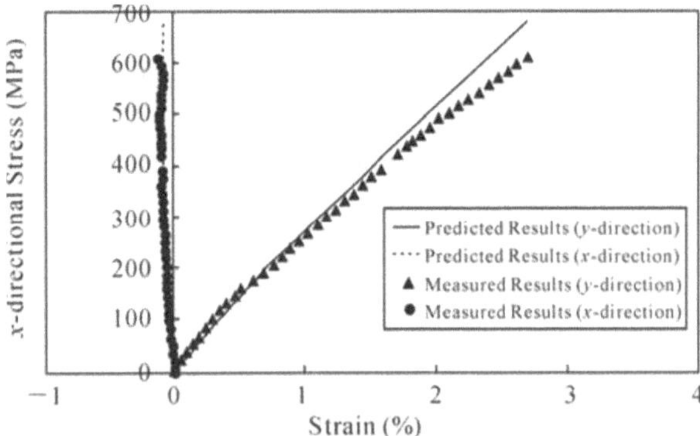

Fig. 5.41 Comparison between predicted and measured stress-strain curves for the exercise Problem 12 ([0°/90°/0°]$_s$ laminate subjected to σ_{xx} load) (from Zhou & Huang, 2008)

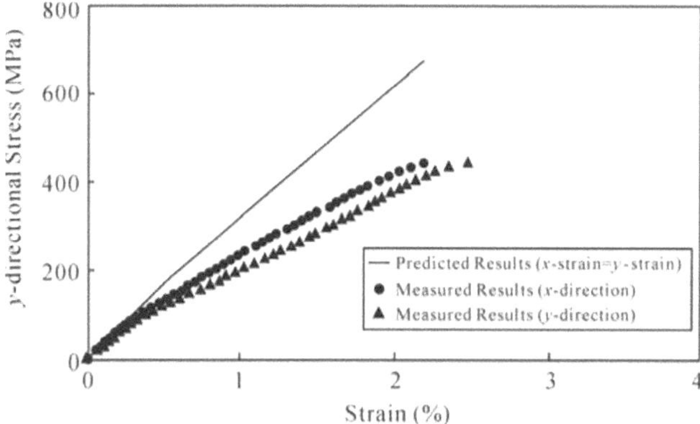

Fig. 5.42 Comparison between predicted and measured stress-strain curves for the exercise Problem 13 ([45°/–45°]$_s$ laminate subjected to σ_{yy} and σ_{xx} loads: $\sigma_{yy}:\sigma_{xx}$ =1:1) (from Zhou & Huang, 2008)

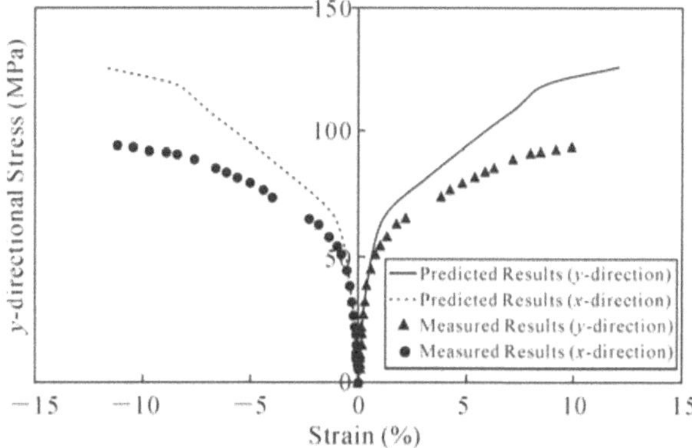

Fig. 5.43 Comparison between predicted and measured stress-strain curves for the exercise Problem 14 ([45°/–45°]$_s$ laminate subjected to σ_{yy} and σ_{xx} loads: $\sigma_{yy}:\sigma_{xx}$ =1:–1) (from Zhou & Huang, 2008)

Remark 5.4

The bridging model was among the 19 theories that took part in the WWFE-I (Huang, 2004a), but the blind predictions made there were only ranked as moderate (Hinton et al., 2004), with a total score of 80 Grade A's plus Grade B's achieved. That was mainly due to two reasons. Firstly, the last-ply failure was used as an ultimate failure criterion in the predictions (Huang, 2004a). Namely, regardless of what laminate layup was subjected to whatever kind of load combination, its ultimate strength was considered to be attained only when all of the laminae in the laminate had failed. Secondly, a total stiffness discount scheme was used in the previous predictions. In other words, as long as a lamina in the laminate attained a failure status, no matter whether failure was caused by a fiber or by a matrix failure, the stiffness of the lamina was completely disregarded in a subsequent analysis, i.e., was not included in the laminate stiffness evaluation through Eq. (5.9.2). It has been shown that the ultimate failure criteria designated by Table 5.3, together with a partial stiffness discount scheme given by Eq. (5.22), is more pertinent for laminate failure analysis and gives overall much better predictions for the ultimate strengths of the laminates. The input parameters listed in Tables 5.15 − 5.19 are exactly the same as those used by Huang (2004a). A further improvement in the failure analysis and ultimate strength prediction is achieved by the introduction of the pure matrix inter-layers in the laminate.

Remark 5.5

The results shown in this section (Figs. 5.30 – 5.43) have been obtained based on the 2D bridging model formulae combined with the classical laminate theory (Section 5.3), see Zhou & Huang (2008) for more details. However, the computer routine presented in Chapter 6 is developed based on the 3D bridging model formulae together with the pseudo 3D laminate theory. The input data on a CD-ROM included with this book are prepared for running the 3D theory computer routine. The thus obtained results can be seen to have slight or negligible differences from those shown in Figs. 5.30 – 5.43.

5.15 Prediction for the WWFE-II Problems

After the WWEF-I, which was designed to assess the predictive capacity of currently available theories for laminate strengths subjected to in-plane (2D) loads, the exercise organizers launched the second worldwide failure exercise (WWFE-II) with a number of typical laminae and laminates subjected to triaxial (3D) load combinations (Kaddour & Hinton, 2011). For these latter problems, the 3D bridging model designated by Eq. (3.70) incorporated with the pseudo 3D laminate theory developed in Section 5.8 is applicable.

There are five material systems used in the WWFE-II, which are E-glass/MY750, S2-glass/epoxy, AS carbon/epoxy, IM7/8551-7 carbon/epoxy and T300/PR319 carbon/ epoxy laminae. Curing conditions (stress-free temperatures) for these laminae together with necessary thermal properties are summarized in Table 5.21. Determination of the input parameters for these material systems is done in a similar way to those used in the WWFE-I. The fiber materials are assumed to be linearly elastic until rupture, whereas the matrices are regarded as elastic-plastic. The elastic properties of the fibers and matrices are directly taken from the provided data, as shown in Table 5.22. It is noted that the elastic properties of a polymer matrix under compression may be different from those under tension. The exercise organizers have provided transverse compressive behaviors of the UD laminae made from the material systems under consideration, which are used to retrieve the compressive elastic properties of the matrices. Furthermore, the retrieved properties with the thermal residual stresses incorporated are different from those that are examined without consideration of the thermal residual stresses. This is because the thermal residual stresses result in a tensile state, which affects the retrieved values. Both of the compressive properties of the matrix materials, with and without consideration of the thermal residual stresses, are listed in Table 5.23. The elastic properties of the fibers under tension and compression are considered to be the same.

Table 5.21 Thermal parameters of the fiber and matrix materials used in the WWFE-II

Properties	E-glass/MY750		S2-Glass /Epoxy		AS /Epoxy		IM7/8551-7 /Epoxy		T300/PR319 /Epoxy	
	Fiber	Matrix	Fiber	Matrix	Fiber	Matrix	Fiber	Matrix	Fiber	Matrix
α_1 (10^{-6}/°C)	4.9	58	5	58	-0.7	58	-0.4	46.7	-0.7	60
α_2 (10^{-6}/°C)	4.9	58	5	58	12	58	5.6	46.7	12	60
α_3 (10^{-6}/°C)	4.9	58	5	58	12	58	5.6	46.7	12	60
Stress-free temperature	120°C		120°C		120°C		177°C		120°C	
Working temperature	25°C		25°C		25°C		25°C		25°C	
$\sigma_{11}^{(T)*}$(MPa)	−14.18	21.27	−14.55	21.82	−13.57	20.36	−23.85	35.78	−4.51	6.76
$\sigma_{22}^{(T)*}$(MPa)	−9.09	13.64	−7.84	11.77	−8.67	13.01	−11.95	17.92	−2.47	3.71
$\sigma_{12}^{(T)*}$(MPa)	0	0	0	0	0	0	0	0	0	0

* Thermal residual stresses in the constituent materials of a unidirectional lamina. As mentioned in Chapter 3, only 2D thermal residual stresses are considered for a unidirectional lamina.

Table 5.22 Elastic Properties of the fiber and matrix materials used in the WWFE-II

Properties	E-glass/MY750		S2-Glass /Epoxy		AS /Epoxy		IM7/8551-7 /Epoxy		T300/PR319 /Epoxy	
	Fiber	Matrix	Fiber	Matrix	Fiber	Matrix	Fiber	Matrix	Fiber	Matrix
E_{11} (GPa)	74	3.35	87	3.2	231	3.2	276	4.08	231	0.95
E_{22} (GPa)	74	3.35	87	3.2	15	3.2	19	4.08	15	0.95
E_{33} (GPa)	74	3.35	87	3.2	15	3.2	19	4.08	15	0.95
G_{12} (GPa)	30.8	1.24	36	1.2	15	1.2	27	1.478	15	0.35
G_{13} (GPa)	30.8	1.24	36	1.2	15	1.2	27	1.478	15	0.35
G_{23} (GPa)	30.8	1.24	36	1.2	7	1.2	7	1.478	7	0.35
ν_{12}	0.2	0.35	0.2	0.35	0.2	0.35	0.2	0.38	0.2	0.35
ν_{13}	0.2	0.35	0.2	0.35	0.2	0.35	0.2	0.38	0.2	0.35
ν_{23}	0.2	0.35	0.2	0.35	0.0714	0.35	0.357	0.38	0.0714	0.35

Table 5.23 Retrieved compressive properties of the matrix materials

(a) With thermal residual stresses incorporated

Properties	MY750	Epoxy1[(1)]	Epoxy2[(2)]	PR-319	8551-7
E(GPa)	3.93	5.75	6.84	1.86	2.36
ν	0.35	0.35	0.35	0.35	0.38

(b) Without considering thermal residual stresses

Properties	MY750	Epoxy1[(1)]	Epoxy2[(2)]	PR-319	8551-7
E(GPa)	3.52	4.16	4.60	1.503	3.14
ν	0.35	0.35	0.35	0.35	0.38

(1) Epoxy1 refers to the matrix in the material system, S2-glass/epoxy;
(2) Epoxy2 refers to the matrix in the material system, AS carbon/epoxy.

Although the exercise organizers have provided some nonlinear stress-strain curves of the matrices, the used nonlinear behaviors are recovered from those of the UD laminae at uniaxial loads, because fabrication processing can make the *in situ* matrix properties in a composite different from those measured in a monolithic specimen. The nonlinear behaviors of the matrices under tension and compression are back-calculated from the in-plane shear and the transverse compressive stress-strain curves of the corresponding UD laminae, respectively. A recovered nonlinear stress-strain curve of the matrix is specified using eight linear segments similar to Eqs. (5.32.1) and (5.32.2). The thus obtained yield strengths and hardening moduli are summarized in Table 5.24. It is seen that the material parameters derived with and without consideration of thermal residual stresses are different.

Table 5.24 Retrieved plastic parameters of the matrix materials with thermal residual stresses considered

Properties (MPa)	MY750	Epoxy1[(a)]	Epoxy2[(b)]	PR-319	8551-7
		Material Parameters at Tension			
$(\sigma_Y)_1$	26.59	34.17	32.2	53.77	35.60
$(\sigma_Y)_2$	31.77	37.74	34.96	64.38	39.80
$(\sigma_Y)_3$	37.60	41.33	38.64	74.95	44.60
$(\sigma_Y)_4$	42.84	44.31	41.72	80.25	50.30
$(\sigma_Y)_5$	46.45	46.91	44.64	85.54	56.63
$(\sigma_Y)_6$	50.35	50.14	47.70	90.82	63.93
$(\sigma_Y)_7$	53.06	52.82	51.60	96.10	72.2
$(\sigma_Y)_8$	56.98	57.15	55.91	103.48	81.1
$(E_T)_1$	3350	3,200	3,200	950	4,080
$(E_T)_2$	2570	3,050	3,060	905	2,560
$(E_T)_3$	1650	2,210	2,640	852	2,024
$(E_T)_4$	1200	1,187	1,680	761	1,620
$(E_T)_5$	824	590	890	702	1,130
$(E_T)_6$	546	458	640	620	760
$(E_T)_7$	302	412	353	536	500
$(E_T)_8$	188	296	233	432	325

(continued)

Properties (MPa)	MY750	Epoxy1[(a)]	Epoxy2[(b)]	PR-319	8551-7
		Material Parameters at Compression			
$(\sigma_Y)_1$	32.32	50.83	40.62	32.77	54.80
$(\sigma_Y)_2$	35.62	55.05	43.59	38.09	64.53
$(\sigma_Y)_3$	38.87	59.50	46.64	43.36	73.07
$\sigma_Y)_4$	42.02	63.91	52.75	48.59	76.76
$(\sigma_Y)_5$	45.01	68.28	55.90	53.69	80.21
$(\sigma_Y)_6$	47.93	72.54	59.23	58.67	83.58
$(\sigma_Y)_7$	50.83	74.63	62.78	63.55	87.11
$(\sigma_Y)_8$	52.37	76.69	66.57	65.98	88.99
$(E_T)_1$	3930	6,840	5,750	1,862	2360
$(E_T)_2$	2750	3,410	2,540	1,350	2,290
$(E_T)_3$	2348	3,370	2,260	1,241	1,475
$(E_T)_4$	1915	3,010	1,675	1,084	890
$(E_T)_5$	1450	2,687	1,200	870	630
$(E_T)_6$	1085	2,255	963	657	454
$(E_T)_7$	770	1,970	762	464	350
$(E_T)_8$	655	1,745	610	359	300

Retrieved plastic parameters of the matrix materials without thermal residual stresses incorporated

Properties (MPa)	MY750	Epoxy1[(a)]	Epoxy2[(b)]	PR-319	8551-7
		Material Parameters at Tension			
$(\sigma_Y)_1$	22.79	33.15	30.68	42.77	26.56
$(\sigma_Y)_2$	36.40	37.22	38.22	53.44	35.18
$(\sigma_Y)_3$	44.17	41.03	41.90	64.11	43.68
$(\sigma_Y)_4$	48.45	44.76	45.55	74.76	47.91
$(\sigma_Y)_5$	50.89	48.46	48.46	80.08	56.27
$(\sigma_Y)_6$	52.54	49.93	49.91	85.39	64.57
$(\sigma_Y)_7$	53.94	51.39	51.36	90.70	72.82
$(\sigma_Y)_8$	55.09	55.78	54.26	103.40	76.94
$(E_T)_1$	3350	3,200	3,200	950	4,080
$(E_T)_2$	1980	3,050	2,564	938	2,860
$(E_T)_3$	1103	1,270	1,078	917	2,094
$(E_T)_4$	604	804	692	850	1,820
$(E_T)_5$	340	583	450	763	1,133
$(E_T)_6$	230	460	350	705	692
$(E_T)_7$	195	405	298	619	408
$(E_T)_8$	161	308	222	470	288.5
		Material Parameters at Compression			
$(\sigma_Y)_1$	35.95	53.13	35.96	28.60	40.59
$(\sigma_Y)_2$	40.39	58.90	44.86	33.32	52.11
$(\sigma_Y)_3$	44.75	64.60	53.56	37.99	63.40
$\sigma_Y)_4$	49.05	70.10	61.94	42.57	74.35
$(\sigma_Y)_5$	53.23	75.42	66.01	47.02	82.25
$(\sigma_Y)_6$	57.32	80.51	70.00	51.34	87.39
$(\sigma_Y)_7$	61.32	82.98	73.95	55.54	94.97

(continued)

Properties(MPa)	MY750	Epoxy1[(a)]	Epoxy2[(b)]	PR-319	8551-7
		Material Parameters at Compression			
$(\sigma_Y)_8$	63.30	85.38	77.90	57.60	102.48
$(E_T)_1$	3,520	4,598	4,157	1,503	3,138
$(E_T)_2$	3,160	4,267	3,815	1,420	2,995
$(E_T)_3$	2,736	4,080	3,110	1,304	2,580
$(E_T)_4$	2,338	3,586	2,165	1,139	1,995
$(E_T)_5$	1,833	3,156	1,497	912	1,346
$(E_T)_6$	1,399	2,649	1,140	689	964
$(E_T)_7$	1,009	2,290	845	488	671
$(E_T)_8$	800	2,004	637	373	433.5

(a) Epoxy1 refers to the matrix in the material system, S2-glass/epoxy;
(b) Epoxy2 refers to the matrix in the material system, AS carbon/epoxy.

Proper bridging parameters, α and β, used in the analysis are calibrated against the transverse and in-plane shear moduli of the laminae in such a way that the predicted moduli agree reasonably with the provided data. These parameters together with the fiber volume fractions are listed in Table 5.25. It is noted that the bridging parameters with and without thermal residual stresses are taken to be the same, as a thermal residual stress hardly influences any elastic behavior of a composite. Furthermore, as bi- or tri-axial tensions and compressions can occur in the considered problems, the generalized maximum normal stress criteria, represented by Eqs. (4.5.1) and (4.5.2) with a power-index of q=3 and by Eqs. (4.6.1) and (4.6.2), will be applied to detect failures of both the fibers and the matrix in a composite lamina.

Table 5.25 Bridging parameters and fiber volume fractions of the composite systems used in the WWFE-II

	E-glass/MY750	S2-Glass/Epoxy	AS/Epoxy	IM7/8551-7/Epoxy	T300/PR319/Epoxy
α	0.35	0.35	0.35	0.45	0.35
β	0.3	0.3	0.3	0.35	0.48
V_f	0.60	0.60	0.60	0.60	0.60

Determination of the constituent strength parameters is done in the same way as carried out for the WWFE-I problems. Tensile and compressive strengths of the fibers are back-calculated from the longitudinal strengths of the UD laminae under tension and compression, respectively. Tensile strengths of the matrix materials are determined by comparing the ultimate tensile stresses of the matrices back-calculated from the longitudinal tensile strength, transverse tensile strength and in-plane shear strength of the corresponding laminae, as similarly done in the preceding section. Compressive strengths of the matrices are back-calculated from the transverse compressive strengths of the laminae. However, as the retrieved compressive strengths of the matrix materials are obviously smaller than those provided by the organizers, the values of 10% higher than the retrieved ones are adopted for the matrix compressive strengths. All of the strength parameters are summarized in Table 5.26.

Table 5.26. Strength parameters of the fibers and matrices used in the WWFE-II
(a) With thermal residual stresses

Properties	E-glass/MY750		S2-Glass/Epoxy		AS/Epoxy		IM7/8551-7 /Epoxy		T300/PR319 /Epoxy	
	Fiber	Matrix	Fiber	Matrix	Fiber	Matrix	Fiber	Matrix	Fiber	Matrix
σ_u (MPa)	2,095.5	57.02	2,794.9	57.82	3,282.6	51.3	4,224.3	64.04	2,285.9	44.9
$\sigma_{u,c}$ (MPa)	1,310.8	64.50	1,885.6	84.30	2,491.1	94.71	2,648	110.55	1,583.5	68.9

(b) Without thermal residual stresses

Properties	E-glass/MY750		S2-Glass/Epoxy		AS/Epoxy		IM7/8551-7 /Epoxy		T300/PR319 /Epoxy	
	Fiber	Matrix	Fiber	Matrix	Fiber	Matrix	Fiber	Matrix	Fiber	Matrix
σ_u (MPa)	2,100.9	48.7	2,800	50.03	3,289.7	40.5	4,235.4	47.2	2,290.4	40.2
$\sigma_{u,c}$ (MPa)	1,297.7	79.75	1,871.8	98.56	2,467.3	107.14	2,630.1	130	1,576.5	73.21

Using the material parameters defined above, thermal residual stresses of the five UD laminae when cooled down from the respective fabrication (stress-free) to room temperatures are evaluated, and are listed in Table 5.21.

It is worth mentioning that the most demanding theory is the one where the mechanical properties of a fibrous composite, not only the effective elastic moduli but also nonlinear stress-strain behaviors, progressive failure strengths and ultimate load carrying capacity under arbitrary load condition, can be well determined by using constituent fiber and matrix properties independently measured from monolithic specimens or taken from a material database. So far, most correlations shown in the book for nonlinear stress-strain behaviors and ultimate strengths of the composites between the bridging model predictions and the measurements, although reasonably accurate in most cases, have been obtained by using constituent properties retrieved from the experimental data of some composites. Although it is a widely accepted assumption that the constituent properties in a composite are different from those measured using monolithic specimens, no one can precisely tell how great this difference will be. Considering that constituent properties are much more easily obtainable than the composite properties and are, in some cases, the only ones available, a thorough assessment of the effectiveness of the bridging model prediction for composite strengths using measured constituent properties is highly valuable. As such, predictions for the WWFE-II problems will also be made by directly employing the constituent properties provided by the organizers (Kaddour & Hinton, 2011) as long as the required properties are available by applying the bridging model.

Essentially, the fiber materials used are considered as linearly elastic until rupture, and only elastic properties and ultimate strengths of the fibers are required.

Although all of the required fiber parameters were already provided by the organizers (Kaddour & Hinton, 2011), only the elastic property parameters are kept unchanged. This is because most fiber properties are difficult to measure directly from monolithic fiber specimens. With the possible exception of the tensile properties (modulus and strength) that may be measured directly but with large deviations, all of the other fiber properties are generally determined by retrieving the composite properties using some micromechanical methods. As many micromechanical models in the literature can be used to accurately calculate elastic properties of a composite, the elastic property parameters of the fibers provided by the organizers can be well regarded as pertinent. The ultimate strengths of the fibers, however, need to be checked before applying the bridging model. This means that all of the fiber parameters as listed in Tables 5.22 and 5.26 are kept unchanged, with only slight amendments to the fiber strengths, if any. The amendments were made due to the use of the provided, instead of the retrieved, matrix properties. The thus obtained fiber strengths are named as 'provided fiber strengths' and are given in Table 5.27.

Table 5.27 Provided strength parameters of the fibers in the WWFE-II

(a) With thermal residual stresses

Properties	E-Glass	S2-Glass	AS Carbon	IM7 Carbon	T300 Carbon
σ_u (MPa)	2,068.4	2,762.6	3,282.6	4,362.2	2,285.9
$\sigma_{u,c}$ (MPa)	1,308.4	1,884.5	2,491.1	2,736.3	1,583.5

(b) Without thermal residual stresses

Properties	E-Glass	S2-Glass	AS Carbon	IM7 Carbon	T300 Carbon
σ_u (MPa)	2,078.6	2,773.8	3,289.7	4,377.5	2,290.4
$\sigma_{u,c}$ (MPa)	1,294.4	1,871.0	2,467.3	2,714.5	1,576.5

On the other hand, the matrix properties given by the organizers should have been obtained through measurements on monolithic specimens, which are directly employed. There are only two exceptions. One is that when a tensile or compressive stress-strain curve of a matrix was not provided by the organizers, the given in-plane shear or the transverse compressive stress-strain data of the corresponding composite are used to retrieve the tensile or the compressive elastic-plastic parameters of the matrix. Another is that when a unidirectional composite is subjected to longitudinal tension or compression up to the given composite strength, and the calculated first or third principal stress in the matrix is greater than the provided matrix tensile strength or smaller than the negative of the provided matrix compressive strength, then the used matrix strength is adjusted accordingly. The provided elastic properties (some are defined according to the provided stress-strain curves) together with tensile and compressive strengths of the five matrix materials are summarized in Table 5.28, whereas the provided plastic behavior parameters are given in Table 5.29.

Table 5.28 Provided elastic and strength properties of the matrix materials in the WWFE-II

Properties	MY750	Epoxy1[(a)]	Epoxy2[(b)]	8551-7	PR-319
		Material Parameters at Tension			
E (GPa)	3.361	3.2	3.191	4.00	0.95
ν	0.35	0.35	0.35	0.38	0.35
σ_u (MPa)	96	85	106	99	70
		Material Parameters at Compression			
E (GPa)	3.356	6.84	3.205	4.00	1.86
ν	0.35	0.35	0.35	0.38	0.35
$\sigma_{u,c}$ (MPa)	120	120	120	130	130

(a) Epoxy1 refers to the matrix in the material system, AS-glass/epoxy;
(b) Epoxy2 refers to the matrix in the material system, S2 carbon/epoxy.

Table 5.29 Provided plastic parameters of the matrix materials in the WWFE-II

Properties (MPa)	MY750	Epoxy1[(a)]	Epoxy2[(b)]	8551-7	PR-319
		Material Parameters at Tension			
$(\sigma_Y)_1$	40	34.17	30	10	53.77
$(\sigma_Y)_2$	50	37.74	40	20	64.38
$(\sigma_Y)_3$	55	41.33	45	30	74.95
$(\sigma_Y)_4$	60	44.31	50	40	80.25
$(\sigma_Y)_5$	65	46.91	55	50	85.54
$(\sigma_Y)_6$	70	50.14	60	60	90.82
$(\sigma_Y)_7$	75	52.82	65	80	96.10
$(\sigma_Y)_8$	80	57.15	73	99	103.48
$(E_T)_1$	3,361	3,200	3,191	4,000	950
$(E_T)_2$	3,333	3,050	2,941	3,571	905
$(E_T)_3$	3,333	2,210	2,941	3,226	852
$(E_T)_4$	3,125	1,187	2,857	2,857	761
$(E_T)_5$	2,778	590	2,778	2,564	702
$(E_T)_6$	2,632	458	2,778	2,222	620
$(E_T)_7$	2,174	412	2,500	1,835	536
$(E_T)_8$	1,724	296	2,424	1,484	432
		Material Parameters at Compression			
$(\sigma_Y)_1$	50	50.83	50	20	32.77
$(\sigma_Y)_2$	60	55.05	60	30	38.09
$(\sigma_Y)_3$	70	59.50	70	40	43.36
$\sigma_Y)_4$	80	63.91	80	60	48.59
$(\sigma_Y)_5$	90	68.28	90	80	53.69
$(\sigma_Y)_6$	100	72.54	100	100	58.67
$(\sigma_Y)_7$	110	74.63	110	120	63.55
$(\sigma_Y)_8$	120	76.69	120	130	65.98
$(E_T)_1$	3,356	6,840	3,205	4,000	1,862
$(E_T)_2$	3,226	3,410	3,125	3,704	1,350
$(E_T)_3$	3,226	3,370	3,125	3,333	1,241
$(E_T)_4$	3,030	3,010	2,857	2,532	1,084
$(E_T)_5$	2,632	2,687	2,632	1,724	870
$(E_T)_6$	2,041	2,255	2,083	1,143	657
$(E_T)_7$	1,493	1,970	1,538	781	464
$(E_T)_8$	980	1,745	1,064	599	359

A total number of 12 problems have been set forth for the WWFE-II. These problems cover a wide range of laminate types and load combinations, including pure matrices, UD laminae and multidirectional laminates subjected to tri-axial loads. Detailed descriptions of the laminate types, materials used, loading conditions and prediction requirements are summarized in Table 5.30.

Table 5.30 Summary of laminate types, material types, loads and plots required (Kaddour & Hinton, 2011)

Case	Laminate Lay-up	Material	Description of Required Prediction
1	Matrix	MY750 epoxy	σ_{xx} versus σ_{zz} (with $\sigma_{yy}=\sigma_{zz}$) envelope
2	0°	T300/PR319	σ_{12} versus σ_{22} (with $\sigma_{11}=\sigma_{22}=\sigma_{33}$) envelope
3	0°	T300/PR319	$2\varepsilon_{12}$ versus σ_{22} (with $\sigma_{11}=\sigma_{22}=\sigma_{33}$) envelope
4	0°	T300/PR319	Shear stress strain curves ($\sigma_{12}-2\varepsilon_{12}$) (for $\sigma_{11}=\sigma_{22}=\sigma_{33}=-600$ MPa)
5	90°	E-glass/MY750 epoxy	σ_{22} versus σ_{33} (with $\sigma_{11}=\sigma_{33}$) envelope
6	0°	S-glass/epoxy	σ_{11} versus σ_{33} (with $\sigma_{22}=\sigma_{33}$) envelope
7	0°	A-S carbon/epoxy	σ_{11} versus σ_{33} (with $\sigma_{22}=\sigma_{33}$) envelope
8	±35°	E-glass/MY750 epoxy	σ_{yy} versus σ_{zz} (with $\sigma_{xx}=\sigma_{zz}$) envelope
9	±35°	E-glass/MY750 epoxy	Stress-strain curves ($\sigma_{yy}-\varepsilon_{xx}$ and $\sigma_{yy}-\varepsilon_{yy}$) at $\sigma_{yy}=\sigma_{zz}=\sigma_{xx}=-100$ MPa
10	(0°/90°/±45°)s	IM7/8551-7	σ_{yz} versus σ_{zz} (with $\sigma_{yy}=\sigma_{xx}=0$) envelope
11	(0°/90°)s	IM7/8551-7	σ_{yz} versus σ_{zz} (with $\sigma_{yy}=\sigma_{xx}=0$) envelope
12	(0°/90°)s	IM7/8551-7	Stress-strain curves ($\sigma_{zz}-\varepsilon_{zz}$, $\sigma_{zz}-\varepsilon_{xx}$ and $\sigma_{zz}-\varepsilon_{yy}$) for $\sigma_{yy}=\sigma_{xx}=0$

The bridging model combined with the pseudo 3D laminate theory is applicable to all of the exercise problems. The predictions made using the retrieved and the provided constituent properties are plotted in Fig. 5.44 through Fig. 5.55 designated, respectively, with "Prediction (retrieved data)" and "Prediction (provided data)" for problem 1 to problem 12. The experimental data provided by Hinton & Kaddour (2011) are also shown in the corresponding figures for comparison. Additional comments are given below for each of the exercise problems.

Problem 1 MY750 epoxy subjected to combined σ_{xx} and σ_{zz} loads (with $\sigma_{yy}=\sigma_{zz}$)

The predicted failure envelopes of the pure matrix under triaxial load combinations using both the retrieved and the provided material properties agree reasonably well with the experimental results (Fig. 5.44). However, the predictions directly using the provided data have a better correlation with the measurements. This is understandable. Although *in situ* properties of a constituent in a composite might be different from those measured from bulk form specimens, they are, possibly, more applicable only to composite analysis. In the present case, the experimental data were obtained using pure matrix specimens, and thus the predictions for the matrix failure envelope based on the measured matrix strength parameters should be, and indeed have been shown to be, more accurate. The comparison between the predictions and the experiments for this problem has confirmed that Eq. (4.5) and, especially, Eq. (4.6) are well applicable when a constituent, specifically matrix material, is subjected to a multiaxial stress state.

The predicted failure mode information for this problem is indicated in Table 5.31. Under the given load combinations, the matrix material failed due to either a tensile or a compressive failure. There was only one exception. When the material was subjected to an equally triaxial compression, $\sigma_{yy}=\sigma_{zz}=\sigma_{xx}=-\sigma$, no failure was detected no matter how great the stress σ could be. This means that an isotropic matrix material will not fail at all if it is subjected to three axial compressive stresses of an equal magnitude. This is consistent with experimental visualizations. A very simple example is that an isotropic material sample, such as a metal ball, will not fail at all when it is thrown into the sea no matter how deep is the sea. Thus, the open envelope predicted as shown in Fig. 5.44 is realistic and, in fact, must be. Furthermore, it is highly possible that a pure matrix material may behave more strongly than a composite under hydrostatic compression. This is because, due to anisotropy of the composite, the matrix or the fiber material may not be subjected to equally hydrostatic compressive stresses even though the composite is under an equally three axial compression. Thus, the fibers or the matrix may still fail under such a load condition.

In addition to the openness, the predicted failure envelopes are discontinuous at some points. This is due to a variation in the equations used to calculate a critical stress (Eqs. (4.5.1) and (4.6.1)).

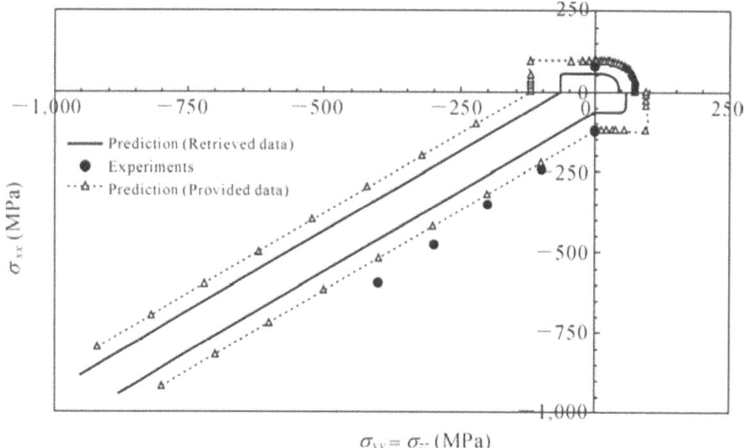

Fig. 5.44 Predicted failure envelope for Problem 1
(A pure matrix material subjected to σ_{xx} *vs* σ_{zz} (with $\sigma_{yy}=\sigma_{zz}$) loads)

Table 5.31 Failure mode information for Problem 1

Load Application Manner: $\Delta\sigma_{yy}=\Delta\sigma_{zz}=\Delta\sigma\cos(\theta)$, $\Delta\sigma_{xx}=\Delta\sigma\sin(\theta)$			
Prediction with Retrieved Data		Prediction with Provided Data	
θ Assumed	Ultimate Failure mode	θ Assumed	Ultimate Failure mode
$-48.5° \le \theta < 138.5°$	MT	$-51.3° \le \theta < 141.3°$	MT
$138.5° \le \theta < 225°$	MC	$141.3° \le \theta < 225°$	MC
$\theta = 225°$	-	$\theta = 225°$	-
$225° < \theta < 311.5°$	MC	$225° < \theta < 308.7°$	MC

(a) MT=matrix tensile failure;
(b) MC=matrix compressive failure.

Problem 2. UD T300/PR319 lamina subjected to combined σ_{12} and σ_{22} (with $\sigma_{11}=\sigma_{22}=\sigma_{33}$) loads

Problem 2 and the subsequent Problem 3 consist of the same composite subjected to the same load condition but with different presentations. For Problem 2, the failure envelopes of the UD composite in an in-plane shear (σ_{12}) versus transverse (σ_{22}) stress plane predicted with different constituent data inputted together with experimental results are shown in Fig. 5.45. The predicted results based on the back-calculated plastic and strength parameters of the matrix seem to be conservative compared to the measurements. However, the predictions obtained using the provided monolithic matrix properties exhibit more accuracy overall. The reason is also understandable. Comparing Table 5.28 with Table 5.26, one can see that the provided matrix tensile and compressive strengths are greater than the corresponding retrieved ones. As the composite failures under the considered load combinations are mainly due to matrix failures at tension and compression, the predicted composite strengths are undoubtedly higher.

Detailed failure mode information for this problem is summarized in Table 5.32. When a tri-axial compressive load applied to this composite is large, the lamina failure is caused by a compression although not necessarily a tri-axially compressive failure of the matrix. Some points on the failure envelope such as those on the right-hand side of a sharp angle (a kink) result from matrix tensile failure. Only under a limited number of load combinations is an excessively large strain failure mode detected in the predictions with the retrieved constituent data, as seen from Table 5.32. However, no matrix compressive failure is detected in the predictions with the provided data. When a combined tri-axial compressive load is in between 10 MPa and 1,085 MPa, the lamina failure is caused by an excessively large shear strain. If a combined tri-axial compressive load exceeds 1,085 MPa but is lower than 1,269 MPa, an extreme through-thickness compressive strain failure mode is recognized. When the load is near to an equally tri-axial compression, the failure mode is changed to an excessively large transverse compressive strain. No matter whether using the retrieved or the provided data in the predictions, the sharp angle occurring on the failure envelope is the result of a transfer of a matrix tensile failure (on the right-hand side) to an excessively large strain failure (on the left-hand side) mode.

It should be pointed out that a load application procedure may possibly have an influence on a predicted strength. All of the predictions for failure envelopes shown in this book have been made based on a proportional load application procedure. For instance, for the present problem, the load application procedure is defined as per

$$\Delta\sigma_{11}=\Delta\sigma_{22}=\Delta\sigma_{33}=\Delta\sigma\cos(\theta) \tag{5.34.1}$$

$$\Delta\sigma_{12}=\Delta\sigma\sin(\theta) \tag{5.34.2}$$

where $\Delta\sigma$ is a load increment. For a chosen θ in the range of $0°\leq\theta<360°$, the above loads are applied to the composite step by step until a composite failure takes place (only half of the complete envelopes were shown in Fig. 5.45 due to symmetry). However, according to Hinton & Kaddour (2011), the experiments for this composite were performed by first applying $\sigma_{11}=\sigma_{22}=\sigma_{33}$ to a pre-specified level, followed by an application of σ_{12} up to composite failure. In order to identify how great the influence is, predictions with the actual load procedure are also carried out using the retrieved and the provided constituent data respectively, and are shown in Fig. 5.45 designated with "Prediction (retrieved data) with actual loads" and "Prediction (provided data) with actual loads" accordingly. It is seen that a different load application procedure does not have any significant influence on the predicted results for this problem.

Table 5.32 Failure mode information for Problems 2 and 3

Load application manner: $\Delta\sigma_{11}=\Delta\sigma_{22}=\Delta\sigma_{33}=\Delta\sigma\cos(\theta)$, $\Delta\sigma_{12}=\Delta\sigma\sin(\theta)$			
Prediction with retrieved data		Prediction with provided data	
θ Assumed	Ultimate failure mode	θ Assumed	Ultimate failure mode
$0°\leq\theta<105°$	MT	$0°\leq\theta<94°$	MT
$105°\leq\theta<111°$	SL	$94°\leq\theta<180°$	SL
$111°\leq\theta<180°$	MC		

SL=a 12% strain limitation condition assumed.

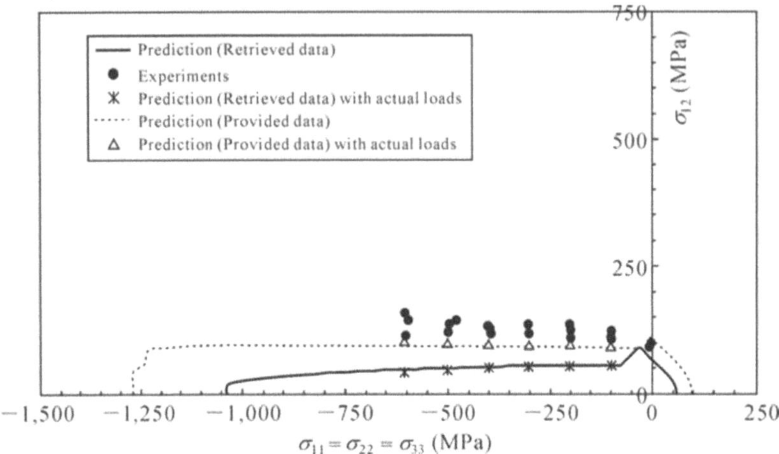

Fig. 5.45 Predicted failure envelopes for Problem 2 (UD lamina subjected to σ_{12} *vs* σ_{22} (with $\sigma_{11}=\sigma_{22}=\sigma_{33}$) loads)

Problem 3. Predicted failure envelope of $2\varepsilon_{12}$ *vs* σ_{22} (with $\sigma_{11}=\sigma_{22}=\sigma_{33}$) for UD T300/PR319 lamina

The predictions for this problem are actually done in the same way as for the previous one, i.e., by applying the stress loads to the composite up to failures, and the failure modes for this problem are exactly the same as those for Problem 2 (Table 5.32). The recorded pair of $2\varepsilon_{12}$ and σ_{22} at each failure is shown in Fig. 5.46. It is seen that the curves in Fig. 5.46 are quite similar to those in Fig. 5.45 and the discussions about Problem 2 are also applicable to this problem. However, two features seem to be more distinct in the present case. One is that the correlation of the predicted results using the monolithic matrix strength data in the experiments is even better than that using the back-calculated counterparts. Another is that the predictions using the provided data with the actual loads seem to differ from the experiments more than the ones based on the proportional load procedure, but those using the provided data do not show any essential difference. The differences must have been in the ultimate strains. As in Problem 2, the predicted ultimate strengths under the different load procedures are essentially the same. Thus, we may draw the conclusion that a different load application procedure may possibly have an influence on a predicted ultimate strain or deformation of a composite by means of the bridging model but have no significant effect on a predicted ultimate strength.

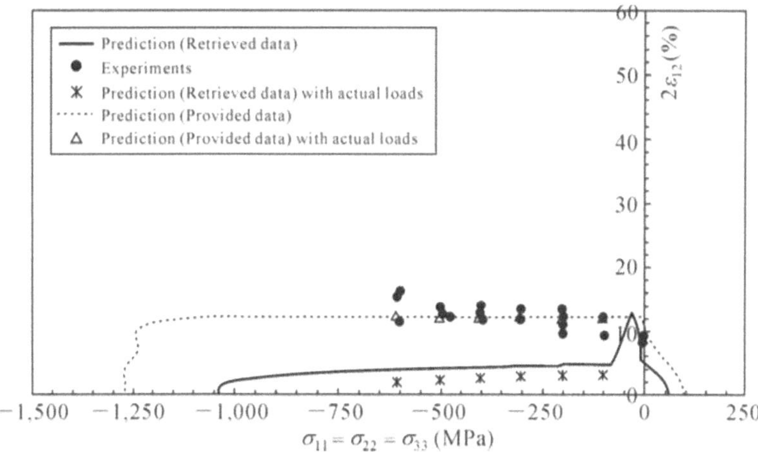

Fig. 5.46 Predicted failure envelopes for Problem 3 (UD lamina subjected to γ_{12} *vs* σ_{22} (with $\sigma_{11}=\sigma_{22}=\sigma_{33}$) loads)

Problem 4. Predicted σ_{12} (in addition to $\sigma_{11}=\sigma_{22}=\sigma_{33}=-600$ MPa) *vs* $2\varepsilon_{12}$ curve for UD T300/PR319 lamina

The internal stresses in the constituent materials generated with the application of $\sigma_{11}=\sigma_{22}=\sigma_{33}=-600$ MPa are shown in Table 5.33. Then a shear load is applied, which will result in additional compression on the matrix in a principal plane (which is a material plane with zero shear stress). The predicted results and the experimental measurements are plotted in Fig. 5.47. The failure mode information is shown in Table 5.34. The predicted stress-strain curve with the retrieved matrix properties agrees reasonably well with the initial portion of the measured data but seems to be much too conservative. The measured composite shear strength (161.1 MPa) is more than three times larger than the predicted one (50.8 MPa) and the measured ultimate shear strain (15.37%) is even five times greater than the predicted one (2.81%). The main reason for this is due to the different failure modes detected with the different input data in the predictions. As shown in Table 5.34, the predicted composite failure is due to a compressive failure occurring in the matrix if the retrieved data are used, whereas this is controlled by a 12% strain limitation when the provided data are employed. As the monolithic matrix compressive strength is higher than the retrieved one, no matrix compressive failure occurs before a predicted shear strain, $2\varepsilon_{12}$, reaches 12%. Thus, the predicted stress-strain curve using the provided data extends a great deal, as shown in Fig. 5.47. The difference between the predicted and measured strengths (102.9 MPa *vs* 161.1 MPa) is reduced to about 36.1%, whereas that between the strains (11.95% *vs* 15.37%) is lowered to 22.2%. Another message we learn from this example is that the modified maximum compressive stress failure criterion, Eqs. (4.6.1) and (4.6.2), is pertinent.

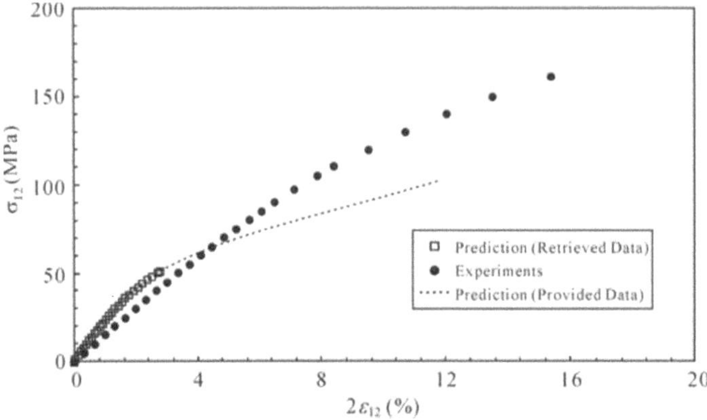

Fig. 5.47 Predicted stress-strain curve for Problem 4 (UD lamina subjected to σ_{12} loads in addition to a triaxial compression)

Table 5.33 Residual stresses of UD T300/PR319 lamina with an initial application of $\sigma_{11}=\sigma_{22}=\sigma_{33}$=-600MPa for Problem 4

	σ_{11}	σ_{22}	σ_{33}	σ_{23}	σ_{13}	σ_{12}
Fiber Stresses (MPa)	−830.46	−795.10	−792.63	0	0	0
matrix Stresses (MPa)	−254.31	−307.35	−311.06	0	0	0
Lamina Strain (%)	−0.1831	−5.7010	−5.8082	0	0	0

Table 5.34 Failure mode information for Problem 4

Load Application Manner: Applying $\sigma_{11}=\sigma_{22}=\sigma_{33}=-600$MPa first, followed by the application of $\Delta\sigma_{12}=\Delta\sigma$ up to failure	
Prediction with Retrieved Data	Prediction with Provided Data
Ultimate Failure Mode	Ultimate Failure Mode
MC	SL

Problem 5. UD E-glass/MY750 epoxy lamina subjected to combined σ_{22} and σ_{33} (with $\sigma_{11}=\sigma_{33}$) loads

In Problems 6 and 7, unidirectional composites are subjected to combined triaixal tensions or compressions, under which both the fibers and the matrix in the composites may be under triaxial tension or compression. As mentioned in Chapter 4, the load sustaining ability of a material under a triaxial compression can be enhanced, and thus the use of the modified maximum compressive stress failure criterion represented by Eqs. (4.6.1) and (4.6.2) can compensate this enhancement. The predictions with the retrieved and provided data for this problem are graphed in Fig. 5.48, respectively. Experimental results are also shown in the figure. To better understand the failure states occurring in this composite, the predicted failure modes versus the load combinations are listed in Table 5.35. The table shows that the present composite failed essentially due to a matrix tensile or compressive failure, very different from the composites in Problems 6 and 7 (Figs. 5.49 and 5.50 and Tables 5.36 and 5.37). The reason is that the E-glass fibers in this composite are not stiff enough, so that with most of the applied load combinations the fiber material is not subjected to an excessively high compressive stress state before a matrix compressive failure occurs. This situation does not change much even when the used matrix compressive strength is increased, as can be seen from the predictions using the provided monolithic matrix plastic and strength parameters shown in Fig. 5.48. The increased matrix strength extended the predicted failure envelope. However, its shape is similar to that obtained with the retrieved data and no fiber failure is detected (Table 5.35).

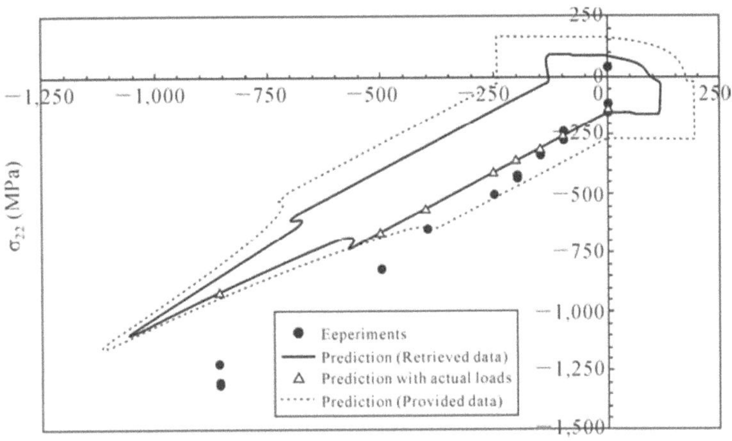

Fig. 5.48 Predicted failure envelopes for Problem 5 (UD lamina subjected to σ_{22} *vs* σ_{33} (with σ_{11}=σ_{33}) loads)

Table 5.35 Failure mode information for Problem 5

Load Application Manner: $\Delta\sigma_{11}=\Delta\sigma_{33}=\Delta\sigma\cos(\theta)$, $\Delta\sigma_{22}=\Delta\sigma\sin(\theta)$			
Prediction with Retrieved Data		Prediction with provided data	
θ Assumed	Ultimate Failure Mode	θ Assumed	Ultimate Failure Mode
$-55°\leq\theta<145°$	MT	$-53°\leq\theta<143°$	MT
$145°\leq\theta<305°$	MC	$143°\leq\theta<307°$	MC

It is interesting to notice that the predicted failure envelopes using both the retrieved and the provided data are closed. This means that the predicted load carrying capacity of the UD composite is limited, even under equal tri-axial compression (i.e., $\sigma_{11}=\sigma_{33}=\sigma_{22}=-0.5\sigma$). Comparing Fig. 5.48 with Fig. 5.44, we can draw a conclusion that a UD composite is much weaker than a pure matrix material when each of them is sustaining a tri-axial compression state with equal magnitude.

Finally, predictions for the actual load procedure are also made using the retrieved input data and are shown in Fig. 5.48. From (Hinton & Kaddour, 2011), the experiments for this composite were carried out by applying $\sigma_{11}=\sigma_{33}$ to a chosen level, followed by the application of a compressive load σ_{22} until the composite attained a failure status. Similarly, as for Problem 2, the effect of a load application procedure on the composite strength prediction is insignificant.

Problem 6. UD S-glass/epoxy lamina subjected to combined σ_{11} and σ_{33} (with $\sigma_{22}=\sigma_{33}$) loads

Ultimate strengths of the UD lamina under varied triaxial load combinations are investigated for this composite. The predicted failure modes are summarized in Table 5.36. The predictions using the retrieved and the provided constituent material properties are compared with the experimental results, as shown in Fig. 5.49. The overall agreement between the predicted results, especially between those based on the retrieved data and the experiments, is good, especially when an applied load combination on the lamina contains a longitudinal tensile stress.

On the other hand, the predictions using monolithic matrix properties for this problem are not as good as those using the retrieved data, although both are coincident with each other when all of the applied stresses on the lamina are compressive. In the latter case, the lamina failure is essentially due to a fiber compressive failure. It seems that when a load combination applied on the lamina contains a transverse tension combined with a longitudinal tension or a relatively large longitudinal compression, the prediction using the provided data differs from the experiment more obviously than that using the retrieved data. In general, the predicted lamina load carrying capacity increased with the application of an increased compression of $\sigma_{22}=\sigma_{33}$ when a longitudinal compressive load was also applied to the lamina. This is consistent with the experimental measurements.

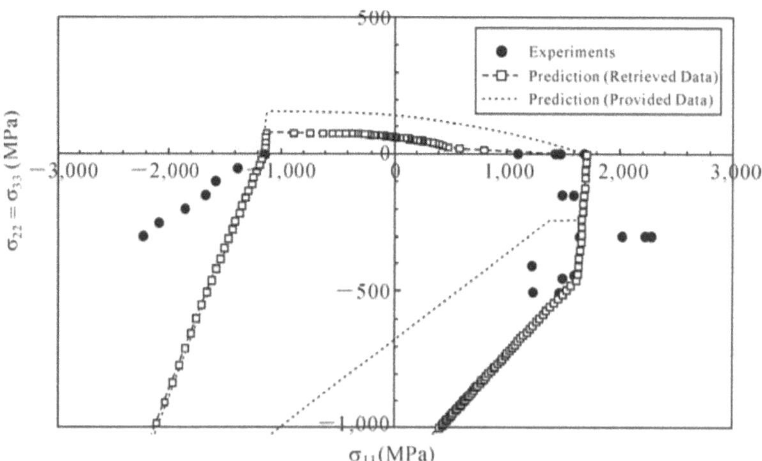

Fig. 5.49 Predicted failure envelope for Problem 6
(UD lamina, glass fiber reinforced composite, subjected to σ_{11} *vs* σ_{33} (with $\sigma_{22}=\sigma_{33}$) loads)

Table 5.36 Failure mode information for Problem 6

Load Application Manner: $\Delta\sigma_{11}=\Delta\sigma\cos(\theta)$, $\Delta\sigma_{22}=\Delta\sigma_{33}=\Delta\sigma\sin(\theta)$			
Prediction with Retrieved Data		Prediction with Provided Data	
θ Assumed	Ultimate Failure Mode	θ assumed	Ultimate Failure Mode
0°<θ<176°	MT	0°<θ<173°	MT
176°≤θ<217°	FC	173°≤θ<210°	FC
217°≤θ<345°	MC	210°≤θ<352°	MC
345°≤θ≤360°	FT	352°≤θ≤360°	FT

* FT=fiber tensile failure, FC=fiber compressive failure.

Problem 7. UD A-S carbon/epoxy lamina subjected to combined σ_{11} and σ_{33} (with $\sigma_{22}=\sigma_{33}$) loads

This problem is similar to Problem 6. The only difference is in the material system used. The predicted failure envelopes based on different definitions of input data together with the experimental results are shown in Fig. 5.50. Detailed failure mode information detected by different prediction schemes is summarized in Table 5.37. Similar conclusions for this problem, as for Problem 6, can be made. Specifically, the prediction using the retrieved data improves the correlation accuracy with the measured results remarkably when the compressive loads along the three axial directions become large. Comparisons between the theoretical results and the experiments for Problems 6 and 7 clearly show that the modified maximum compressive stress failure criterion, represented by Eqs. (4.6.1) and (4.6.2), applied to detect not only the matrix but also the fiber failures in a composite, is necessary.

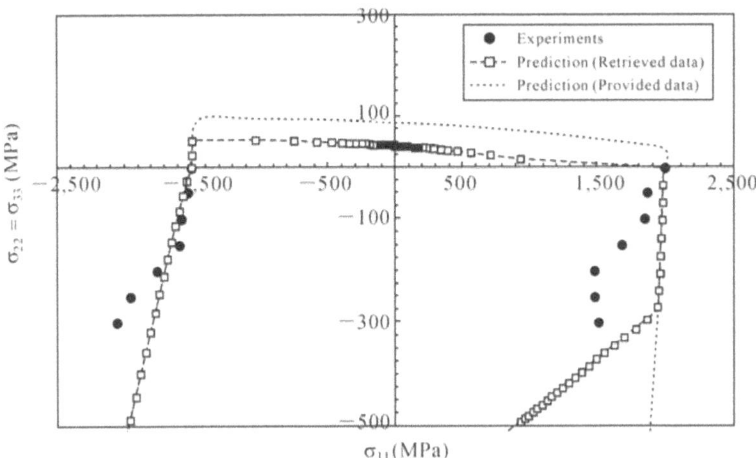

Fig. 5.50 Predicted failure envelope for Problem 7
(UD AS/epoxy lamina subjected to σ_{11} *vs* σ_{33} (with $\sigma_{22}=\sigma_{33}$) loads)

Table 5.37 Failure mode information for Problem 7

Load Application Manner: $\Delta\sigma_{11}=\Delta\sigma\cos(\theta)$, $\Delta\sigma_{22}=\Delta\sigma_{33}=\Delta\sigma\sin(\theta)$			
Prediction with Retrieved Data		Prediction with Provided Data	
θ Assumed	Ultimate Failure Mode	θ Assumed	Ultimate Failure Mode
0°<θ<178°	MT	0°<θ<177°	MT
178°≤θ<205°	FC	177°≤θ<210°	FC
205°≤θ<352°	MC	210°≤θ<338°	MC
352°≤θ≤360°	FT	338°≤θ≤360°	FT

Problem 8. (±35°)$_s$ E-glass/MY750 epoxy laminate subjected to combined σ_{yy} and σ_{zz} (with σ_{xx}=σ_{zz}) loads

Comparison between the predicted failure envelopes and the experimental data for this composite is shown in Fig. 5.51. The predicted failure modes varied with different load combinations are indicated in Table 5.38. Excessively large strains or matrix compressive failures are the predicted two failure modes for this problem, corresponding to the cases where the composite is subjected to tensile or compressive loads respectively. It is seen from Fig. 5.51 that most of the measured data are located in between the predicted failure envelopes using the retrieved and the provided data. In general, the predictions fit well with the experiments although only limited amounts of experimental data are available.

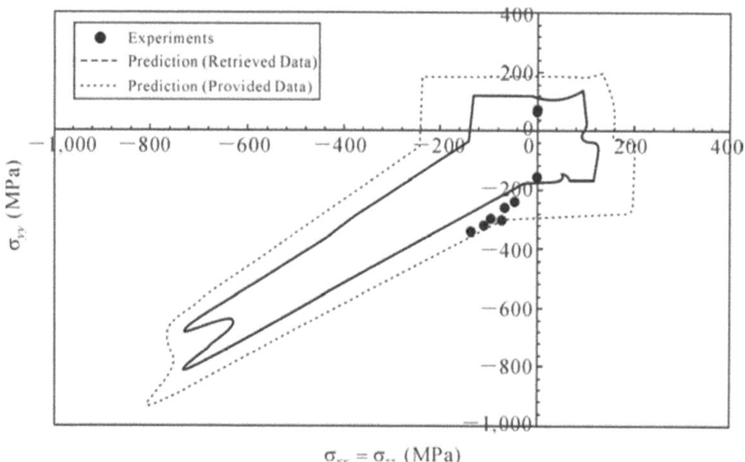

Fig. 5.51 Predicted failure envelopes for Problem 8 ([±35°] laminate subjected to σ_{yy} *vs* σ_{zz} (with σ_{xx}= σ_{zz}) loads)

Table 5.38 Failure mode information for Problem 8

Load Application Manner: $\Delta\sigma_{xx}=\Delta\sigma_{zz}=\Delta\sigma\cos(\theta)$, $\Delta\sigma_{yy}=\Delta\sigma\sin(\theta)$			
Prediction with Retrieved Data		Prediction with Provided Data	
θ Assumed	Ultimate Failure Mode	θ Assumed	Ultimate Failure Mode
$-62°\leq\theta<145°$	SL	$-54°\leq\theta<142°$	SL
$145°\leq\theta<298°$	MC	$142°\leq\theta<306°$	MC

It deserves mentioning that predictions following an actual load application procedure, in which a bi-axial load $\sigma_{xx}=\sigma_{zz}$ was applied to a prescribed value before a third stress σ_{yy} was applied to the composite up to an ultimate failure (Hinton & Kaddour, 2011), were also carried out for this composite. The effect of the different load application procedure on the predictions is so limited that the predicted failure load combinations are essentially located in the same failure envelope. Thus, those combinations are not marked out in the envelope.

Problem 9. (±35°)$_s$ E-glass/MY750 laminate subjected to compression σ_{yy}: in the presence of $\sigma_{zz}=\sigma_{yy}=\sigma_{xx}=-100$ MPa

For this composite, an initial triaxial compression load, $\sigma_{zz}=\sigma_{yy}=\sigma_{xx}=-100$ MPa, is applied before the continuous application of a uniaxial compression load, σ_{yy}, up to failure. The residual stresses in the constituents of each layer and the laminate strains corresponding to the initial triaxial compression are listed in Table 5.39. The predicted stress-strain curves as well as the experimentally measured data are plotted in Fig. 5.52. It is seen that the predicted curves using the retrieved data up to ultimate failure fit well with the measurements. For example, the predicted strains of the laminate, $\varepsilon_y=-0.1723$ and $\varepsilon_x=-0.1093$ as listed in Table 5.39, at a load level of $\sigma_{zz}=\sigma_{yy}=\sigma_{xx}=-100$MPa, are close to the measured ones, $\varepsilon_y=-0.1845$ and $\varepsilon_x=-0.110$. The majority of the predictions using the provided data also agree well with the corresponding experimental data. However, when the compressive load σ_{yy} is increased to over −244 MPa, the latter predictions seem to be stiffer. The nonlinear feature of the measured stress-strain curve is very significant when the stress is over −250 MPa. From the experimental data, the measured y-directional strain is increased from about 2% to almost 7.5% when the compressive load σ_{yy} varies from −250 MPa to −299 MPa. In terms of the present theory, this might be possible if the hardening modulus of the matrix material near to failure is significantly smaller than the preceding modulus. All of the predictions indicated that the ultimate failure of the composite is due to a matrix compressive failure.

Table 5.39 Residual stresses of $(\pm 35°)_s$ E-glass/MY750 laminate with an initial application of $\sigma_{zz}=\sigma_{yy}=\sigma_{xx}=-100$ MPa for Problem 9

(a) Prediction with Retrieved Data

	σ_{11}	σ_{22}	σ_{33}	σ_{23}	σ_{13}	σ_{12}
35°-ply fiber stresses (MPa)	−194.91	−108.93	−131.12	0	0	−11.86
35°-ply matrix stresses (MPa)	−15.25	−24.64	−50.21	0	0	−3.92
−35°-ply fiber stresses (MPa)	−193.48	−108.04	−129.43	0	0	12.26
−35°-ply matrix stresses (Mpa)	−14.98	−24.52	−49.56	0	0	4.05
Inter-layer stresses (MPa)	−51.96	−54.68	−100	0	0	0
Laminate strain (%)	−0.1151	−0.1996	−0.3368	0	0	0

(b) Prediction with Provided Data

	σ_{11}	σ_{22}	σ_{33}	σ_{23}	σ_{13}	σ_{12}
35°-ply fiber stresses (MPa)	−198.26	−108.18	−131.35	0	0	−12.45
35°-ply matrix stresses (MPa)	−14.50	−23.91	−49.85	0	0	−4.09
−35°-ply fiber stresses (MPa)	−196.87	−107.33	−129.64	0	0	12.87
−35°-ply matrix stresses (Mpa)	−14.24	−23.81	−49.20	0	0	4.23
Inter-layer stresses (MPa)	−37.34	−41.32	−100.0	0	0	0
Laminate strain (%)	−0.1093	−0.1723	−0.2907	0	0	0

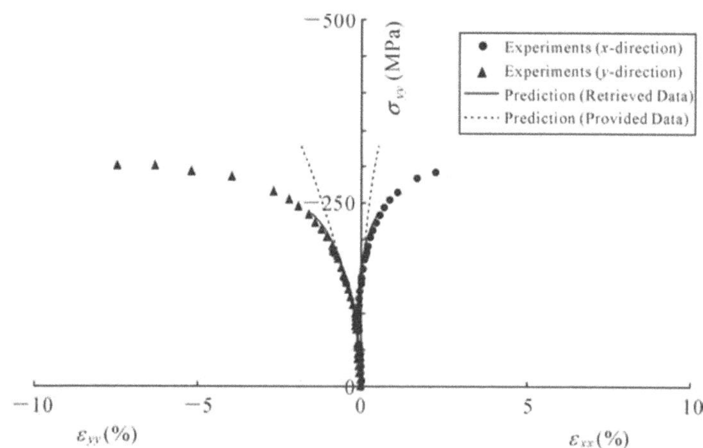

Fig. 5.52 Predicted stress-strain curves for Problem 9 ($[\pm 35°]_s$ laminate subjected to σ_{yy} loads after $\sigma_{yy}=\sigma_{zz}=\sigma_{xx}=-100$ MPa)

Problem 10. $(0°/90°/\pm 45°)_s$ IM7/8551-7 laminate subjected to combined σ_{yz} and σ_{zz} (with $\sigma_{yy}=\sigma_{xx}=0$) loads

For this problem, the experimental measurements were carried out by first applying a constant load along the thickness direction followed by the application of another load up to failure (Hinton & Kaddour, 2011). On the other hand, a proportional load application procedure similar to Eqs. (5.34.1) and (5.34.2) is assumed in the predictions (Table 5.40). However, by comparing the predicted results based on the actual experimental loading condition with those based on the proportional load application procedure, no significant difference is recognized. The predicted results together with the experimental data are shown in Fig. 5.54 for comparison. Detailed failure mode information from the predictions with the retrieved and the provided data is summarized in Table 5.40.

It is noted that for Problems 1 – 9, the difference between the predicted failure envelopes with and without consideration of the thermal residual stresses is insignificant, and hence all of the results shown in Fig. 5.44 to Fig. 5.53 have incorporated the influence of the thermal residual stresses. For the present problem as well as Problem 11, however, the situation is different. The influence of the thermal residual stresses is relatively large. Thus, both predictions with and without considering thermal residual stresses have been made and are shown in Fig. 5.54. It is seen that when the through-thickness compression is not large, the bridging model predictions agree well with the experiments. As the through-thickness compression is increased, the predictions seem to be conservative. Generally speaking, the predictions without considering the thermal residual stresses perform better than those with the thermal stress effect. Even under the extreme load condition, i.e., only a pure through-thickness compressive load is applied, the predicted strength without considering the thermal stress effect is close to the measured one from a "cylindrical" specimen (Hinton & Kaddour, 2011). As pointed out by Hinton and Kaddour (2011) and Kaddour et al. (2011), a strong dependency of the measured strength on the shape of a specimen was recognized in the experiments. In the present predictions, only 2D (in-plane) thermal stresses are calculated, as in general the thickness of a laminate is much smaller than the other two planar dimensions. This calculation, however, may not be quite suitable when the thickness is relatively large and when a through- thickness load is the main cause for the composite to fail, such as in the present case.

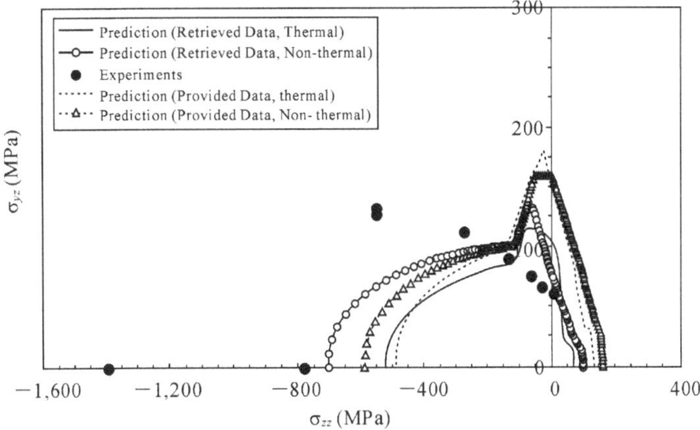

Fig. 5.53 Predicted failure envelopes for Problem 10 ([0°/90°/±45°] laminate subjected to σ_{yz} *vs* σ_{zz} (with $\sigma_{yy}=\sigma_{xx}=0$) loads)

It is noticed that for almost all of the previous problems, i.e., Problems 1 – 9, the predictions using monolithic matrix properties gave higher strength magnitudes or enlarged failure envelopes than those using the retrieved data. In the present and the next problems, however, the situations are somewhat different. In other words, the predicted failure envelopes using the provided data are smaller overall, see Figs. 5.53 and 5.54. This is because the provided matrix stress-strain curves are stiffer than the retrieved ones. The monolithic matrix moduli (elastic and hardening moduli) are larger than the retrieved ones. In the present case as well as in Problem 11, the composite failures are essentially caused by the matrix failures. Although the monolithic matrix strengths are greater than the retrieved counterparts affording the matrix to be failed at a higher stress level, the stiffer monolithic matrix moduli result in even higher stresses sustained by the matrix materials when the composite is subjected to the same load combinations. Thus, the predicted composite load carrying ability is reduced.

Table 5.40 Failure mode information for Problem 10

Load application manner: $\Delta\sigma_{zz}=\Delta\sigma\cos(\theta)$, $\Delta\sigma_{yz}=\Delta\sigma\sin(\theta)$							
Prediction with Retrieved Data				Prediction with Provided Data			
With Thermal Residual Stresses		Without Thermal Residual Stresses		With Thermal Residual Stresses		Without Thermal Residual Stresses	
θ Assumed	Ultimate Failure Mode	θ Assumed	Ultimate Failure Mode	θ Assumed	Ultimate Failure Mode	θ Assumed	Ultimate Failure Mode
$0°\le\theta<124°$	SL	$0°\le\theta<122°$	SL	$0°\le\theta<109°$	SL	$0°\le\theta<109°$	SL
$124°\le\theta\le180°$	MC	$122°\le\theta\le180°$	MC	$109°\le\theta\le180°$	MC	$109°\le\theta\le180°$	MC

Problem 11. (0°/90°) IM7/8551-7 laminate subjected to combined σ_{yz} and σ_{zz} (with $\sigma_{yy}=\sigma_{xx}=0$) loads

The predicted failure envelopes (Fig. 5.54) for this problem are quite similar to those for Problem 10. Failure mode information (Table 5.41) for this problem is comparable to that for Problem 10. This means that different lay-ups for a quasi-isotropic laminate may have only a limited effect on its through-thickness behaviors. All discussions about Problem 10 are applicable to this problem as well.

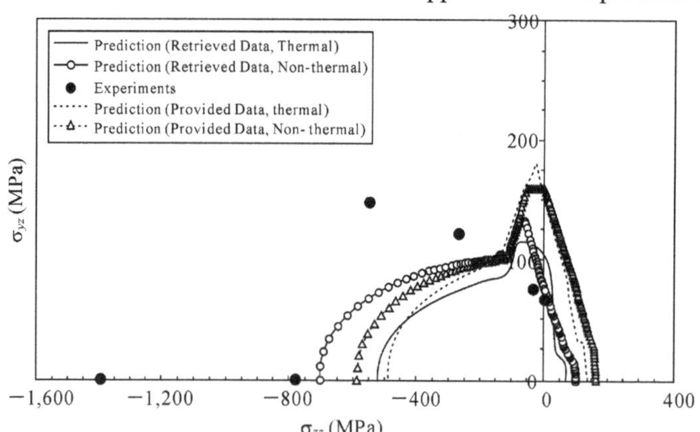

Fig. 5.54 Predicted failure envelopes for Problem 11 ([0°/90°] laminate subjected to σ_{yz} *vs* σ_{zz} (with $\sigma_{yy}=\sigma_{xx}=0$) loads)

Table 5.41 Failure mode information for Problem 11

Load application manner: $\Delta\sigma_{zz}=\Delta\sigma\cos(\theta)$, $\Delta\sigma_{yz}=\Delta\sigma\sin(\theta)$							
Prediction with Retrieved Data				Prediction with Provided Data			
With Thermal Residual Stresses		Without Thermal Residual Stresses		With Thermal Residual Stresses		Without Thermal Residual Stresses	
θ assumed	Ultimate failure mode	θ assumed	Ultimate failure mode	θ assumed	Ultimate failure mode	θ assumed	Ultimate failure mode
$0°\leq\theta<125°$	SL	$0°\leq\theta<146°$	SL	$0°\leq\theta<109°$	SL	$0°\leq\theta<109°$	SL
$122°\leq\theta\leq180°$	MC	$146°\leq\theta\leq180°$	MC	$109°\leq\theta\leq180°$	MC	$109°\leq\theta\leq180°$	MC

Problem 12. (0°/90°) IM7/8551-7 laminate subjected to σ_{zz} with $\sigma_{yy}=\sigma_{xx}=0$

As pointed out by the exercise organizers, no measured data were available for this composite (Hinton & Kaddour, 2011), and the provided experimental results were actually based on a $[45°/0°/-45°/90°]_s$ lay-up laminate. The stress-strain curves of the latter laminate subjected to through-thickness load σ_{zz} are predicted using both the retrieved and the monolithic constituent properties, and are plotted in Fig. 5.55. Also shown in the figure are the experimental results. Similarly as in Problems 10 and 11, the predicted through-thickness strength of the quasi-isotropic laminate is lower than the measured one. Moreover, the predicted stress-strain curves up to the ultimate failure are almost linear whereas the measured one in the thickness direction is concave when the through-thickness compression load becomes large. Currently, the latter phenomenon, i.e., an instantaneous tangential modulus increasing with an increase in the applied loads, is not likely to be explained in terms of the bridging model together with the laminate theory used.

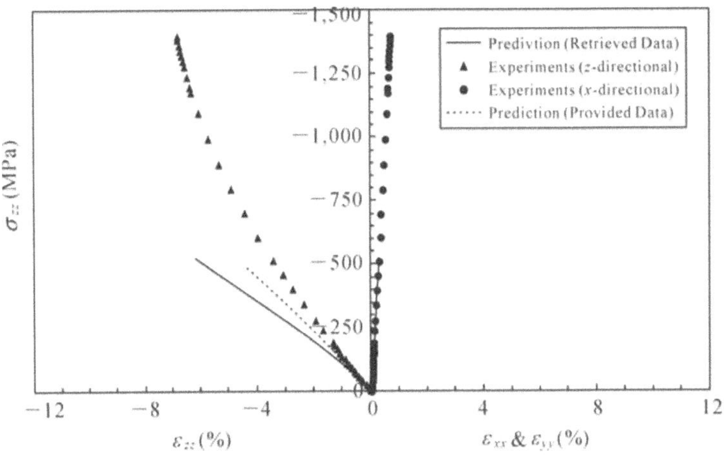

Fig. 5.55 Predicted stress-strain curves for Problem 12 ([0°/90°] laminate subjected to σ_{zz} load with $\sigma_{yy}=\sigma_{xx}=0$)

5.16 Concluding Remarks

In the previous sections of this chapter, analyzing procedures based on the bridging model (Chapter 3) for the mechanical properties of laminated composites subjected to various load conditions are presented. Correlations of the model predictions with experimental data of different sources have been shown. Before ending the chapter, some additional comments are given in the following.

5.16.1 Laminate Theory

The bridging model presented in Chapter 3 is a unified constitutive theory for unidirectional composites and a tool for stress calculation. The most important and, perhaps, the unique feature of this theory is that the internal stresses in the constituent fiber and matrix materials of the composite subjected to arbitrary load condition including a temperature variation can be explicitly determined using fiber and matrix properties only. For the analysis of a laminate consisting of multidirectional layers as shown in this chapter, however, a laminate theory must be incorporated so that the loads shared by each layer in its local coordinate system can be obtained. Although the classical laminate theory is the most well known and is efficient in most cases where the laminate is subjected to in-plane loads, the pseudo 3D laminate theory shown in Section 5.8 is universally applicable, because the former is a reduced form of the latter. The only drawback with the use of the 3D theory is that a total number of $3n+6$ linear algebraic equations must be solved simultaneously at each load increment, whereas the classical laminate theory only results in 6 such equations, where n is the number of the final layers constituting the laminate to be analyzed. It is noted that after introducing pure matrix interface layers, the original N layers of the laminate become $n=2N-1$ layers.

5.16.2 Effect of the Load Application Manner

In principle, the manner of the load application will have an influence on the mechanical behavior of a composite beyond a linearly elastic deformation range. This is because the mechanical responses (constitutive equations) of an elasto-plastic material, at a plastic deformation range, to loading and unloading are different. As a matrix material is generally able to deform plastically, the resulting composite can exhibit different behavior at loading or unloading. Another influence may possibly occur due to a different progressive failure process involved. If a pre-load application already causes some layers in the laminate to have failed before a subsequent load application takes place, a different load application procedure such as a proportional load application manner adopted in this book may probably affect the prediction of the ultimate behavior of the laminate. However, the influences are likely to be more effective when a strain or deformation calculation for the composite is required. Prediction of an ultimate strength of the composite seems to be little influenced by a different load application procedure, as seen from the examples shown in the book. This might be attributed to the fact that changing a failure mode from a matrix failure to a fiber failure or vice versa generally needs a significant variation in the loads applied to a composite, whereas the variation in a load application procedure can seldom cause a change in the composite failure mode. The stress states in the constituents mainly depend on the final load magnitudes applied to the composite rather than on how the loads are applied to the composite.

5.16.3 Definition of Fiber and Matrix Properties

It has been shown in the preceding section that mechanical properties of the constituent fiber and matrix materials of a composite can be defined independently, without retrieval from the composite behaviors, although the back-calculated constituent properties may sometimes result in a better prediction for composite properties. However, there is a sacrifice in cost and time in preparing composite specimens and doing mechanical testing for a prediction which is not always better. In general, the fiber elastic properties from a reliable source are directly applicable, whereas the fiber strength parameters may be calibrated in terms of the bridging model. On the other hand, much more attention should be paid to determination of the matrix properties. Although uniaxial tension and compression failure tests are generally enough to define matrix properties, several points should be taken into account. Firstly, when a bounding range for a matrix modulus (elastic or hardening modulus) exists, a better choice is likely to be use of the lower bound to define that quantity. The experimental verifications have shown that most of the discrepancies from the bridging model based predictions, if any, for the examples considered in this book are conservative overall. A smaller matrix modulus would result in a larger portion of the applied loads on the composite to be sustained by the fiber material, and thus less conservative predictions can be expected. Secondly, whenever possible the matrix tensile strength should be calibrated against the transverse tensile and in-plane shear strengths of a UD composite, as described previously. Alternatively, the lower bound of the measured tensile strength for the matrix should be used to define that quantity. This is because the predicted transverse tensile and in-plane shear strengths of the UD composite using the measured tensile strength of the pure matrix are often higher than the measured counterparts of the composite. Hence, the *in-situ* tensile strength of a matrix must be smaller than the measured one of a monolithic specimen. On the other hand, it has already been recognized that using measured uniaxial strengths of a lamina can result in a much more conservative prediction for a laminate strength compared with the measured one. Thus, some researchers already suggested using enlarged uniaxial strength parameters for the prediction of a laminate strength (Rotem, 1998; Sun & Tao, 1998). Similarly, an enlarged matrix tensile strength, where the predicted transverse tensile and in-plane shear strengths of a UD composite are in a certain range, such as 20% greater than the measured counterparts of the composite, may be better to use to predict a laminate strength.

Thirdly, the measured compressive strength of the matrix using monolithic specimens can be reasonably well applied. Finally, predicted longitudinal strengths of a UD composite should be equal or close to the measured ones and hence the matrix strength parameters should be adjusted if necessary.

It is well known that stress concentration occurs when an isotropic plate contains a circular hole. The plate load carrying capacity lowers down a lot due to the occurrence of the hole. Similarly, the matrix *in situ* strength in a composite must be decreased as a result of the introduction of the fibers. This is why the transverse tensile strength of a UD composite is generally smaller than the matrix tensile strength in a bulk form. It is highly possible that the *in situ* strength of a constituent especially matrix material can be defined from its monolithic strength measured independently divided by a stress concentration factor. This factor must depend on the fiber and matrix properties and on the fiber volume fraction, among others.

With the above points in mind, properties of different constituents can be determined and documented in a material database. The thus obtained property parameters are likely to be applicable for any composites made of the corresponding constituents, or at least can be used as input data for a preliminary design of a structure containing composites under development.

5.16.4 Failure Criteria

Although satisfactorily good predictions for most of the examples have been achieved in this book, based on the failure criteria for lamina and laminate summarized in Chapters 4 and 5 respectively, the most challenging work for more efficient application of the bridging model to determine composite load carrying capacity remains with the development of even more powerful failure criteria to detect a lamina failure without using composite strength parameters. There are at least two pieces of evidence to show that the used failure criterion for detecting a matrix failure needs to be further developed. The first is that when a measured tensile strength of a monolithic matrix specimen is used for lamina strength prediction, much higher strengths than the measured transverse tensile and shear strengths may be recognized. However, the same matrix tensile strength is able to result in a better correlation between predicted and measured strengths of a laminate. Another piece of evidence is that the measured tensile and shear strengths of a matrix are generally different from each other, whereas the predicted ones using the failure criterion, Eqs. (4.5.1) and (4.5.2), are the same. However, simply adding a shear stress failure criterion such as $0.5(\sigma^{(1)}-\sigma^{(3)})\geq\tau_u$ to detect a matrix failure, where τ_u is a shear strength, does not result in better predictions overall for the problems considered in this book. Thus, more thorough investigation should be made in this regard.

Before a more powerful failure criterion is available, a way of compensation is to calibrate the matrix tensile strength upon the transverse tensile and in-plane shear strengths of a UD composite, as done in most examples in this book.

5.16.5 Analysis of Composite Structures

One more important advantage of the bridging model in practical applications is that it can be easily incorporated into a commercial finite element software package such as ABAQUS to perform structural failure analysis for a complicated structure containing laminated composites (Huang, 2007). The minimum required input data, the explicit and closed-form formula expressions, a universal applicability and a satisfactorily high prediction accuracy are among several outstanding features of the bridging model. In (Huang, 2007), a user subroutine UGENS has been developed based on all of the bridging model formulae together with the classical laminate theory to define an elemental instantaneous stiffness matrix of a shell element for a composite structure. Failure detection and a total stiffness discount scheme were also implemented into the subroutine so that at each load increment each composite element would be checked. Further development can be expected by incorporating, for example, the 3D theories into the UGENS subroutine with which a constituents-based optimal design for a critical structure can be achieved by virtue of ABAQUS.

5.16.6 Application to Other Kinds of Composites

Although only unidirectional laminae and multidirectional laminates have been dealt with in this book, no restriction exists on the applicability of the bridging model to the mechanical property simulation of a different kind of continuous fiber reinforced composite such as a woven, braided or knitted fabric reinforced composite. Essentially, a laminate global analysis and a lamina local analysis are necessary, with an incremental solution strategy used. A schematic diagram shown in Fig. 5.56 illustrates the global analysis procedure, whereas that in Fig. 5.57 or Fig. 5.58 explains how a local analysis is carried out for a braided or knitted fabric reinforced composite lamina. The purpose of the global analysis is to determine the stress increments shared by each lamina in the laminate, given in the laminate global coordinate system. For this, the classical or the 3D laminate theory based on the instantaneous stiffness matrix of each fibrous lamina is applied. On the other hand, the purpose of the local analysis is three-fold. First, the instantaneous stiffness matrix of the lamina is updated. Second, the internal stress increments induced in the fiber and matrix of the fibrous composite are calculated. Third, total stresses in the fiber and matrix are updated and any failure status of the fibrous lamina as well as an ultimate failure of the laminated composite is assessed. The failure criteria for a single layer lamina represented by Eqs. (4.5) and (4.6) and by Table 5.3 for a laminate are all applicable. For more details refer to (Huang, 2000a, 2002b, 2002c, 2004b, 2005; Huang et al., 2003; and Huang & Ramakrishna, 2002a, 2002b, 2002c, 2003).

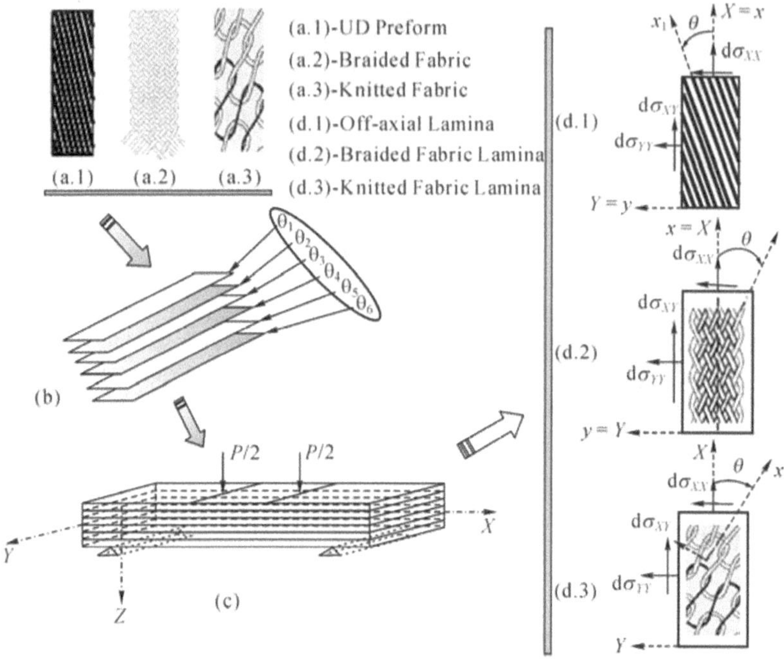

Fig. 5.56 Schematic analysis procedure for a laminated composite: (a) Single layer fibrous architecture; (b) Lamination; (c) Under arbitrary load condition; (d) Loads sustained by a single layer lamina can be determined using a laminate theory based on the instantaneous stiffness matrix of all the lamina layers (from Huang & Fujihara, 2005)

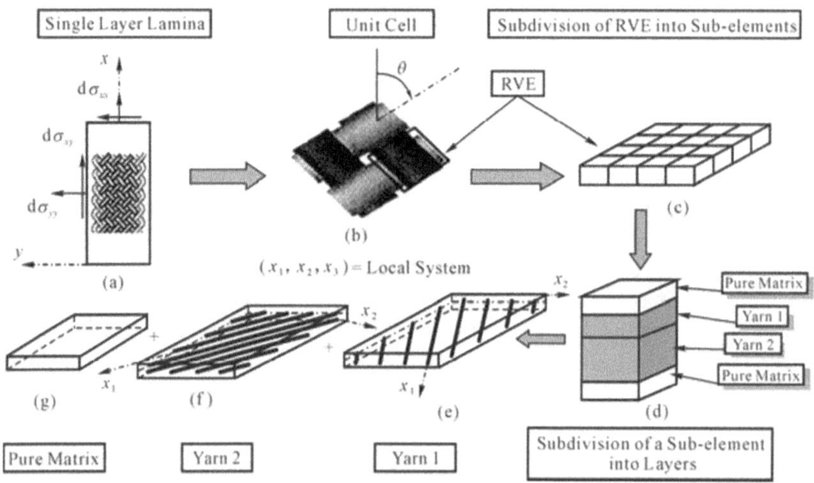

Fig. 5.57 Schematic analysis procedure for a single layer braided fabric reinforced lamina: (a) The single layer lamina with known stress increments; (b) A unit cell for the lamina; (c) Subdivision for an RVE of the lamina; (d) A sub-element consisting of at most four material layers; (e), (f) and (g) UD laminae to constitute the sub-element (pure matrix layer is a UD lamina with zero fiber volume fraction) (from Huang & Fujihara, 2005)

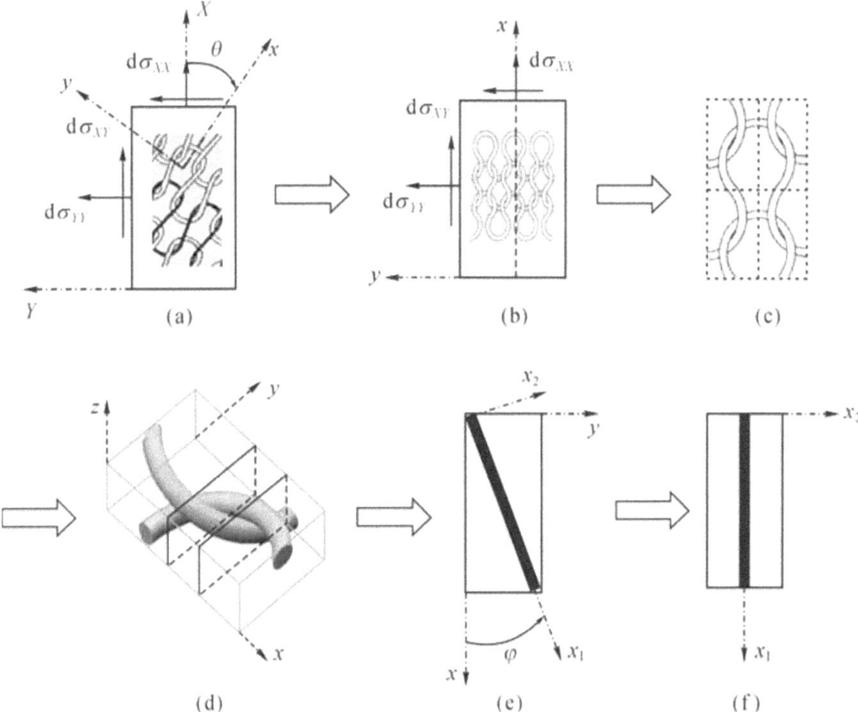

Fig. 5.58 Schematic analysis procedure for a single layer knitted fabric reinforced lamina: (a) The lamina with known stress increments in the global coordinate system; (b) In the lamina coordinate system; (c) A unit cell for lamina; (d) Subdivision of an RVE into segments; (e) A segment can be regarded as a UD composite in an off-axial angle; (f) Analysis of the segment in its local coordinate system(from Huang & Fujihara, 2005)

References

Bigelow, C.A. (1993) Thermal residual stresses in a silicon-carbide/ titanium [0/90] laminate. ASTM J. Comp. Tech. Res. 15, 304-310.

Hashion, Z. & Rotem, A. (1973) A fatigue failure criterion for fiber reinforced materials. J. Compos. Mater. 7, 448-464.

Hinton, M.J. & Soden, P.D. (1998) Predicting failure in composite laminates: the background to the exercise. Composite Science and Technology 58, 1001-1010.

Hinton, M.J., Kaddour, A.S., Soden, P.D. (2002) A comparison of the predictive capabilities of current failure theories for composite laminates, judged against experimental evidence. Composite Science and Technology 62, 1725-1797.

Hinton, M.J. & Kaddour, A.S. (2011) Triaxial Test Results for Fiber Reinforced Composites: Second World-Wide Failure Exercise Benchmark Data.

Composite Science and Technology (to appear).

Hinton, M.J., Kaddour, A.S., Soden, P.D. (2004) A further assessment of the predictive capabilities of current failure theories for composite laminates: comparison with experimental evidence. Composite Science and Technology 64, 549-588.

Huang, Z.M. (2000a) The mechanical properties of composites reinforced with woven and braided fabrics. Comp. Sci. Tech. 60(4), 479-498.

Huang, Z.M. (2000b) Simulation of inelastic response of multidirectional laminates based on stress failure criteria. Mater. Sci. Tech. 16(6), 692-698.

Huang, Z.M. (2001) Micromechanical life prediction for composite laminates. Mechanics of Materials 33, 185-199.

Huang, Z.M. (2002a) Cyclic response of metal matrix composite laminates subjected to thermomechanical loads. International Journal of Fatigue 24, 463-475.

Huang, Z.M. (2002b) Modeling and characterization of bending strength of braided fabric reinforced Laminates. J. Composite Mater. 36(22), 2537-2566.

Huang, Z.M. (2002c) Fatigue life prediction of a woven fabric composite subjected to biaxial cyclic loads. Composites Part A, 33(2), 253-266.

Huang, Z.M. (2004a) A bridging model prediction of the ultimate strength of composite laminates subjected to biaxial loads. Composite Science and Technology 64, 395-448.

Huang, Z.M. (2004b) Progressive flexural failure analysis of laminated composites with knitted fabric reinforcement. Mechanics of Materials, 36(3), 239-260.

Huang, Z.M. (2004c). Correlation of the bridging model predictions of the biaxial failure strengths of fibrous laminates with experiments. Comp. Sci. & Tech., 64, 529-548.

Huang, Z.M. (2005) Efficient approach to the structure-property relationship of woven and braided fabric reinforced composites up to failure. J. Reinf. Plastics Comp. 2005, 24(12), 1289-1309.

Huang, Z.M. (2007) Inelastic and failure analysis of laminate structures by ABAQUS incorporated with a general constitutive relationship. Journal of Reinforced Plastics and Composites 26, 1135-1181.

Huang, Z.M., Fujihara, K. & Ramakrishna, S. (2003) Tensile stiffness and strength of regular braid composites: Correlation of theory with experiments. J. Comp. Tech. Res., ASTM 25(1), 35-49.

Huang, Z.M. & Fujihara, K. (2005) Stiffness and Strength Design of Composite Bone Plates. Comp. Sci. & Tech., 65(1), 73-85.

Huang, Z.M. & Ramakrishna, S. (2002a) Modeling mechanical properties of knitted fabric composites—Part I: Overview and Geometric Description. Science & Engineering of Composite Materials 2002, 10(3), 163-188.

Huang, Z.M. & Ramakrishna, S. (2002b) Modeling Mechanical Properties of Knitted Fabric Composites—Part II: Theoretical Description, Science & Engineering of Composite Materials 10(3), 189-212.

Huang, Z.M. & Ramakrishna, S. (2002c) Modeling Mechanical Properties of Knitted Fabric Composites—Part III: Applications, Science & Engineering of Composite Materials 10(3), 213-240.

Huang, Z.M. & Ramakrishna, S. (2003) Modeling inelastic and strength properties of textile composites: A unified approach. Comp. Sci. & Tech. 2003, 63(3-4), 445-466.

Kaddour, A.S. & Hinton, M.J. (2011) Instructions to contributors of the Second World-Wide Failure Exercise (WWFE-II): Part (A), Composite Science and Technology (to appear).

Kaddour, A.S., Thompson, L., Li, S. & Hinton, M.J. (2011) Through- thickness compressive behaviour of carbon/epoxy laminates. Composite Science and Technology (to appear).

Kroupa, J.L., Neu, R.W., Nicholas, T., Coker, D., Robertson, D.D. & Mall, S. (1996) A comparison of analysis tools for predicting the inelastic cyclic response of cross-ply titanium matrix composites. In: Johnson W.S. & Cox. B.N. (eds.), Life Prediction Methodology for Titanium Matrix Composites. ASTM STP 1253 ASTM, 297-327.

Lakes, R.S., Lee, T., Bersie, A. & Wang, Y.C. (2001) Extreme damping in composite materials with negative-stiffness inclusions. Nature 410(6828), 565-567.

Lekhnitskii, S.G. (1968) Anisotropic Plates. Trans. from the 2nd Russian ed. by S.W. Tsai, and T. Cheron. New York: Gordon and Breach Science Publishers.

Liu, K.S. & Tsai, S.W. (1998) A progressive quadratic failure criterion of a laminate. Composite Science and Technology 58, 1123-1132.

Mall, S. & Nicholas, T. (eds.). (1997) Titanium Matrix Composites- Mechanical Behavior. Lancaster, Basel: Technomic Publishing Co., Inc.

NASA (1975) Aircraft Field Conservation Technology, Task Force Report. Office of Aeronautics and Space Technology.

Newaz, G.M. & Majumdar, B.S. (1994) A comparison of mechanical response of MMC at room and elevated temperatures. Comp. Sci. Tech. 50, 85-90.

Pister, K.S. & Dong, S.B. (1959) Elastic bending of layered plates, in Proceedings of the American Society of Civil Engineers. J. Eng. Mech. Division 85 (EM4), 1-10.

Reissner, E. & Stavsky, Y. (1961) Bending and stretching of certain types of heterogeneous aeolotropic elastic plates. J. Appl. Mech. 28, 402-408.

Robertson, D.D. & Mall, S. (1996) Incorporating fiber damage in a micromechanical analysis of metal matrix composite laminates. ASTM J. of Comp. Tech. Res. 18, 265-273.

Robertson, D.D. & Mall, S. (1998) Micromechanical analysis and modeling. In: Mall S. & Nicholas T. (eds.), Titanium Matrix Composites—Mechanical Behavior. Lancaster: Technomic 397-464

Rotem, A. (1998) Prediction of laminate failure with Rotem failure criterion. Comp. Sci. Tech. 58, 1083-1094.

Rotem, A. & Hashion, Z. (1976) Fatigue failure of angle ply laminates. AIAA J

14(7), 867-872.

Rotem, A. & Nelson, H.G. (1981) Fatigue Behavior of Graphite-Epoxy Laminates at Elevated Temperatures, Fatigue of Fibrous Composite Materials. ASTM STP 723, American Society for Testing Materials 152-173.

Sendeckyj, G.P., Richardson, M.D. & Pappas, J.E. (1975) Fracture Behavior of Thornel 300/5208 Graphite-Epoxy Laminates—Part 1: Unnotched Laminates, Composite Reliability. ASTM STP 580, 528-546.

Smith, C.B. (1953) Some new types of orthotropic plates laminated of orthotropic materials. J. Appl. Mech. 20, 286-288.

Soden, P.D., Kitching, R., Tse, P.C., Tsavalas, Y. & Hinton, M.J. (1993) Influence on winding angle of the strength and deformation of filament-wound composite tubes subjected to uniaxial and biaxial loads. Comp. Sci. Tech 46, 363-378.

Soden, P.D., Hinton, M.J. & Kaddour, A.S. (1998a) Lamina properties, lay-up configurations and loading conditions for a range of fiber- reinforced composite laminates. Comp. Sci. Tech 58, 1011-1022.

Soden, P.D., Hinton, M.J. & Kaddour, A.S. (1998b) A comparison of the predictive capabilities of current failure theories for composite laminates. Comp. Sci. Tech 58, 1225-1254.

Soden, P.D., Hinton, M.J. & Kaddour, A.S. (2002) Biaxial test results for strength and deformation of a range of E-glass and carbon fibre reinforced composite laminates: failure exercise benchmark data. Comp. Sci. & Tech 62, 1489-1514.

Stavsky, Y. (1964) On the general theory of heterogeneous aeolotropic plates, Aeronautical Quarterly 15, 29-38.

Sun, C.T. & Tao, J.X. (1998) Prediction of failure envelopes and stress/strain behaviour of composite laminates. Comp. Sci. & Tech. 58, 1125-1136.

Timoshenko, S.P. (1940) Theory of Plates and Shells. New York: McGraw Hill.

Wang, Z.M. (1992a) Advanced in Mechanics of Composite Materials and Structures (First Part). Guangzhou: South China University of Technology Press.

Wang, Z.M. (1992b) Advanced in Mechanics of Composite Materials and Structures (Second Part). Guangzhou: South China University of Technology Press.

Watts, A.A. (ed.) (1984) Commercial Opportunities for Advanced Composites, ASTM Special Technical Publication 704. American Society for Testing and Materials.

Webber, R.S., Alderson, K.L. & Evans, K.E. (2000) Novel variations in the microstructure of the auxetic microporous ultra-high molecular weight polyethylene, Part I: Processing and microstructure, Polymer Engineering & Science 40(8), 1894-1905.

Zhang, R.G., Yeh, H.L. & Yeh, H.Y. (1999) A discussion of negative Poisson's ratio design for composites. J. Reinforced Plastics & Composites 18(17), 1546-1556.

Zhang, Y.Z., Bini, T.B., Huang, Z.M. & Ramakrishna, S. (2000) Fracture Characteristics of Knitted Fabric Composites under Tensile Load. Advanced Composites Letters 9(2), 133-137.

Zhang, Y.Z., Huang, Z.M. & Ramakrishna, S. (2001) Tensile Behavior of Multilayer Knitted Fabric Composites with Different Stacking Configuration. Appl. Comp. Mater. 8(4), 279-295.

Zhou, Y.X. & Huang, Z.M. (2008) A modified ultimate failure criterion and material degradation scheme in bridging model prediction for biaxial strength of laminates. Journal of Composite Materials 42(20), 2123-2141.

Zinoviev, P.A., Grigoriev, S.V., Lebedeva, O.V. & Tairova, L.P. (1998) The strength of multilayered composites under a plane-stress state. Composite Science and Technology 58, 1209-1223.

6 Computer Routine Implementation

6.1 Introduction

In this chapter, the analyzing formulae presented in the previous chapters are programmed into a computer routine called CABM (Composite Analysis by Bridging Model) by using the FORTRAN-77 language. The computer program developed here can be used to simulate mechanical properties of a multidirectional laminate (a UD lamina is a special laminate) made of fiber and matrix materials with any but given elastic-plastic properties subjected to arbitrary load conditions. The simulated properties include effective elastic moduli, an overall instantaneous stiffness matrix, a stress-strain curve along any but a specified direction, internal stresses in the fibers and matrix of every layer, the thermal stresses in the fibers and the matrix at a given temperature and each progressive failure strength of the laminate. For the sake of generality, the 3D bridging model formulation and the pseudo 3D laminate theory are employed in the developed program. It is noted that although all of the curvature increments of the laminate are available, integration of them to obtain a deflection of the laminate has not been dealt with in this book. Hence, a bending load-deflection curve of a laminate cannot be directly calculated with the present computer program.

Essentially, the computer routine consists of three modules, i.e., data input module, solution module and result output module. In Section 6.2, all of the three modules together with necessary subroutines are described in as much detail as possible. Explanation of the input data as well as the input format required by running the routine is shown in Section 6.3. A complete list of the original computer code is given in Section 6.4. Finally, in Section 6.5, several examples containing input files and output results are illustrated for easy reference.

6.2 Description of the Computer Routine

The computer routine is programmed in FORTRAN 77 language, consisting of 31 subroutines most of which are self-contained, and can be compiled in most of the current FORTRAN compilers. However, it should be noted that in the subroutines COUPLE and STRTH, the external subroutines DNEQNF and DZPLRC from the IMSL FORTRAN 90 MP library are employed, respectively, to perform standard mathematical computations. The user should make sure that the mathematical library (the IMSL FORTRAN 90 MP library or a higher version) has been correctly installed on his/her computer before compiling this routine. For more information on this library, the user can refer to the website http://www.vni.com. A compiled executable routine (.exe) is provided on the attached CD-ROM. Using it or transferring it into a local folder, the reader, even without installing a FORTRAN compiler, can directly run the computer routine to do his analysis. A short introduction in PDF format telling how to work with the executable routine can be found on the CD-ROM, and the original code of the computer routine is also included.

A flowchart for the routine structure is indicated in Fig. 6.1. As can be seen from the figure, the routine is separated into an input, a solution and an output module, which are described one by one in the following.

6.2.1 Main Routine and Data Input Module

In the main routine, the input and output files are defined and a number of characterizing parameters, such as the maximum number of layers to identify a laminate, the maximum temperature points to specify different sets of the fiber and matrix properties and the maximum linear segments to approximate a matrix stress/strain curve, are pre-set. The whole analysis procedure then starts by calling the other subroutines.

The Data Input Module contains five subroutines, i.e., SOLVE, PARAM, GLOBAL, RESID and INITIL0. All of the input data are read in this module. In the subroutine SOLVE, the controlling parameters for the problem to be analyzed, including the number of layers in the laminate, the number of incremental solution steps used, the parameter to control whether any residual stresses are included, and so on, are defined (for more details refer to Section 6.3). The subroutine PARAM is used to generate the elastic-plastic properties of the fiber and matrix materials for the problem, whereas the subroutine GLOBAL is used to form the coordinate transformation matrices based on the lay-up angles as illustrated in Section 1.5. To assign all of the working arrays at the beginning to zero is done by the subroutine INITIL0. If there are any residual stresses occurring in the fiber and matrix materials of each layer, the subroutine RESID can be called to achieve the purpose.

Once the input data are completed, the subroutine PROSS is called to start the solution module.

6.2.2 Solution Module

This module is the core of this computer routine, and is represented by the subroutine PROSS, which works by calling a number of other subroutines. A flowchart for the solution module is shown in Fig. 6.2, where for each solution process the corresponding subroutines are listed using "Capital Bold" characters. Before going into any details of the subroutines, it deserves mentioning that when the working temperature is different from the stress free temperature, a solution process should be carried out firstly to obtain the thermal residual stresses as shown in Fig. 6.1. The analysis procedure to obtain the thermal residual stresses is the same as that for the internal stresses to be illustrated in this module. The only difference is that only the 'thermal load' induced by the temperature variation is involved. An illustration of the flowchart shown in Fig. 6.2 is given as follows.

(1) First of all, as the fiber and matrix properties can be dependent on the stress and temperature states, the input material parameters must be adjusted to define the fiber and matrix properties corresponding to the current stress state and the current temperature. This will be done by the subroutines FIBERS and MATRIX, respectively. Furthermore, the instantaneous compliance matrices of the fiber and matrix materials are defined by calling the subroutines ELASC, PLASC, STATUS and ELAPS if necessary. The functions of these subroutines, respectively, are:

    ELASC: to form an elastic compliance matrix;

    PLASC: to form the plastic part of an instantaneous compliance matrix;

    STATUS: to judge whether the material is under an essential tension or essential compression;

    ELAPS: to assemble the elastic and plastic parts to obtain the instantaneous compliance matrix.

(2) After determination of the constituent material properties, the instantaneous bridging matrix and the local compliance matrix as well as the thermal expansion coefficients of each lamina layer can be calculated by using the bridging model. This process is done in the subroutine UDTAPE. The other subroutines to be called in UDTAPE are BRIDG2, COUPLE, BRIDGE, ZERO, DNEQNF, FCN, BRIGA, BRIDGE and THERM.

    BRIDGE2: a subroutine to obtain a compliance matrix of a UD lamina in its local coordinate system by using the bridging model;

    COUPLE: a subroutine to determine a 3D bridging matrix by calling other subroutines, i.e., ZERO, DNEQNF and BRIDGE;

    ZERO: a subroutine to provide an initial guess for the bridging matrix. The initial guess at the current load step is always taken to be that at the preceding step (except for the first load step at which the initial guess is determined by the subroutine BRIGA);

    DNEQNF: a standard nonlinear equation system solver subroutine provided in the IMSL FORTRAN 90 MP library to solve Eq. (3.71) to determine the dependent elements of the bridging matrix;

FCN: a subroutine to form an external function required in DNEQNF;
BRIGA: a subroutine to calculate a 2D bridging matrix for the initial guess of a 3D bridging matrix at the first load step and thermal stress analysis;
BRIDGE: a subroutine to assign the known elements to a 3D bridging matrix;
THERM: a subroutine to define the thermal expansion coefficient vectors and the thermal stress concentration factors of the fiber and matrix materials of a lamina.

(3) When the compliance matrix of each layer is available, the subroutine OVERSTIF is used to calculate the lamina global stiffness matrix and a global thermal stress concentration vector. In this subroutine, the subroutine INVER is called to evaluate a stiffness matrix from a given compliance matrix.

(4) The subroutine STRESS is used to solve the strain and curvature increments of the laminate and then to obtain the stress increments in each layer. To do this, the subroutines MECHF and TEMPF are employed first to obtain the applied mechanical and thermal load increments, respectively. Then the equilibrium equations, Eqs. (5.26), (5.27) and (5.29), are formed and solved to give the strain and curvature increments. This latter work is done by means of the subroutines LAMINATE and EQUA. A Gaussian elimination scheme, implemented into a subroutine GAUSS, is used to solve the simultaneous linear algebraic equations. Finally, the stress increments shared by each layer of the laminate are calculated using the strain and curvature increments and the global stiffness matrix of the layer.

(5) The subroutine UPDATE is used to derive the internal stress increments in the constituent fiber and matrix materials of each layer. Then the total internal stresses in the constituents are updated.

(6) Failure detection for each lamina as well as for the laminate is performed in the subroutine STRTH. The micromechanical failure criteria for detecting a lamina failure presented in Chapter 4 are implemented in this subroutine. Two steps of failure detection are performed. In the first step, an ultimate failure for the laminate is assessed. Namely, it is checked whether a fiber tensile or compressive failure, a matrix compressive failure or an extreme strain limitation is attained. If any of these ultimate failure conditions occur, the solution module is terminated and the computer routine will go to the next module, i.e., results module. If an ultimate failure does not occur, the second step failure detection is carried out by assessing whether there is a matrix tensile failure in each layer. Should any such (matrix tensile) failure occur, a stiffness degradation as described in Chapter 5 is applied to the failed lamina layer. When the detection for all of the layers in the laminate is completed, go to (1) for a next load increment analysis.

A relationship interconnecting different subroutines used in the solution module is shown in Fig. 6.3.

6.2.3 Results Module

As aforementioned, a variety of mechanical property parameters of a laminate can be simulated with this computer routine. The output of these parameters depends on the choice of proper controlling parameters specified in the input data. The output process of the simulated results is performed by using the subroutine WRTE. Definition for the input data is illustrated in the next section.

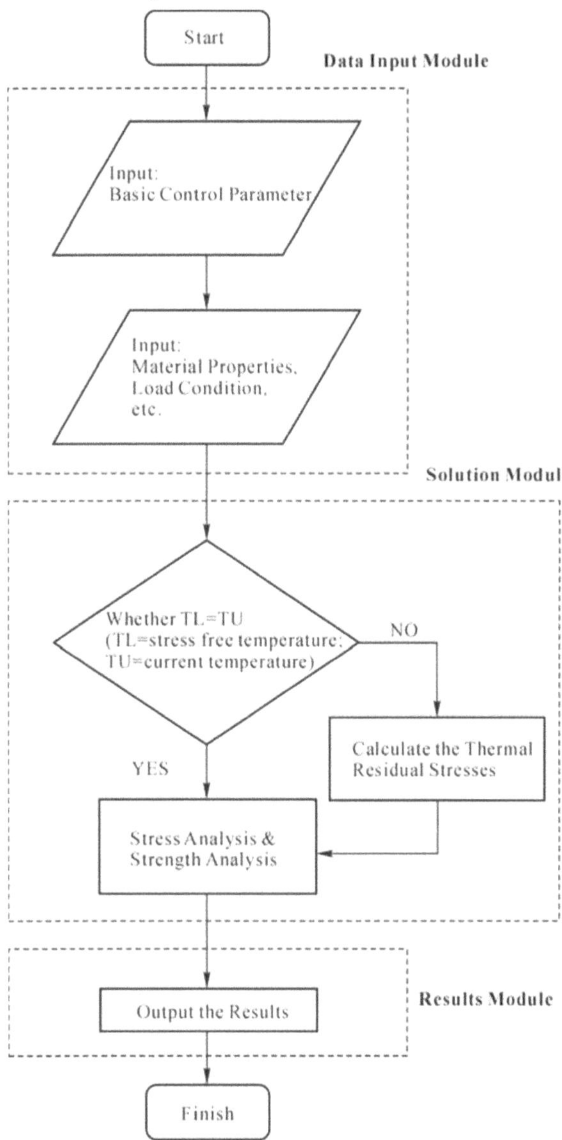

Fig 6.1 Flow Chart for the Computer Routine

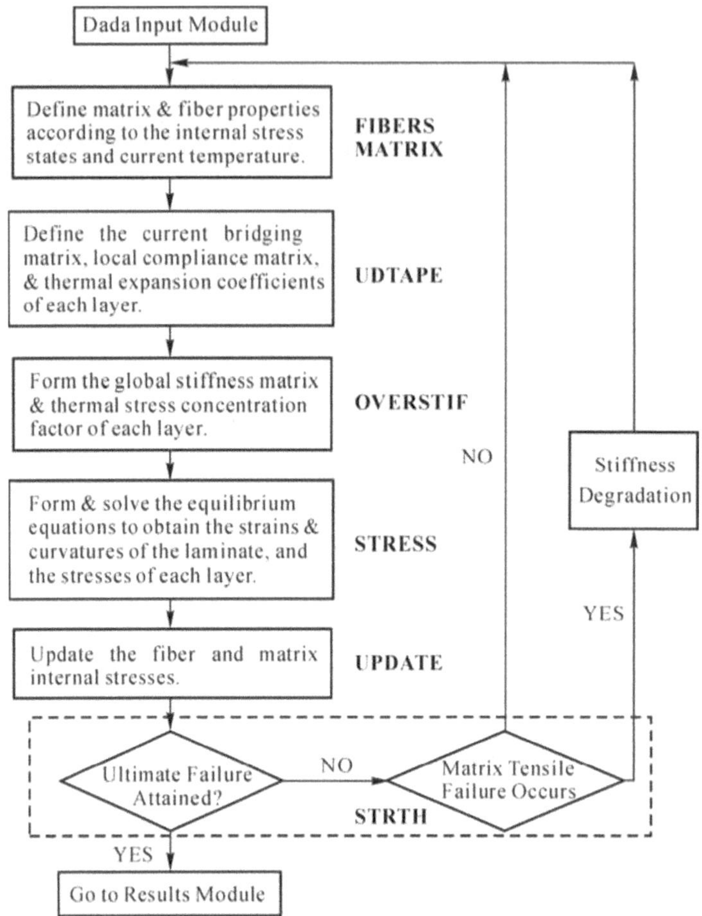

Fig 6.2 Flow Chart for the solution module

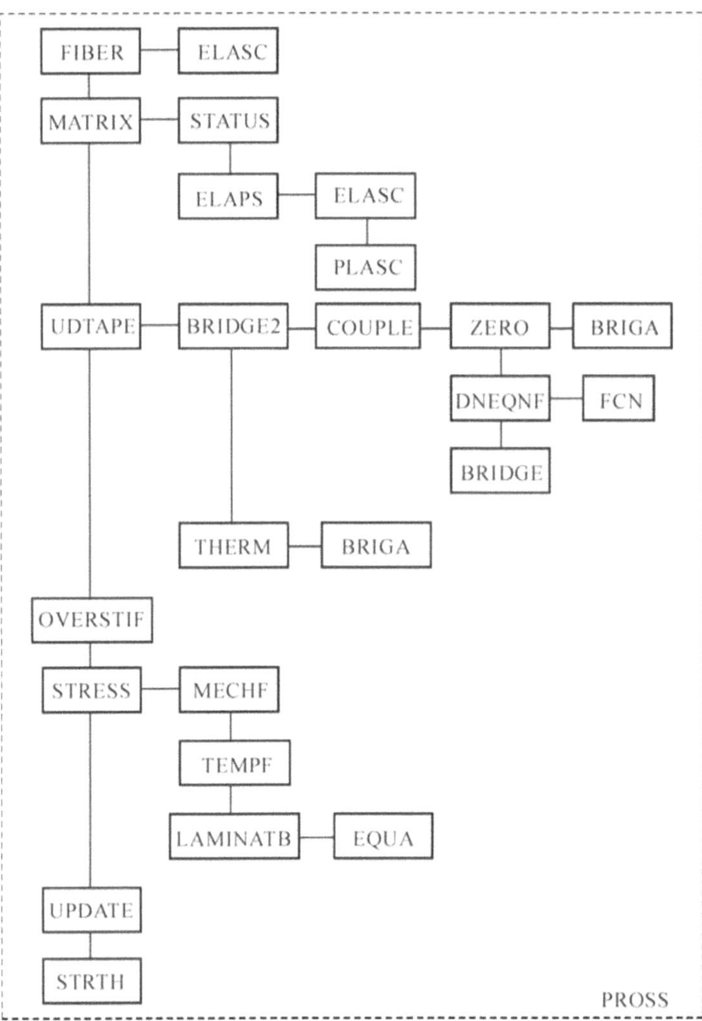

Fig. 6.3 Interrelationship between subroutines used in the solution module (subroutine PROSS)

6.3 Explanation of Input Data

Before compiling or running the computer routine, four data files named "moduli_input.dat", "moduli_output.dat", "deflec1.dat" and "deflec2.dat" should be set ready in the folder where the routine is loaded. All of the input data required for running the routine must be written and stored in the data file "moduli_input.dat" sequentially. Most of the simulated results, such as an overall stiffness matrix of the laminate, effective elastic moduli of each lamina and the laminate, the ultimate failure strength of the laminate, progressive failure information, internal stresses in the constituent materials, are output into the file "moduli_output.dat". The in-plane and the through-thickness stress/strain curves, however, are printed out into the files "deflec1.dat" and "deflec2.dat", respectively.

Units

The **CABM** uses the following units throughout the routine:

MPa for a stress;
N (Newton) for a force or a moment per unit length;
mm (millimeter) for a length;
N/mm for a force per unit length;
degree (not radian) for an angle;
% for a strain;
°C for a temperature;
10^{-6}/°C for a thermal expansion coefficient (e.g., if the thermal expansion coefficient of a material is α=58×10^{-6}/°C, only 58 is provided)

In the routine, the characteristic of a variable is implicitly defined. This means that the variables starting with "I", "J", "K", "L", "M" and "N" stand for integers or integer arrays, whereas those starting with all of the other characters represent real variables or arrays.

All of the following parameters (except for possibly selective ones) much be inputted sequentially and stored in the data file "moduli_input.dat". It is noticed that all of the data are inputted in a free format. Namely, each number (either an integer or a real number) ends with a comma ",".

(1) NQ

q = NQ, the power-index in the generalized maximum normal stress theory, see Eq. (4.5.2).

(2) BETA, ALFA

BETA = β, the bridging parameter in defining the bridging element a_{22}, a_{33} and a_{44}.

ALFA = α, the bridging parameter in defining the bridging elements a_{55} and a_{66}.

(3) NL, NSTP, JPRIT, IDSOL, JSTP, IDRES, IDPLY, IDSEC, ISPLY, NINT

NL = number of layers of the laminate under consideration;
NSTP = incremental steps for the considered problem;
JPRIT = 0, no print;
= –1, print stress/strain curve plus the compliance matrices at the first step;
= –2, print stress/strain curve plus the compliance matrices at the first and the last steps;
= –3, only print stress/strain curve;
= 1, print the compliance matrices at the first step;
= 2, print the compliance matrices at the first and the last steps;
= 4, print internal stresses at the last step;

Note: when the stress/strain curves are printed (JPRIT = –2 or –3), all of the in-plane stresses and strains are output into the file 'deflec1.dat' and the outputs are carried out once every JSTP steps. Whether or not the out-of-plane stresses and strains are output into the file 'deflec2.dat' depends on another controlling parameter ISPLY. Namely, the in-plane strains and the out-of-plane strains are output independently.

IDSOL = 1, first apply thermal and then isothermal mechanical loads;
= 2, coupled thermo-mechanical analysis;

JSTP > 0, print stress/strain curve once after 'JSTP' steps

IDRES = 0, no additional residual stresses;
= 1, additional residual stresses to be inputted;

IDPLY = 0, each layer has the same fiber volume fraction;
= 1, each layer has a different fiber volume fraction;

IDSEC = 0, each layer has the same width;
= 2, each layer has a different width to be inputted independently;

Note: if each layer has a different width, stress transfers, especially those of the through-thickness shear stresses between the layers in the framework of the 3D theory, are quite complicated. Thus, if a through-thickness load is considered, IDSEC must be set to 0.

ISPLY = the number of the layer of which the out-of-plane shear strain/stress curves together with the out-of-plane normal strain/stress curve of the laminate are output into the file 'deflec2.dat';
= 0, no print for the through-thickness strain/stress curves;

Note: when ISPLY=J>0, the three out-of-plane stresses, the out-of-plane normal strains of the laminate and the out-of-plane shear strains of the J-th layer are to be output into the file 'deflec2.dat' at every JSTP load step.

NINT =0, pure matrix inter-layers are not introduced;
=1, pure matrix inter-layers are incorporated;

(4) Z(1, 1), Z(1, 2), …, Z(1, NL+1)

Z(1, I) = Z-coordinate (through-thickness direction) of the I-th layer's lower surface, see Section 5.2 for definition;

Z(1, I+1) = Z-coordinate (through-thickness direction) of the I-th layer's upper surface;

(No matter whether or not the pure matrix inter-layers are introduced, NL always is the number of the primary layers. Only the Z-coordinates of the primary layers are specified in the input data)

(5) Z(2, 1), Z(2, 1), Z(2, 2), …, Z(2, NL)

Z(2,1)= width of the laminate if IDSEC=0, i.e., if each layer has the same width;

Z(2, I) = width of the I-th layer if IDSEC=2, i.e., if each layer has a different width;

(6) ANGEL(1), ANGEL(2), …, ANGEL(NL)

ANGEL(I)= inclined angle of the fiber axis of the I-th layer with the global x-axis, see Section 5.2 for definition;

(7) TL, TU, T0, T1

TL = stress free temperature;

TU = the temperature at which the external loads are applied;

T0 = initial working temperature (usually T0 = TU);

T1 = final working temperature;

(8) NSUB

NSUB= subintervals to cover [TL, TU]

(9) SSMIN(1), SSMIN(2), … , SSMIN(6), SSMAX(1),SSMAX(2), … , SSMAX(6)

SSMIN(1), SSMIN(2), … , SSMIN(6)= $N_{xx}^{\min}, N_{yy}^{\min}, N_{xy}^{\min}, \sigma_{yz}^{\min}, \sigma_{xz}^{\min}, \sigma_{zz}^{\min}$ the minimum applied loads;

SSMAX(1),SSMAX(2), … , SSMAX(6)= $N_{xx}^{\max}, N_{yy}^{\max}, N_{xy}^{\max}, \sigma_{yz}^{\max}, \sigma_{xz}^{\max}, \sigma_{zz}^{\max}$ the maximum applied loads.

The actual applied loads on the laminate are (SSMAX(I)-SSMIN(I)) for I=1, 2, …, 6. If IDSEC=0, i.e., if each layer has the same width, the in-plane force loads applied, $N_{xx}^{\min}, N_{yy}^{\min}, N_{xy}^{\min}, N_{xx}^{\max}, N_{yy}^{\max}, N_{xy}^{\max}$, should be replaced by the stress loads, $\sigma_{xx}^{\min}, \sigma_{yy}^{\min}, \sigma_{xy}^{\min}, \sigma_{xx}^{\max}, \sigma_{yy}^{\max}$, $\sigma_{xy}^{\max}$.

(10) NEF

NEF = number of different temperatures at which different fiber properties (longitudinal modulus, tensile and compressive strengths, longitudinal and transverse thermal expansion coefficients) are provided. If the fiber properties are independent of temperature, set *NEF* = 2.

(11) Fiber properties: input the following 6×*NEF*+5 quantities sequentially.

$E_{11}^f, \nu_{12}^f, E_{22}^f, \nu_{23}^f, G_{12}^f$, fiber elastic properties;

T_1, …, T_{NEF} = temperatures at which the fiber assumes different properties;

$E_{11,1}^f$, …, $E_{11,NEF}^f$ = the longitudinal moduli of the fiber assumed at the *NEF* temperatures, respectively;

$\sigma_{u,1}^f$, …, $\sigma_{u,NEF}^f$ = the longitudinal tensile strengths of the fiber assumed at the *NEF* temperatures, respectively;

$\sigma_{u,c,1}^f$, …, $\sigma_{u,c,NEF}^f$ = the longitudinal compressive strengths of the fiber assumed at the *NEF* temperatures, respectively;

$\alpha_{1,1}^f$, …, $\alpha_{1,NEF}^f$ = the longitudinal thermal expansion coefficients of the fiber assumed at the *NEF* temperatures, respectively;

$\alpha_{2,1}^f$, …, $\alpha_{2,NEF}^f$ = the transverse thermal expansion coefficients of the fiber assumed at the *NEF* temperatures, respectively.

If the fiber properties are independent of temperature, the temperature range $[T_1, T_{NEF}]$ is assigned as large enough and the fiber properties can be set to be the same at the temperatures T_1 and T_{NEF}.

(12) MSEG, NEM

MSEG= number of linear segments to form a stress-strain curve of the matrix;

NEM= number of different temperatures at which different matrix properties (Poisson's ratio, thermal expansion coefficients, tensile and compressive strengths as well as the stress-strain curves) are provided.

(13) Matrix properties: input the following $(4\times MSEG+5)\times NEM$ quantities sequentially:

1st line: $T_1, \ldots, T_{NEM}$ = temperatures at which the matrix assumes different properties;

2nd line: $\alpha_1^m, \ldots, \alpha_{NEM}^m$ = thermal expansion coefficients of the matrix assumed at the *NEM* temperatures, respectively;

3rd line: $\nu_1^m, \ldots, \nu_{NEM}^m$ = Poisson's ratios of the matrix assumed at the *NEM* temperatures, respectively;

4th line: $\sigma_{u,1}^m, \ldots, \sigma_{u,NEM}^m$ = tensile strengths of the matrix assumed at the *NEM* temperatures, respectively;

5th line: $\sigma_{u,c,1}^m, \ldots, \sigma_{u,c,NEM}^m$ = compressive strengths of the matrix assumed at the *NEM* temperatures, respectively;

6th line: $\sigma_{Y,L,1}^{m,1}, \ldots, \sigma_{Y,L,NEM}^{m,1}$ = the first end stresses (i.e., yield strengths) of the matrix tensile stress-strain curves (linear segments) at the *NEM* temperatures, respectively;

……………………………………..

(*MSEG*+5)th line: $\sigma_{Y,L,1}^{m,MSEG}, \ldots, \sigma_{Y,L,NEM}^{m,MSEG}$ = the last end stresses of the matrix tensile stress-strain curves (linear segments) at the *NEM* temperatures, respectively;

(*MSEG*+6)th line: $E_{T,L,1}^{m,1}, \ldots, E_{T,L,NEM}^{m,1}$ = the tangent moduli (i.e. Young's moduli) of the first segments of the matrix tensile stress-strain curves (linear segments) at the *NEM* temperatures, respectively;

…………………………………….

(2×*MSEG*+5)th line: $E_{T,L,1}^{m,MSEG}, \ldots, E_{T,L,NEM}^{m,MSEG}$ = the tangent moduli of the last segment of the matrix tensile stress-strain curves (linear segments) at the *NEM* temperatures, respectively;

(2×*MSEG*+6)th line: $\sigma_{Y,C,1}^{m,1}, \ldots, \sigma_{Y,C,NEM}^{m,1}$ = the first end stresses (i.e., yield strengths) of the matrix compressive stress-strain curves (linear segments) at the *NEM* temperatures, respectively;

……………………………………

(3×*MSEG*+5)th line: $\sigma_{Y,C,1}^{m,MSEG}$, …, $\sigma_{Y,C,NEM}^{m,MSEG}$ = the last end stresses of the matrix compressive stress-strain curves (linear segments) at the *NEM* temperatures, respectively;

(3×*MSEG*+6)th line: $E_{T,C,1}^{m,1}$, …, $E_{T,C,NEM}^{m,1}$ = the tangent moduli (i.e., Young's moduli) of the first segments of the matrix compressive stress-strain curves (linear segments) at the *NEM* temperatures, respectively;

……………………………………

(4×*MSEG*+5)th line: $E_{T,C,1}^{m,MSEG}$, …, $E_{T,C,NEM}^{m,MSEG}$ = the tangent moduli of the last segments of the matrix compressive stress-strain curves (linear segments) at the *NEM* temperatures, respectively.

If the matrix properties are independent of temperature, the same treatment as for the fiber material can be done. Namely, set *NEM* = 2 and the temperature range [T_1, T_{NEM}] to be large enough. The matrix properties can be assigned to be the same at the temperatures of T_1 and T_{NEM}.

(14) VF(1), VF(2), … ,VF(NL)

VF(1) = fiber volume fraction of the laminate (if IDPLY=0, i.e., if all of the layers have the same fiber volume fraction);

VF(I) = fiber volume fraction of the *I*-th layer (if IDPLY=1, i.e., if each layer has a different fiber volume fraction)

(15) SSF(1,1), SSF(1,2), …, SSF(1,6), SSM(1,1), SSM(1,2), …, SSM(1,6), SSF(2,1), SSF(2,2), …, SSF(2,6), SSM(2,1), SSM(2,2), …, SSM(2,6),

…………………………

SSF(NL*,1), SSF(NL*,2), …, SSF(NL*,6), SSM(NL*,1), SSM(NL*,2), …, SSM(NL*,6),

These are residual stresses in the constituent materials (to be inputted only if IDRES=1). It should be noted that if NINT=1, i.e., if the pure matrix inter-layers are introduced, the residual stresses in all of the primary layers and the pure matrix inter-layers should be inputted, i.e., NL^*=2×NL–1. If NINT=0, i.e., if the pure matrix inter-layers are not incorporated, NL^*=NL.

6.4 Original Code of the Computer Routine

```
C---------------------------------------------------------------------------
C ******* A PROGRAM TO ANALYSE MULTIPLIED LAMINATE    *******-
C      PROGRESSIVE FAILURE STRENGTHS ARE PREDICTED          -
```

```
C      FIBER IS TRAINSVERSELY ISOTROPIC, LINEARLY ELASTIC     -
C      MATRIX IS ISOTROPIC, MULTI-SEGMENT ELASTO-PLASTIC     -
C      BOTH FIBER & MATRIX CAN BE TEMPERATURE DEPENDENT
C***    THE LAMINATE CAN BE SUBJECTED TO 3D LOADS          -
C***    Unit--Stress(Strength) & Moduli:MPa, Temperature:C,       ** -
C**    Angle:degree, length(coordinate):mm, Strain:%, CTE:10**(-6),     **-
C*******************************************************************
       PARAMETER (ML=100, MUM=25, MUF=7, MEM=10, MEF=10)
C
C***ML–MAXIMUM NUMBER OF PLIES
C***MUM = 5+2*Mseg,
C***Mseg = LINEAR SEGMENTS OF THE MATRIXSTRESS/STRAIN CURVE
C***MAXIMUM Mseg = (MUM–5)/2
C***MEM = MAXIMUM TEMP. POINTS TO SPECIFY MATRIX PROPERTIES
C***MEF = MAXIMUM TEMP. POINTS TO SPECIFY FIBER PROPERTIES
C
       IMPLICIT DOUBLE PRECISION (A-H, O-Z)
C
         COMMON /DATA2/WK(100)
         COMMON /DATA3/WK1(320)
       OPEN(10, FILE = 'moduli_input.dat')
       OPEN(20, FILE = 'moduli_output.dat')
       OPEN(15, FILE = 'deflec1.dat')
        OPEN(25, FILE = 'deflec2.dat')
       CALL SOLVE(ML, MUM, MUF, MEM, MEF)
       STOP
       END
C***********************************************************
       SUBROUTINE SOLVE(ML, MUM, MUF, MEM, MEF)
       IMPLICIT DOUBLE PRECISION (A-H, O-Z)
       DIMENSION RESIN(2, MUM, MEM), FIBER(MUF, MEF), SSF(ML, 6),
      &              SSM(ML, 6), SSP(ML, 6), VF(ML), LAYER(ML), ZT(ML+1),
      &              Z(2, ML+1), ANGLE(ML), SS(6), SA(6), CG(ML, 6, 6),
      &    TS(ML, 6, 6), TC(ML, 6, 6), ALFT(ML, 6), FAIL(ML, 12), ANGLE1(ML),
      &   SSMIN(6), SSMAX(6), SA1(ML, 3), DSS(6), ALFF(ML, 3), ALFM(ML, 3)
       DATA LFAIL/0/, DSS/6*0.0/
       READ(10, *) NQ
      IF(NQ.EQ.0) NQ = 3
       WRITE(20, 250) NQ
      READ(10, *) A_BETA, A_ALFA
       IF(A_ALFA.GE.1..OR.A_ALFA.LE.0.) A_ALFA = 0.5
       IF(A_BETA.GE.1..OR.A_BETA.LE.0.) A_BETA = 0.5
      WRITE(20, 260) A_BETA, A_ALFA
C*** NQ>0,
C*** INDEX IN MODIFIED MAXIMUM NORMAL STRESS CRITERION
C*** A_BETA & A_ALFA = Bridging parameters
       READ(10, *) NL, NSTP, JPRIT, IDSOL, JSTP, IDRES, IDPLY,
      &               IDSEC, ISPLY, NINT
      IF(NL.GT.ML) THEN
          WRITE(20, 2030) NL, ML
2030      FORMAT(15X, 'ACTUAL LAYERS', I3, 2X,
```

```
     &      'MORE THAN ALLOWABLE', I3)
            STOP
      ENDIF
C
C*** NL = LAYERS IN THE LAMINATE; NSTP = INCREAMENTAL STEPS
C*** JPRIT = 1, PRINT COMPLIANCE & MODULUS AT FIRST STEP
C*** = 2, PRINT COMPLIANCE & MODULUS AT FIRST AND LAST STEP
C*** = –1, PRINT STRESS/STRAIN CURVE PLUS FIRST STEP MODULI
C** = –2, PRINT STRESS/STRAIN CURVEPLUS FIRST&LAST STEP MODULI
C*** = –3, ONLY PRINT STRESS(idss)/STRAIN(idss) CURVE
C*** = 4, ONLY PRINT INTERNAL STRESSES AT THE LAST STEP
C*** = 0, NO PRINT
C*** = –4, PRINIT INTERNAL STRESSES
C*** IDSOL = 1, FIRSTLY APPLY THERMAL,
C***                THEN ISOTHERMAL MECHANICAL LOADS;
C***           = 2, COUPLED THERMO-MECHANICAL ANALYSIS
C***               (e.g., TMF problem)
C***               (THERMAL RESIDUAL STRESSES INCLUDED
C***                IF "TL" NOT EQUAL TO "TU")
C*** IDRES = 0, NO ADDITIONAL RESIDUAL STRESS
C***         = 1, ADDITIONAL RESIDUAL STRESSES TO BE INPUTED
C*** JSTP   > 0, PRINT STRESS(idss)/STRAIN(idss) CURVE,
C***              ONCE AFTER "JSTP" STEPS
C***         = 0, NOT PRINT
C*** IDPLY = 0, EACH PLY HAS SAME GEOMETRIC PARAMETERS
C***        = 1, EACH PLY HAS DIFFERENT GEOMETRIC PARAMETERS
C*** IDSEC = 0, EACH PLY HAS THE SAME WIDTH
C***         = 1, EACH PLY HAS DIFFERENT WIDTH,
C***              BEING INPUTED INDEPENDENTLY
C*** NINT = 1, INTER–LAYER IS CONSIDERED
C***       = 0, INTER–LAYER IS NOT CONSIDERED
C
      READ(10, *) (Z(1, I), I = 1, NL+1)
            IF(NINT.EQ.1.AND.NL.NE.1) THEN
               DO 3020 I = 1, NL+1
3020                 ZT(I) = Z(1, I)
                 DO I = 1, NL–1
                     Z(1, 2*I) = ZT(I+1)–(ZT(I+1)–ZT(I))/40.
                     Z(1, 2*I+1) = ZT(I+1)+(ZT(I+2)-ZT(I+1))/40.
               END DO
                     Z(1, 2*NL) = ZT(NL+1)
            ENDIF
      IF(IDSEC.NE.1) THEN
            READ(10, *) Z(2, 1)
          DO I = 2, NL
                   Z(2, I) = Z(2, 1)
          END DO
      ELSE
            READ(10, *) (Z(2, I), I = 1, NL)
      ENDIF
            IF(NINT.EQ.1.AND.NL.NE.1) THEN
```

```
                DO 3030 I = 1, NL
3030                ZT(I) = Z(2, I)
                  DO 3031 I = 1, NL–1
                    Z(2, 2*I–1) = ZT(I)
3031                Z(2, 2*I) = (ZT(I)+ZT(I+1))/2.
                    Z(2, 2*NL–1) = ZT(NL)
           ENDIF
C
C*** Z(1, i)–Z-COORDINATE OF THE i-th LAYER;
C*** Z(2, i) –WIDTH OF THE i-th LAYER
C
      READ(10, *) (ANGLE1(I), I = 1, NL)
      READ(10, *) TL, TU, T0, T1
      READ(10, *) NTSUB
      READ(10, *) (SSMIN(I), I = 1, 6), (SSMAX(I), I = 1, 6)
     IF(NINT.EQ.1.AND.NL.NE.1) THEN
     IF(NINT.EQ.1) THEN
     DO 3000 I = 1, NL–1
     ANGLE(2*I–1) = ANGLE1(I)
3000 ANGLE(2*I) = 0.
     ANGLE(2*NL–1) = ANGLE1(NL)
     ENDIF
     ELSE
     DO 3001 I = 1, NL
3001  ANGLE(I) = ANGLE1(I)
      ENDIF
C*** ANGLE(i)–INCLINED ANGLE OF i-th PLY X-AXIS
C***                        WITH THE GLOBAL X-AXIS
C*** TL–STRESS FREE TEMPERATURE;
C*** TU–PRESENT TEMPERATURE (TO APPLY LOAD)
C*** T0–INITIAL WORKING TEMP. (USUALLY = TU);
C*** T1–FINIAL WORKING TEMP.
C*** NTSUB–SUBINTERVALS TO COVER [TL, TU]
C*** SSMIN–MINIMUM APPLED STRESSES ON THE LAMINATE
c*** SSMAX–MAXIMUM APPLED STRESSES ON THE LAMINATE
      CALL PARAM(ML, MUM, MUF, MEM, MEF, NEM,
     &             NEF, MSEG, RESIN, FIBER)
     IF(IDPLY.NE.0) THEN
           READ(10, *) (VF(IL), IL = 1, NL)
     ELSE
           READ(10, *) VF(1)
           DO 100 IL = 1, NL
                 VF(IL) = VF(1)
 100       continue
     ENDIF
     IF(NINT.EQ.1.AND.NL.NE.1) THEN
     VF1 = VF(1)
     VF(1) = VF1*100./97.5
     DO 3010 I = 1, NL–1
     VF(2*I+1) = VF1*100./95.0
3010 VF(2*I) = 0.00000001
```

```
      VF(2*NL-1) = VF1*100./97.5
      ENDIF
C
      NL2 = NL
      IF(NINT.EQ.1) NL = 2*NL2-1
C
      DO 25 IL = 1, NL
      LAYER(IL) = 1
      DO 25 J = 1, 3
25    SSP(IL, J) = 0.
c
       CALL GLOBAL(ML, NL, ANGLE, TS, TC)
c
       WRITE(20, 70)
       DO 80 IL = 1, NL
       ALFF(IL, 3) = 0.
       ALFM(IL, 3) = 0.
 80    WRITE(20, 90) IL, ANGLE(IL), VF(IL), Z(1, IL), Z(1, IL+1)
c
       CALL INITIL0(ML, NL, SSF, SSM, SA, SA1, SS)
       IF(IDRES.GT.0) CALL RESID(ML, NL, SSF, SSM)
       IF(TL.NE.TU) THEN
       WRITE(20, 10) TL, TU
       CALL PROSS(ML, NL2, MUM, MUF, MEM, MEF, NL, NEM, NEF, MSEG, 4,
      &     LFAIL, NTSUB, NQ, TL, TU, SS, SA, DSS, A_BETA, A_ALFA, dis_z,
      &     RESIN, FIBER, SSF, SSM, SSP, ISPLY, VF, LAYER, Z, CG, TS,
      &     TC, ALFT, FAIL, SA1, NINT, JSTP)
      do 35 I = 1, 3
       DO 351 J = 1, NL
351   SA1(J, I) = 0.
 35    SA(I) = 0.
       dis_z = 0.
       ENDIF
       IF(IDSOL.EQ.1) THEN
      IF(ABS(T1-T0).GT.0.) THEN
       WRITE(20, 20) T0, T1
       CALL PROSS(ML, NL2, MUM, MUF, MEM, MEF, NL, NEM, NEF, MSEG, 4,
      &     LFAIL, NTSUB, NQ, T0, T1, SS, SA, DSS, A_BETA, A_ALFA, dis_z,
      &     RESIN, FIBER, SSF, SSM, SSP, ISPLY, VF, LAYER, Z, CG, TS,
      &     TC, ALFT, FAIL, SA1, NINT, JSTP)
       WRITE(20, 40)
      ENDIF
       do 302 i = 1, 6
302    DSS(I) = (SSMAX(I)-SSMIN(I))/REAL(NSTP)
       CALL PROSS(ML, NL2, MUM, MUF, MEM, MEF, NL, NEM, NEF, MSEG,
      &     JPRIT, LFAIL, NSTP, NQ, T1, T1, SS, SA, DSS, A_BETA,
      &     A_ALFA, dis_z, RESIN, FIBER, SSF, SSM, SSP, ISPLY, VF,
      &     LAYER, Z, CG, TS, TC, ALFT, FAIL, SA1, NINT, JSTP)
       ELSE
       WRITE(20, 50) T0, T1
       do 602 i = 1, 6
```

```
602   DSS(I) = (SSMAX(I)–SSMIN(I))/REAL(NSTP)
      CALL PROSS(ML, NL2, MUM, MUF, MEM, MEF, NL, NEM, NEF, MSEG,
     &       JPRIT, LFAIL, NSTP, NQ, T0, T1, SS, SA, DSS, A_BETA,
     &       A_ALFA, dis_z, RESIN, FIBER, SSF, SSM, SSP, ISPLY, VF,
     &       LAYER, Z, CG, TS, TC, ALFT, FAIL, SA1, NINT, JSTP)
      ENDIF
      RETURN
 10   FORMAT(/10X, 'THERMAL RESIDUAL STRESSES COOLING
     &          FROM', 1X, F8.2, 3X, 'TO', 1X, F8.2/)
 20   FORMAT(/10X, 'DECOUPLED SOLUTION BY FIRST APPLYING',
     &         ' TEMPERATURE FROM', F6.1, 2X, 'TO', 2X, F6.1/)
 40   FORMAT(3X, 'ABOVE STRESSES ARE TOTAL RESIDUAL
     &  STRESSES', ' BEFORE APPLYING ANY MECHANICAL LOAD'/)
 50   FORMAT(/'COUPLED THERMO-MECHANICAL SOLUTION FROM',
     &          ' INITIAL TEMPERATURE', F6.1, 2X, 'TO FINAL',
     &          ' TEMPERATURE', F6.1/)
 70   FORMAT(/5X, 'PLY', 3X, 'PLY-ANGLE', 5X, 'Vf', 8X, 'Z(i)', 8X, 'Z(i+1)')
 90   FORMAT(4X, I3, 4X, F6.2, 7X, F5.3, 5X, F6.3, 7X, F6.3, 4X, F7.2)
 250  FORMAT(1X, 'INDEX IN MODIFIED MAXIMUM NORMAL
     &         STRESS CRITERION = ', I3)
 260  FORMAT(1X, 'Beta(in defining bridging elements a2) = ', F5.3,
     &          2X, 'Alfa(in defining a3) = ', F5.3/)
 2000 FORMAT(15X, 'ACTUAL LAYERS', I3, 2X,
     &         'MORE THAN ALLOWABLE', I3)
      END
C*********************************************************
      SUBROUTINE PARAM(ML, MUM, MUF, MEM, MEF, NEM, NEF,
     &                     MSEG, RESIN, FIBER)
C*** TO DETERMINE FIBER & MATRIX EFFECTIVE MODULI OF
C*** EACH PLY, FIBER & MATRIX COMPLIANCES, ULTIMATE
C*** STRENGTHS, AND THERMAL EXPANSION COEFFICIENTS
C*** USING CURRENT STRESS STATES, TEMPERATURE, STRAIN RATES
C*** SUPPOSE THE LAMINATE CONSISTS OF TWO PHASE MATERIALS
C*** MUM–MAXIMUM SETS OF MATRIX PARAMETERS
C*** MUF–MAXIMUM SETS OF FIBER PARAMETERS
C*** MEM–MAXIMUM NUMBER OF ONE PARAMETER SET FOR MATRIX
C*** MEF–MAXIMUM NUMBER OF ONE PARAMETER SET FOR FIBER
C*** NEM–ACTUAL NUMBER OF ONE PARAMETER SET FOR MATRIX
C*** NEF–ACTUAL NUMBER OF ONE PARAMETER SET FOR FIBER
C*** MSEG–NUMBER OF SEGMENTS FOR MATRIX
C***             STRESS/STRAIN CURVE
      IMPLICIT DOUBLE PRECISION (A-H, O-Z)
      DIMENSION RESIN(2, MUM, MEM), FIBER(MUF, MEF), ID(5)
     ID(1) = 1
     ID(2) = 2
      READ(10, *) NEF
      IF(NEF.GT.MEF.OR.MUF.LT.7) THEN
      NUF = 7
      GOTO 2000
      ENDIF
      READ(10, *) (FIBER(1, I), I = 1, 5)
```

```
      WRITE(20, 200) (FIBER(1, I), I = 1, 5), NEF
      DO 20 J = 2, 7
      READ(10, *) (FIBER(J, I), I = 1, NEF)
 20   CONTINUE
      WRITE(20, 110) (FIBER(2, I), I = 1, NEF)
      WRITE(20, 120) (FIBER(3, I), I = 1, NEF)
      WRITE(20, 130) (FIBER(4, I), I = 1, NEF)
      WRITE(20, 140) (FIBER(5, I), I = 1, NEF)
      WRITE(20, 150) (FIBER(6, I), I = 1, NEF)
      WRITE(20, 160) (FIBER(7, I), I = 1, NEF)

      READ(10, *) MSEG, NEM
      WRITE(20, 1010) MSEG
      NUM = 5+2*MSEG
      IF(NEM.GT.MEM.OR.NUM.GT.MUM) GOTO 2000
     ID(3) = 2*MSEG+5
     ID(4) = 2*MSEG+3
     ID(5) = 2*MSEG+4
     DO 500 J = 1, 5
     READ(10, *) (RESIN(1, ID(J), I), I = 1, NEM)
     DO 500 I = 1, NEM
 500 RESIN(2, ID(J), I) = RESIN(1, ID(J), I)
     DO 490 K = 1, 2
     IF(K.EQ.1) THEN
     WRITE(20, 470)
     ELSE
     WRITE(20, 480)
     ENDIF
      DO 50 J = 3, 2*MSEG+2
      READ(10, *) (RESIN(K, J, I), I = 1, NEM)
 50   CONTINUE
      WRITE(20, 410) NEM
      DO 55 I = 1, NEM
 55   WRITE(20, 420) (RESIN(K, J, I), J = 1, 2), RESIN(K, 2*MSEG+5, I),
     &                 RESIN(K, 2*MSEG+3, I), RESIN(K, 2*MSEG+4, I)
      WRITE(20, 415)
      DO 60 I = 1, NEM
 60   WRITE(20, 430) RESIN(K, 1, I), (RESIN(K, J, I), J = 3, 2+MSEG)
      WRITE(20, 460)
      DO 80 I = 1, NEM
 80   WRITE(20, 430) RESIN(K, 1, I), (RESIN(K, J, I), J = 3+MSEG, 2*MSEG+2)
 490 CONTINUE
      RETURN
1010  FORMAT(/2X, 'SEGMENTS (Mseg) OF THE MATRIX
     &           STRESS-STRAIN', 1X, 'CURVE = ', I4)
 200  FORMAT('FIBER MODULI(E1, U12, E2, U23, G12) = ', E10.4, 1X, F5.3, 1X,
     &           E10.4, 1X, F5.3, 1X, E10.4//1X, 'TEMPERATURE POINTS
     &    (Nef) = ', I4, 2X, 'ON WHICH FIBER PROPERITES ARE CHANGED')
 110  FORMAT(2X, 'T(1-Nef) = ', 8(F7.1, 1X))
 120  FORMAT(1X, 'E1(1-Nef) = ', 8(F7.0, 1X))
 130  FORMAT(1X, 'Su(1-Nef) = ', 8(F7.1, 1X))
```

```
 140   FORMAT('Suc(1-Nef) = ', 8(F7.1, 1X))
 150   FORMAT('Af1(1-Nef) = ', 8(F7.3, 1X))
 160   FORMAT('Af2(1-Nef) = ', 8(F7.3, 1X)/)
 300   FORMAT(5X, 'POISSONS RATIO = ', F5.3, 2X, 'TENSILE & COMPR.',
     &              1X, 'STRENGTHS = ', 2(F7.2, 2X)/)
 410   FORMAT(5X, 'TEMPERATURE POINTS (ON WHICH MATRIX
     &          PROPERTIES', 1X, 'ARE VARIED) = ', I4/2X, 'TEMP.'
     &          , 4X, 'Alfm', 4X, 'Poissons Ratio', 2x,
     &          'Tensile Strength', 2x, 'Com.', 1x, 'Strength')
 420   FORMAT(F7.1, 2X, F7.2, 6X, F5.3, 10X, F7.1, 9X, F7.1)
 415   FORMAT(2X, 'TEMP.', 19X, 'YIELD STRENGTH, Ys(1--Nem)')
 430   FORMAT(F7.1, 2X, 9(F8.1, 1X)/9X, 9(F8.1, 1X))
 460   FORMAT(2X, 'TEMP.', 17X, 'TANGENTENT MODULI, ET(1--Nem)')
 470  FORMAT(/20X, 'MATRIX PROPERTIES UNDER TENSION:')
 480  FORMAT(/18X, 'MATRIX PROPERTIES UNDER COMPRESSION:')
 2000 WRITE(20, 2100) MUM, MEM, MUF, MEF, NUM, NEM, NUF, NEF
 2100 FORMAT(/'ALLOWABLE (MUM, MEM, MUF, MEF) = ', 4(I4, 1X),
     &           'ACTUAL (NUM, NEM, NUF, NEF) = ', 4(I4, 1X))
       STOP
       END
C***********************************************************
       SUBROUTINE RESID(ML, NL, SSF, SSM)
       IMPLICIT DOUBLE PRECISION (A-H, O-Z)
       DIMENSION SSF(ML, 6), SSM(ML, 6), SF(6), SM(6)
       DO 10 IL = 1, NL
       READ(10, *) (SF(I), I = 1, 6), (SM(I), I = 1, 6)
       DO 10 I = 1, 6
       SSF(IL, I) = SSF(IL, I)+SF(I)
       SSM(IL, I) = SSM(IL, I)+SM(I)
 10    CONTINUE
       RETURN
       END
C***********************************************************
       SUBROUTINE FIBERS(ML, NL, MUF, MEF, NEF, TEMP,
     &                       FIBER, EGF, SF, SUF, ALFF)
       IMPLICIT DOUBLE PRECISION(A-H, O-Z)
       DIMENSION FIBER(MUF, MEF), EGF(ML, 5), SF(ML, 6, 6),
     &              SUF(ML, 2), ALFF(ML, 6)
C
       DO 40 I = 1, NEF-1
       IF(TEMP.GE.FIBER(2, I).AND.TEMP.LE.FIBER(2, I+1)) THEN
       I0 = I
       GOTO 50
       ENDIF
 40    CONTINUE
       WRITE(20, 25) TEMP, FIBER(2, 1), FIBER(2, NEF)
       STOP
 50    CONTINUE
       DO 60 IL = 1, NL
       DO 70 I = 1, 5
 70    EGF(IL, I) = FIBER(1, I)
```

```
      DO 80 J = 6, 7
      ALFF(IL, J–5) = FIBER(J, I0)+(TEMP-FIBER(2, I0))*
     &          (FIBER(J, I0+1)–FIBER(J, I0))/(FIBER(2, I0+1)–FIBER(2, I0))
 80   SUF(IL, J–5) = FIBER(J–2, I0)+(TEMP-FIBER(2, I0))*
     &               (FIBER(J–2, I0+1)–FIBER(J–2, I0))/
     &                  (FIBER(2, I0+1)–FIBER(2, I0))
     ALFF(IL, 3) = ALFF(IL, 2)
     ALFF(IL, 4) = 0.
      ALFF(IL, 5) = 0.
      ALFF(IL, 6) = 0.
      EGF(IL, 1) = FIBER(3, I0)+(TEMP–FIBER(2, I0))*
     &        (FIBER(3, I0+1)–FIBER(3, I0))/(FIBER(2, I0+1)–FIBER(2, I0))
 60  CONTINUE
      DO 110 IL = 1, NL
     CALL ELASC(ML, IL, EGF(IL, 1), EGF(IL, 3), EGF(IL, 2), EGF(IL, 5),
     &              EGF(IL, 4), SF)
 110  CONTINUE
      RETURN
 25   FORMAT(/'GIVEN TEMPERATURE T = ', E11.4, 1X, 'IS OUT OF
     &          RANGE(', E11.4, 1X, E11.4, ') OF THE FIBERS'/)
      END
C*****************************************************************
     SUBROUTINE ELASC(ML, IL, E1, E2, U, G, V, SE)
     IMPLICIT DOUBLE PRECISION (A-H, O-Z)
     DIMENSION SE(ML, 6, 6)
C***TO CALCULATE PLANE ELASTIC COMPLIANCE MATRIX SE
     DO 10 I = 1, 6
     DO 10 J = 1, 6
 10  SE(IL, I, J) = 0.
     SE(IL, 1, 1) = 1./E1
     SE(IL, 1, 2) = –U/E1
     SE(IL, 2, 1) = SE(IL, 1, 2)
     SE(IL, 2, 2) = 1./E2
     SE(IL, 1, 3) = SE(IL, 1, 2)
     SE(IL, 3, 1) = SE(IL, 1, 2)
     SE(IL, 2, 3) = –V/E2
     SE(IL, 3, 2) = SE(IL, 2, 3)
     SE(IL, 3, 3) = SE(IL, 2, 2)
     G23 = E2/(2*(1+V))
     SE(IL, 4, 4) = 1./G23
     SE(IL, 5, 5) = 1./G
     SE(IL, 6, 6) = SE(IL, 5, 5)
     RETURN
     END
C*****************************************************************
      SUBROUTINE PLASC(ML, IL, MSEG, ETM, SSM, EGM, SP, ID,
     &                    LAYER, ID_M, SM_Y)
     IMPLICIT DOUBLE PRECISION (A-H, O-Z)
      DIMENSION ETM(2, 20), SSM(ML, 6), EGM(ML, 5), SP(6, 6),
     &          SA(6), LAYER(ML), SM_Y(ML)
      ID = 0
```

```
      ID_M = 0
      S1 = SQRT(SSM(IL, 1)**2+SSM(IL, 2)**2+SSM(IL, 3)**2-SSM(IL, 1)
      &          *SSM(IL, 2)–SSM(IL, 1)*SSM(IL, 3)–SSM(IL, 2)
      &          *SSM(IL, 3)+3.*SSM(IL, 4)**2+3.*SSM(IL, 5)**2
      &          +3.*SSM(IL, 6)**2)
       IF(S1.LE.ETM(1, 1)) RETURN
       ID = 1
      IF(S1.GE.SM_Y(IL)) SM_Y(IL) = S1
      IF(S1.LT.SM_Y(IL)) THEN
      ID_M = 1
      RETURN
      ENDIF
       DO 10 I = 1, MSEG–1
       IF(S1.GT.ETM(1, I).AND.S1.LE.ETM(1, I+1)) THEN
       ET1 = ETM(2, I+1)
       GOTO 50
       ENDIF
10     CONTINUE
       ET1 = ETM(2, MSEG)
50     IF(ET1.LE.0.) ET = 0.01
       ET = ET1
      E = ETM(2, 1)
      IF(LAYER(IL).LT.0) ET = 0.01*ET1
      SA(1) = SSM(IL, 1)–(SSM(IL, 1)+SSM(IL, 2)+SSM(IL, 3))/3.
      SA(2) = SSM(IL, 2)–(SSM(IL, 1)+SSM(IL, 2)+SSM(IL, 3))/3.
      SA(3) = SSM(IL, 3)–(SSM(IL, 1)+SSM(IL, 2)+SSM(IL, 3))/3.
      SA(4) = SSM(IL, 4)
      SA(5) = SSM(IL, 5)
      SA(6) = SSM(IL, 6)
      C = 9.*(E-ET)/(4.*E*ET*S1*S1)
      SP(1, 1) = C*SA(1)*SA(1)
      SP(1, 2) = C*SA(1)*SA(2)
      SP(1, 3) = C*SA(1)*SA(3)
      SP(1, 4) = 2.*C*SA(1)*SA(4)
      SP(1, 5) = 2.*C*SA(1)*SA(5)
      SP(1, 6) = 2.*C*SA(1)*SA(6)
      SP(2, 2) = C*SA(2)*SA(2)
      SP(2, 3) = C*SA(2)*SA(3)
      SP(2, 4) = 2.*C*SA(2)*SA(4)
      SP(2, 5) = 2.*C*SA(2)*SA(5)
      SP(2, 6) = 2.*C*SA(2)*SA(6)
      SP(3, 3) = C*SA(3)*SA(3)
      SP(3, 4) = 2.*C*SA(3)*SA(4)
      SP(3, 5) = 2.*C*SA(3)*SA(5)
      SP(3, 6) = 2.*C*SA(3)*SA(6)
      SP(4, 4) = 4.*C*SA(4)*SA(4)
      SP(4, 5) = 4.*C*SA(4)*SA(5)
      SP(4, 6) = 4.*C*SA(4)*SA(6)
      SP(5, 5) = 4.*C*SA(5)*SA(5)
      SP(5, 6) = 4.*C*SA(5)*SA(6)
      SP(6, 6) = 4.*C*SA(6)*SA(6)
```

```
      DO 20 I = 1, 6
      DO 20 J = I, 6
 20   SP(J, I) = SP(I, J)
C*** CHANGE MODULUS TO DEFINE BRIDGING MATRIX
       EGM(IL, 1) = ET1
       EGM(IL, 2) = 0.5
       EGM(IL, 3) = ET1
       EGM(IL, 4) = 0.5
       EGM(IL, 5) = ET1/3.
      RETURN
      END
C*****************************************************************
      SUBROUTINE STATUS(ML, IL, SSM, L)
      IMPLICIT DOUBLE PRECISION (A-H, O-Z)
      DIMENSION SSM(ML, 6)
      IF((SSM(IL, 1)+SSM(IL, 2)+SSM(IL, 3)).GE.0.) THEN
      L = 1
      ELSE
      L = 2
      ENDIF
      RETURN
      END
C*****************************************************************
       SUBROUTINE MATRIX(ML, NL, MUM, MEM, NEM, MSEG, LAYER,
      &                   TEMP, RESIN, SSM, EGM, SM, SUUM, ALFM, SM_Y)
       IMPLICIT DOUBLE PRECISION (A-H, O-Z)
       DIMENSION RESIN(2, MUM, MEM), SSM(ML, 6), EGM(ML, 5),
      &    SM(ML, 6, 6), SUUM(ML, 2), ALFM(ML, 6), ETM(2, 20),
      &    LAYER(ML), EGM1(ML, 5), SM_Y(ML)
       IF(MSEG.GT.20) STOP
       DO 40 I = 1, NEM-1
       IF(TEMP.GE.RESIN(1, 1, I).AND.TEMP.LE.RESIN(1, 1, I+1)) THEN
       I0 = I
       GOTO 50
       ENDIF
 40    CONTINUE
       WRITE(20, 25) TEMP, RESIN(1, 1, 1), RESIN(1, 1, NEM)
       STOP
 50    CONTINUE
       DO 60 IL = 1, NL
      ID_M = 0
      CALL STATUS(ML, IL, SSM, L)
       ALFM(IL, 1) = RESIN(L, 2, I0)+(TEMP-RESIN(L, 1, I0))*
      &  (RESIN(L, 2, I0+1)-RESIN(L, 2, I0))/(RESIN(L, 1, I0+1)-RESIN(L, 1, I0))
       ALFM(IL, 2) = ALFM(IL, 1)
       ALFM(IL, 3) = ALFM(IL, 2)
       ALFM(IL, 4) = 0.
       ALFM(IL, 5) = 0.
       ALFM(IL, 6) = 0.
       DO 80 J = 1, 2
       J1 = 2*MSEG+2+J
```

```
80   SUUM(IL, J) = RESIN(L, J1, I0)+(TEMP–RESIN(L, 1, I0))*
    &                  (RESIN(L, J1, I0+1)–RESIN(L, J1, I0))/
    &                  (RESIN(L, 1, I0+1)–RESIN(L, 1, I0))
     EGM(IL, 1) = RESIN(L, MSEG+3, I0)+(TEMP–RESIN(L, 1, I0))*
    &                (RESIN(L, MSEG+3, I0+1)–RESIN(L, MSEG+3, I0))/
    &                (RESIN(L, 1, I0+1)–RESIN(L, 1, I0))
     EGM(IL, 2) = RESIN(L, 2*MSEG+5, I0)+(TEMP–RESIN(L, 1, I0))*
    &                (RESIN(L, 2*MSEG+5, I0+1)–RESIN(L, 2*MSEG+5, I0))/
    &                (RESIN(L, 1, I0+1)–RESIN(L, 1, I0))
     EGM(IL, 3) = EGM(IL, 1)
     EGM(IL, 4) = EGM(IL, 2)
     EGM(IL, 5) = 0.5*EGM(IL, 1)/(1.+EGM(IL, 2))
     DO 90 K = 1, MSEG
     ETM(1, K) = RESIN(L, K+2, I0)+(TEMP–RESIN(L, 1, I0))*
    &               (RESIN(L, K+2, I0+1)–RESIN(L, K+2, I0))/
    &               (RESIN(L, 1, I0+1)–RESIN(L, 1, I0))
90   ETM(2, K) = RESIN(L, K+2+MSEG, I0)+(TEMP–RESIN(L, 1, I0))*
    &               (RESIN(L, K+2+MSEG, I0+1)–RESIN(L, MSEG+K+2, I0))/
    &               (RESIN(L, 1, I0+1)–RESIN(L, 1, I0))
    IF(LAYER(IL).LT.0) THEN
    EGM1(IL, 1) = 0.01*EGM(IL, 1)
    EGM1(IL, 3) = 0.01*EGM(IL, 3)
    EGM1(IL, 5) = 0.01*EGM(IL, 5)
    ELSE
    EGM1(IL, 1) = EGM(IL, 1)
    EGM1(IL, 3) = EGM(IL, 3)
    EGM1(IL, 5) = EGM(IL, 5)
    ENDIF
    EGM1(IL, 2) = EGM(IL, 2)
    EGM1(IL, 4) = EGM(IL, 4)
     CALL ELAPS(ML, IL, MSEG, ETM, SSM, EGM1, SM,
    &               LAYER, ID_M, SM_Y)
    EGM(IL, 1) = EGM1(IL, 1)
    EGM(IL, 3) = EGM1(IL, 3)
    EGM(IL, 5) = EGM1(IL, 5)
    EGM(IL, 2) = EGM1(IL, 2)
    EGM(IL, 4) = EGM1(IL, 4)
60   CONTINUE
     RETURN
25   FORMAT(/'GIVEN TEMPERATURE T = ', E11.4, 1X,
    &             'IS OUT OF RANGE(', E11.4, 1X, E11.4, ')
    &             OF THE MATRIX'/)
     END
C******************************************************************
     SUBROUTINE ELAPS(ML, IL, MSEG, ETM, SSM, EGM, SM, LAYER,
    &                       ID_M, SM_Y)
     IMPLICIT DOUBLE PRECISION (A-H, O-Z)
     DIMENSION ETM(2, 20), SSM(ML, 6), EGM(ML, 5), SM(ML, 6, 6),
    &              SP(6, 6), LEYER(ML), SM_Y(ML)
     CALL ELASC(ML, IL, EGM(IL, 1), EGM(IL, 3), EGM(IL, 2),
    &                EGM(IL, 5), EGM(IL, 4), SM)
```

```
      CALL PLASC(ML, IL, MSEG, ETM, SSM, EGM, SP, ID, LAYER,
     &            ID_M, SM_Y)
     IF(ID.EQ.0.OR.ID_M.EQ.1) RETURN
      DO 10 I = 1, 6
      DO 10 J = 1, 6
 10   SM(IL, I, J) = SM(IL, I, J)+SP(I, J)
      RETURN
      END
C*****************************************************************
      SUBROUTINE GLOBAL(ML, NL, ANGLE, TS, TC)
      IMPLICIT DOUBLE PRECISION (A-H, O-Z)
      DIMENSION ANGLE(ML), TS(ML, 6, 6), TC(ML, 6, 6)
      PAI = 3.14159265359/180.
      DO 10 IL = 1, NL
      RL1 = COS(ANGLE(IL)*PAI)
     RM2 = RL1
      RL2 = –SIN(ANGLE(IL)*PAI)
     RM1 = –RL2

     TC(IL, 1, 1) = RL1*RL1
     TC(IL, 1, 2) = RL2*RL2
     TC(IL, 1, 3) = 0
     TC(IL, 1, 4) = 0
     TC(IL, 1, 5) = 0
     TC(IL, 1, 6) = 2.*RL1*RL2
     TC(IL, 2, 1) = RM1*RM1
     TC(IL, 2, 2) = RM2*RM2
     TC(IL, 2, 3) = 0
     TC(IL, 2, 4) = 0
     TC(IL, 2, 5) = 0
     TC(IL, 2, 6) = 2.*RM1*RM2
     TC(IL, 3, 1) = 0
     TC(IL, 3, 2) = 0
     TC(IL, 3, 3) = 1
     TC(IL, 3, 4) = 0
     TC(IL, 3, 5) = 0
     TC(IL, 3, 6) = 0
     TC(IL, 4, 1) = 0
     TC(IL, 4, 2) = 0
     TC(IL, 4, 3) = 0
     TC(IL, 4, 4) = RL1
     TC(IL, 4, 5) = RM1
     TC(IL, 4, 6) = 0
     TC(IL, 5, 1) = 0
     TC(IL, 5, 2) = 0
     TC(IL, 5, 3) = 0
     TC(IL, 5, 4) = RL2
     TC(IL, 5, 5) = RM2
     TC(IL, 5, 6) = 0
     TC(IL, 6, 1) = RL1*RM1
     TC(IL, 6, 2) = RL2*RM2
```

```
      TC(IL, 6, 3) = 0
      TC(IL, 6, 4) = 0
      TC(IL, 6, 5) = 0
      TC(IL, 6, 6) = RL1*RM2+RL2*RM1

      DO 20 I = 1, 6
      DO 20 J = 1, 6
 20   TS(IL, I, J) = TC(IL, I, J)
      TS(IL, 1, 6) = RL1*RL2
      TS(IL, 2, 6) = RM1*RM2
      TS(IL, 6, 1) = 2.*RL1*RM1
      TS(IL, 6, 2) = 2.*RL2*RM2
 10    CONTINUE
       RETURN
       END
C*************************************************************
      SUBROUTINE INVER(D, A, N, L)
C*** SUBROUTINE TO FIND THE INVERSION S OF
C*** THE POSITIVE DEFINITE MATRIX
C***   N<7, OTHERWISE, A & B SHOULD BE MODIFIED
       IMPLICIT DOUBLE PRECISION (A-H, O-Z)
      DIMENSION D(N, N), ME(N), B(N), C(N), A(N, N)
       DO 5 I = 1, N
      DO 5 j = 1, N
      A(I, J) = D(I, J)
5     CONTINUE
      L = 1
      DE = 1.
      DO 10   J = 1, N
10    ME(J) = J
       DO 20 I = 1, N
      Y = 0.
      DO 30 J = I, N
      IF(ABS(A(I, J)).LE.ABS(Y)) GOTO 30
      K = J
      Y = A(I, J)
30    CONTINUE
       DE = DE*Y
      IF((ABS(Y)+1).EQ.1.0) THEN
      L = 0
      STOP
      ENDIF
      Y = 1./Y
      DO 40 J = 1, N
      C(J) = A(J, K)
      A(J, K) = A(J, I)
      A(J, I) = -C(J)*Y
      B(J) = A(I, J)*Y
40    A(I, J) = A(I, J)*Y
      A(I, I) = Y
      J = ME(I)
```

```
      ME(I) = ME(K)
      ME(K) = J
      DO 11 K = 1, N
      IF(K.EQ.I) GOTO 11
      DO 12 J = 1, N
      IF(J.EQ.I) gOTO 12
      A(K, J) = A(K, J)-B(J)*C(K)
12    CONTINUE
11    CONTINUE
20    CONTINUE
       DO 33 I = 1, N
      DO 44 K = 1, N
      IF(ME(K).EQ.I) GOTO 55
44    CONTINUE
55     IF(K.EQ.I) GOTO 33
       DO 66 J = 1, N
      W = A(I, J)
      A(I, J) = A(K, J)
66    A(K, J) = W
       IW = ME(I)
      ME(I) = ME(K)
      ME(K) = IW
      DE = -DE
33    CONTINUE
       RETURN
      END
C*****************************************************************
       SUBROUTINE GAUSS(A, B, N, X, L, JS)
C*** SUBROUTINE TO SOLVE [A]{X} = {B},
C*** SOLUTION IS BACKED IN {X}
       IMPLICIT DOUBLE PRECISION (A–H, O–Z)
       DIMENSION A(N, N), X(N), B(N), JS(N)
       L = 1
       DO 50 K = 1, N–1
       D = 0.
       DO 210 I = K, N
       DO 210 J = K, N
       IF(ABS(A(I, J)).GT.D) THEN
       D = ABS(A(I, J))
       JS(K) = J
       IS = I
       ENDIF
 210   CONTINUE
       IF((D+1.0).EQ.1.) THEN
       L = 0
       ELSE
       IF(JS(K).NE.K) THEN
       DO 220 I = 1, N
       T = A(I, K)
       A(I, K) = A(I, JS(K))
       A(I, JS(K)) = T
```

```
220  CONTINUE
     ENDIF
     IF(IS.NE.K) THEN
     DO 230 J = K, N
     T = A(K, J)
     A(K, J) = A(IS, J)
     A(IS, J) = T
230  CONTINUE
     T = B(K)
     B(K) = B(IS)
     B(IS) = T
     ENDIF
     ENDIF
     IF(L.EQ.0) THEN
     WRITE(20, 100)
     RETURN
     ENDIF
     DO 10 J = K+1, N
     A(K, J) = A(K, J)/A(K, K)
10   CONTINUE
     B(K) = B(K)/A(K, K)
     DO 30 I = K+1, N
     DO 20 J = K+1, N
     A(I, J) = A(I, J)–A(I, K)*A(K, J)
20   CONTINUE
     B(I) = B(I)–A(I, K)*B(K)
30   CONTINUE
50   CONTINUE
     IF(ABS(A(N, N))+1.EQ.1.) THEN
     L = 0
     WRITE(20, 100)
     RETURN
     ENDIF
     X(N) = B(N)/A(N, N)
     DO 70 I = N–1, 1, –1
     T = 0.
     DO 60 J = I+1, N
     T = T+A(I, J)*X(J)
60   CONTINUE
     X(I) = B(I)–T
70   CONTINUE
100  FORMAT(/10X, 'LINEAR FAIL')
     JS(N) = N
     DO 150 K = N, 1, –1
     IF(JS(K).NE.K) THEN
     T = X(K)
     X(K) = X(JS(K))
     X(JS(K)) = T
     ENDIF
150  CONTINUE
     RETURN
```

```
      END
C****************************************************************
      SUBROUTINE MECHF(ML, NL, Z, DSS, DF, IDSEC)
      IMPLICIT DOUBLE PRECISION (A–H, O–Z)
      DIMENSION Z(2, ML+1), DSS(6), DF(9)
      IF(IDSEC.EQ.0) THEN
     Z0 = Z(1, NL+1)–Z(1, 1)
      IF(Z0.LE.0.) THEN
      WRITE(20, 100) Z0
      STOP
      ENDIF
     ELSE
     Z0 = 1.
     ENDIF
      DO 20 I = 1, 3
      DF(I) = DSS(I)*Z0
      DF(I+3) = DSS(I+3)
20    DF(I+6) = 0.0
 100  FORMAT(/5X, 'THICKNESS OF THE LAMINATE = ', E11.4)
      RETURN
      END
C****************************************************************
      SUBROUTINE TEMPF(ML, NL, LAYER, BTE, Z, DFT, DTEMP)
      IMPLICIT DOUBLE PRECISION (A–H, O–Z)
      DIMENSION BTE(ML, 6), Z(2, ML+1), DFT(9), LAYER(ML)
      DO 10 I = 1, 3
      DFT(I) = 0.
10    DFT(I+3) = 0.
      DO 15 J = 1, NL
      DFT(1) = DFT(1)+BTE(J, 1)*DTEMP*(Z(1, J+1)–Z(1, J))*Z(2, J)
      DFT(2) = DFT(2)+BTE(J, 2)*DTEMP*(Z(1, J+1)–Z(1, J))*Z(2, J)
     DFT(3) = DFT(3)+BTE(J, 6)*DTEMP*(Z(1, J+1)–Z(1, J))*Z(2, J)
      DFT(4) = 0.
     DFT(5) = 0.
     DFT(6) = 0
      DFT(7) = DFT(7)+BTE(J, 1)*DTEMP*(Z(1, J+1)**2–Z(1, J)**2)*Z(2, J)
      DFT(8) = DFT(8)+BTE(J, 2)*DTEMP*(Z(1, J+1)**2–Z(1, J)**2)*Z(2, J)
     DFT(9) = DFT(9)+BTE(J, 6)*DTEMP*(Z(1, J+1)**2–Z(1, J)**2)*Z(2, J)
15    CONTINUE
      DO 20 I = 1, 6
 20   DFT(I) = DFT(I)*1.E–6
      RETURN
      END
C****************************************************************
      SUBROUTINE LAMINATE(DF, DST, DST1, DST2, LAYER, CG, Z,
     &                    ML, NL, CG0, BTE, DTEMP)
C*** TO CALCULATE OVERALL STIFFNESS ELEMENTS OF
C*** A LAMINATE AND TO GET
C*** IN-PLANE STRAIN {dst}
C*** AND OUT-OF-PLANE STRAIN {DST1}.
C*** CG(IL, ., .) = GLOBAL STIFFNESS MATRIX OF IL-TH LAYER
```

```
      IMPLICIT DOUBLE PRECISION (A-H, O-Z)
      DIMENSION DF(9), DST(3), LAYER(ML), CG(ML, 6, 6), Z(2, ML+1),
     &          A(6+3*NL, 6+3*NL), B(6+3*NL), CG0(6, 6), BTE(ML, 6),
     &              DST0(6+3*NL), DST1(NL, 3), DST2(3), A1(3, 3), B1(3, 3)
      B(1) = DF(1)
     B(2) = DF(2)
     B(3) = DF(3)
      B(4) = DF(7)
     B(5) = DF(8)
     B(6) = DF(9)
     DO 5 I = 1, NL
     B(4+3*I) = DF(4)
     B(5+3*I) = DF(5)
     B(6+3*I) = DF(6)
     IF(DTEMP.NE.0) B(6+3*I) = BTE(I, 3)*DTEMP*1.0e-6+DF(6)
5    CONTINUE
      DO 10 I = 1, 3*NL+6
      DO 10 J = 1, 3*NL+6
10    A(I, J) = 0.
!      MFAIL = 0
      DO 100 IL = 1, NL
      A(1, 1) = A(1, 1)+CG(IL, 1, 1)*(Z(1, IL+1)-Z(1, IL))*Z(2, IL)
      A(1, 2) = A(1, 2)+CG(IL, 1, 2)*(Z(1, IL+1)-Z(1, IL))*Z(2, IL)
      A(1, 3) = A(1, 3)+CG(IL, 1, 6)*(Z(1, IL+1)-Z(1, IL))*Z(2, IL)
      A(2, 3) = A(2, 3)+CG(IL, 2, 6)*(Z(1, IL+1)-Z(1, IL))*Z(2, IL)
      A(2, 1) = A(2, 1)+CG(IL, 2, 1)*(Z(1, IL+1)-Z(1, IL))*Z(2, IL)
      A(3, 1) = A(3, 1)+CG(IL, 6, 1)*(Z(1, IL+1)-Z(1, IL))*Z(2, IL)
      A(3, 2) = A(3, 2)+CG(IL, 6, 2)*(Z(1, IL+1)-Z(1, IL))*Z(2, IL)
      A(2, 2) = A(2, 2)+CG(IL, 2, 2)*(Z(1, IL+1)-Z(1, IL))*Z(2, IL)
      A(3, 3) = A(3, 3)+CG(IL, 6, 6)*(Z(1, IL+1)-Z(1, IL))*Z(2, IL)
      A(1, 4+3*IL) = CG(IL, 1, 4)*(Z(1, IL+1)-Z(1, IL))*Z(2, IL)
      A(1, 5+3*IL) = CG(IL, 1, 5)*(Z(1, IL+1)-Z(1, IL))*Z(2, IL)
      A(1, 6+3*IL) = CG(IL, 1, 3)*(Z(1, IL+1)-Z(1, IL))*Z(2, IL)
      A(2, 6+3*IL) = CG(IL, 2, 3)*(Z(1, IL+1)-Z(1, IL))*Z(2, IL)
      A(2, 4+3*IL) = CG(IL, 2, 4)*(Z(1, IL+1)-Z(1, IL))*Z(2, IL)
      A(3, 4+3*IL) = CG(IL, 6, 4)*(Z(1, IL+1)-Z(1, IL))*Z(2, IL)
      A(3, 5+3*IL) = CG(IL, 6, 5)*(Z(1, IL+1)-Z(1, IL))*Z(2, IL)
      A(2, 5+3*IL) = CG(IL, 2, 5)*(Z(1, IL+1)-Z(1, IL))*Z(2, IL)
      A(3, 6+3*IL) = CG(IL, 6, 3)*(Z(1, IL+1)-Z(1, IL))*Z(2, IL)

      A(4, 4+3*IL) = CG(IL, 1, 4)*(Z(1, IL+1)**2-Z(1, IL)**2)*Z(2, IL)/2.0
      A(4, 5+3*IL) = CG(IL, 1, 5)*(Z(1, IL+1)**2-Z(1, IL)**2)*Z(2, IL)/2.0
      A(4, 6+3*IL) = CG(IL, 1, 3)*(Z(1, IL+1)**2-Z(1, IL)**2)*Z(2, IL)/2.0
      A(5, 6+3*IL) = CG(IL, 2, 3)*(Z(1, IL+1)**2-Z(1, IL)**2)*Z(2, IL)/2.0
      A(5, 4+3*IL) = CG(IL, 2, 4)*(Z(1, IL+1)**2-Z(1, IL)**2)*Z(2, IL)/2.0
      A(6, 4+3*IL) = CG(IL, 6, 4)*(Z(1, IL+1)**2-Z(1, IL)**2)*Z(2, IL)/2.0
      A(6, 5+3*IL) = CG(IL, 6, 5)*(Z(1, IL+1)**2-Z(1, IL)**2)*Z(2, IL)/2.0
      A(5, 5+3*IL) = CG(IL, 2, 5)*(Z(1, IL+1)**2-Z(1, IL)**2)*Z(2, IL)/2.0
      A(6, 6+3*IL) = CG(IL, 6, 3)*(Z(1, IL+1)**2-Z(1, IL)**2)*Z(2, IL)/2.0

      A(4, 1) = A(4, 1)+CG(IL, 1, 1)*(Z(1, IL+1)**2-Z(1, IL)**2)*Z(2, IL)/2.0
```

```
A(4, 2) = A(4, 2)+CG(IL, 1, 2)*(Z(1, IL+1)**2-Z(1, IL)**2)*Z(2, IL)/2.0
A(4, 3) = A(4, 3)+CG(IL, 1, 6)*(Z(1, IL+1)**2-Z(1, IL)**2)*Z(2, IL)/2.0
A(5, 3) = A(5, 3)+CG(IL, 2, 6)*(Z(1, IL+1)**2-Z(1, IL)**2)*Z(2, IL)/2.0
A(5, 1) = A(5, 1)+CG(IL, 2, 1)*(Z(1, IL+1)**2-Z(1, IL)**2)*Z(2, IL)/2.0
A(6, 1) = A(6, 1)+CG(IL, 6, 1)*(Z(1, IL+1)**2-Z(1, IL)**2)*Z(2, IL)/2.0
A(6, 2) = A(6, 2)+CG(IL, 6, 2)*(Z(1, IL+1)**2-Z(1, IL)**2)*Z(2, IL)/2.0
A(5, 2) = A(5, 2)+CG(IL, 2, 2)*(Z(1, IL+1)**2-Z(1, IL)**2)*Z(2, IL)/2.0
A(6, 3) = A(6, 3)+CG(IL, 6, 6)*(Z(1, IL+1)**2-Z(1, IL)**2)*Z(2, IL)/2.0
A(1, 4) = A(1, 4)+CG(IL, 1, 1)*(Z(1, IL+1)**2-Z(1, IL)**2)*Z(2, IL)/2.0
A(1, 5) = A(1, 5)+CG(IL, 1, 2)*(Z(1, IL+1)**2-Z(1, IL)**2)*Z(2, IL)/2.0
A(1, 6) = A(1, 6)+CG(IL, 1, 6)*(Z(1, IL+1)**2-Z(1, IL)**2)*Z(2, IL)/2.0
A(2, 6) = A(2, 6)+CG(IL, 2, 6)*(Z(1, IL+1)**2-Z(1, IL)**2)*Z(2, IL)/2.0
A(2, 4) = A(2, 4)+CG(IL, 2, 1)*(Z(1, IL+1)**2-Z(1, IL)**2)*Z(2, IL)/2.0
A(3, 4) = A(3, 4)+CG(IL, 6, 1)*(Z(1, IL+1)**2-Z(1, IL)**2)*Z(2, IL)/2.0
A(3, 5) = A(3, 5)+CG(IL, 6, 2)*(Z(1, IL+1)**2-Z(1, IL)**2)*Z(2, IL)/2.0
A(2, 5) = A(2, 5)+CG(IL, 2, 2)*(Z(1, IL+1)**2-Z(1, IL)**2)*Z(2, IL)/2.0
A(3, 6) = A(3, 6)+CG(IL, 6, 6)*(Z(1, IL+1)**2-Z(1, IL)**2)*Z(2, IL)/2.0
A(4, 4) = A(4, 4)+CG(IL, 1, 1)*(Z(1, IL+1)**3-Z(1, IL)**3)*Z(2, IL)/3.0
A(4, 5) = A(4, 5)+CG(IL, 1, 2)*(Z(1, IL+1)**3-Z(1, IL)**3)*Z(2, IL)/3.0
A(4, 6) = A(4, 6)+CG(IL, 1, 6)*(Z(1, IL+1)**3-Z(1, IL)**3)*Z(2, IL)/3.0
A(5, 6) = A(5, 6)+CG(IL, 2, 6)*(Z(1, IL+1)**3-Z(1, IL)**3)*Z(2, IL)/3.0
A(5, 4) = A(5, 4)+CG(IL, 2, 1)*(Z(1, IL+1)**3-Z(1, IL)**3)*Z(2, IL)/3.0
A(6, 4) = A(6, 4)+CG(IL, 6, 1)*(Z(1, IL+1)**3-Z(1, IL)**3)*Z(2, IL)/3.0
A(6, 5) = A(6, 5)+CG(IL, 6, 2)*(Z(1, IL+1)**3-Z(1, IL)**3)*Z(2, IL)/3.0
A(5, 5) = A(5, 5)+CG(IL, 2, 2)*(Z(1, IL+1)**3-Z(1, IL)**3)*Z(2, IL)/3.0
A(6, 6) = A(6, 6)+CG(IL, 6, 6)*(Z(1, IL+1)**3-Z(1, IL)**3)*Z(2, IL)/3.0

A(4+3*IL, 1) = CG(IL, 4, 1)*Z(2, IL)
A(4+3*IL, 2) = CG(IL, 4, 2)*Z(2, IL)
A(4+3*IL, 3) = CG(IL, 4, 3)*Z(2, IL)
A(5+3*IL, 1) = CG(IL, 5, 1)*Z(2, IL)
A(5+3*IL, 2) = CG(IL, 5, 2)*Z(2, IL)
A(5+3*IL, 3) = CG(IL, 5, 3)*Z(2, IL)
A(6+3*IL, 1) = CG(IL, 3, 1)*Z(2, IL)
A(6+3*IL, 2) = CG(IL, 3, 2)*Z(2, IL)
A(6+3*IL, 3) = CG(IL, 3, 6)*Z(2, IL)
A(4+3*IL, 4) = CG(IL, 4, 1)*Z(2, IL)*(Z(1, IL+1)+Z(1, IL))/2.0
A(4+3*IL, 5) = CG(IL, 4, 2)*Z(2, IL)*(Z(1, IL+1)+Z(1, IL))/2.0
A(4+3*IL, 6) = CG(IL, 4, 3)*Z(2, IL)*(Z(1, IL+1)+Z(1, IL))/2.0
A(5+3*IL, 4) = CG(IL, 5, 1)*Z(2, IL)*(Z(1, IL+1)+Z(1, IL))/2.0
A(5+3*IL, 5) = CG(IL, 5, 2)*Z(2, IL)*(Z(1, IL+1)+Z(1, IL))/2.0
A(5+3*IL, 6) = CG(IL, 5, 3)*Z(2, IL)*(Z(1, IL+1)+Z(1, IL))/2.0
A(6+3*IL, 4) = CG(IL, 3, 1)*Z(2, IL)*(Z(1, IL+1)+Z(1, IL))/2.0
A(6+3*IL, 5) = CG(IL, 3, 2)*Z(2, IL)*(Z(1, IL+1)+Z(1, IL))/2.0
A(6+3*IL, 6) = CG(IL, 3, 6)*Z(2, IL)*(Z(1, IL+1)+Z(1, IL))/2.0
A(3*IL+4, 3*IL+4) = CG(IL, 4, 4)*Z(2, IL)
A(3*IL+4, 3*IL+5) = CG(IL, 4, 5)*Z(2, IL)
A(3*IL+4, 3*IL+6) = CG(IL, 4, 3)*Z(2, IL)
A(3*IL+5, 3*IL+4) = CG(IL, 5, 4)*Z(2, IL)
A(3*IL+5, 3*IL+5) = CG(IL, 5, 5)*Z(2, IL)
A(3*IL+5, 3*IL+6) = CG(IL, 5, 3)*Z(2, IL)
```

```
      A(3*IL+6, 3*IL+4) = CG(IL, 3, 4)*Z(2, IL)
      A(3*IL+6, 3*IL+5) = CG(IL, 3, 5)*Z(2, IL)
      A(3*IL+6, 3*IL+6) = CG(IL, 3, 3)*Z(2, IL)
 100   CONTINUE
       CALL EQUA(A, B, LAYER, NL, DST0, L)
       DO 25 I = 1, 3
      DST(I) = DST0(I)
25    CONTINUE
      DO 27 I = 1, 3
27    DST2(I) = DST0(3+I)
       DO 26 I = 1, NL
      DO 26 J = 1, 3
      DST1(I, J) = DST0(3*I+3+J)
26    CONTINUE
       IF(L.EQ.0) THEN
       DO 30 I = 1, NL
      DO 30 J = 1, 3
       DST1(I, J) = 0.
      DST2(I) = 0.
 30    DST(I) = 0.
       ENDIF
       RETURN
       END
C***************************************************************
       SUBROUTINE EQUA(A, B, LAYER, NL, DST0, L)
C*** SOLVE THE EQUTION OF THE LAMINATE
C*** TO OBTAIN THE IN-PLANE AND OUT-OF-PLANE STRAIN
       IMPLICIT DOUBLE PRECISION (A-H, O-Z)
      DIMENSION A(6+3*NL, 6+3*NL), LAYER(NL),
      &     DST00(6+3*NL), B(6+3*NL), C(6+3*NL, 6+3*NL),
      &     DST0(6+3*NL), E(6+3*NL), JS(6+3*NL)
       J = 1

      DO 30 I = 1, 6+3*NL
           DO 30 J = 1, 6+3*NL
30         C(I, J) = A(I, J)
       DO 40 I = 1, 6+3*NL
40    E(I) = B(I)

       CALL GAUSS(C, E, 3*NL+6, DST00, L, JS)
       DO 50 I = 1, 6+3*NL
50    DST0(I) = DST00(I)
       RETURN
      END
C***************************************************************
       SUBROUTINE STRESS(ML, NL, IDSEC, LAYER, Z, CG, SSP, BTE,
      &                      DSS, DTEMP, SS, SA, CG0, DSA, DSA1,
      &                      SA1, dis_z)
       IMPLICIT DOUBLE PRECISION (A-H, O-Z)
       DIMENSION LAYER(ML), Z(2, ML+1), CG(ML, 6, 6),
      &            SSP(ML, 6), BTE(ML, 6), DSA(3), DF(9),
```

```
     &              DFT(9), DSS(6), SS(6), CG0(6, 6), SA(6),
     &              DSA1(NL, 3), SA1(mL, 3), BF(6), DSA2(3)
C***GIVEN: 1) OVERALL MECHANICAL STRESS-INCREMENTS
C***            ON THE LAMINATE, "DSS"
C***         2) TEMPERATURE VARIATION DTEMP
C***OBTAIN: STRESS-INCREMENTS "SSP",
C***             GIVEN IN LAMINATE GLOBAL SYSTEM
C*** CG = LAMINA INSTANTANEOUS STIFFNESS MATRIX
C***         IN GLOBAL SYSTEM
C*** BTE = LAMINA THERMAL STRESS CONCENTRATION VECTOR
C***         IN GLOBAL SYSTEM
      CALL MECHF(ML, NL, Z, DSS, DF, IDSEC)
      CALL TEMPF(ML, NL, LAYER, BTE, Z, DFT, DTEMP)
      DO 10 I = 1, 9
 10   DF(I) = DF(I)+DFT(I)
      CALL LAMINATE(DF, DSA, DSA1, DSA2, LAYER, CG, Z,
     &                    ML, NL, CG0, BTE, DTEMP)
C*** {DSA} = IN-PLANE STRAIN INCREMENTS
C*** {DSA1} = OUT-OF-PLANE STRAIN INCREMENTS
C*** {SA} = IN-PLANE STRAIN
C*** {SA1} = OUT-OF-PLANE STRAIN
     DO 40 I = 1, 3
     SA(I) = DSA(I)+SA(I)
     SA(3+I) = DSA2(I)+SA(3+I)
      DO 40 J = 1, NL
 40  SA1(J, I) = DSA1(J, I)+SA1(J, I)
      DO I = 1, 6
     SS(I) = SS(I)+DSS(I)
     end do
      DISPLACE = 0.
      DO 42 I = 1, NL
      IF(SA1(I, 3).LT.(-1.)) GOTO 42
     DISPLACE = DISPLACE+DSA1(I, 3)*(Z(1, I+1)-Z(1, I))
 42  CONTINUE
      DIS_Z = DIS_Z+DISPLACE/(Z(1, NL+1)-Z(1, 1))

      DO 100 IL = 1, NL
      DO 20 I = 1, 3
     BF(I) = DSA(I)+DSA2(I)*(Z(1, IL+1)+Z(1, IL))/2.0
 20  BF(I+3) = DSA1(IL, I)
      C1 = BF(6)
     BF(6) = BF(3)
     BF(3) = C1
     DO 25 I = 1, 6
 25  SSP(IL, I) = -BTE(IL, I)*DTEMP*1.E-6
      DO 30 I = 1, 6
     DO 30 J = 1, 6
 30  SSP(IL, I) = SSP(IL, I)+CG(IL, I, J)*BF(J)
 100  CONTINUe
      RETURN
      END
```

```
C****************************************************************
        SUBROUTINE INITIL0(ML, NL, SSF, SSM, SA, SA1, SS)
        IMPLICIT DOUBLE PRECISION (A-H, O-Z)
        DIMENSION SSF(ML, 6), SSM(ML, 6), SA(3), SA1(ML, 3), SS(6)
        DO 20 I = 1, 3
        SA(I) = 0.
        SS(I) = 0.
20      SS(I+3) = 0.
        DO 10 IL = 1, NL
        DO 10 J = 1, 3
       SA1(IL, J) = 0.
        SSF(IL, J) = 0.
       SSF(IL, J+3) = 0.
        SSM(IL, J) = 0.
10      SSM(IL, J+3) = 0.
        RETURN
        END
C****************************************************************
          SUBROUTINE COUPLE(ML, IL, EUF, EUM, SF, SM,
     &                  VF, A, B, S, A_BETA, A_ALFA)
C*** OBTAIN THE BRIDGING MATRIX [A]
C***         AND COMPLIANCE MATRIX [S]
C***         IN THE LOACAL COORDINATE SYSTEM
          USE IMSL
         IMPLICIT DOUBLE PRECISION (A-H, O-Z)
          DIMENSION X(15), XGUESS(15), EUF(ML, 5), EUM(ML, 5), VF(ML),
     &                 SF(ML, 6, 6), SM(ML, 6, 6), A(6, 6), B(6, 6), S(6, 6)
          EXTERNAL FCN
          COMMON /DATA2/WK(100)
          COMMON /DATA3/WK1(320)
          N = 15
C
          A1 = EUM(IL, 1)/EUF(IL, 1)
          A2 = A_BETA+(1.-A_BETA)*EUM(IL, 1)/EUF(IL, 3)
          A5 = A_ALFA+(1.-A_ALFA)*EUM(IL, 5)/EUF(IL, 5)
          I12 = 0
          IF(ABS(0.5-EUM(IL, 2)).GT.0.0001) THEN
          DO 30 I = 3, N
 30       X(I) = 0.
          IF((SF(IL, 1, 1)-SM(IL, 1, 1)).EQ.0.) THEN
          X(1) = 0.
          ELSE
          X(1) = (SF(IL, 1, 2)-SM(IL, 1, 2))*(A1-A2)/(SF(IL, 1, 1)-SM(IL, 1, 1))
          ENDIF
          X(2) = X(1)
          I12 = 100
          ENDIF
          K1 = 0
          K2 = 36
          DO 40 I = 1, 6
          DO 40 J = 1, 6
```

```
          K1 = K1+1
          K2 = K2+1
          WK(K1) = SF(IL, I, J)
 40       WK(K2) = SM(IL, I, J)
          WK(73) = VF(IL)
          WK(74) = A1
          WK(75) = A2
          WK(76) = A5
          IF(I12.EQ.100) GOTO 100
C
         ITMAX = 100000
         ERRREL = 0.0001
          CALL ZERO(ML, IL, VF, SF, SM, N, XGUESS, A1, A2, A5)
C
          CALL DNEQNF(FCN, ERRREL, N, ITMAX, XGUESS, X, FNORM)
C***    DNEQNF IS A STANDARD SUBROUTINE TO FIND OUT
C***    THE ROOTS OF A VECTOR FUNCTION FCN
          DO 50 I = 1, 15
 50       WK1((IL-1)*16+I) = X(I)
          WK1(IL*16) = 1.
 100      CALL BRIDGE(X, N, A, B, S)
        RETURN
        END
C***************************************************************************
***************
          SUBROUTINE BRIDGE(X, N, A, B, S)
         IMPLICIT DOUBLE PRECISION (A-H, O-Z)
          DIMENSION X(N), S1(6, 6), A(6, 6), B(6, 6), S(6, 6), SF(6, 6), SM(6, 6),
     &               C(6, 6)
          COMMON /DATA2/WK(100)
          K1 = 0
          K2 = 36
          DO 100 I = 1, 6
          DO 100 J = 1, 6
          K1 = K1+1
          K2 = K2+1
          SF(I, J) = WK(K1)
 100      SM(I, J) = WK(K2)
          VF = WK(73)
          VM = 1.–VF
          A(1, 1) = WK(74)
          A(2, 2) = WK(75)
          A(3, 3) = WK(75)
          A(4, 4) = WK(75)
          A(5, 5) = WK(76)
          A(6, 6) = WK(76)
        K = 0
        DO 10 I = 2, 6
        DO 10 J = 1, I–1
        K = K+1
        A(J, I) = X(K)
```

```
      A(I, J) = 0.
      B(I, J) = 0.
 10   CONTINUE
      L = 0
       DO 20 I = 1, 6
      C(I, I) = VF+VM*A(I, I)
       DO 30 K = I+1, 6
      C(I, K) = VM*A(I, K)
30    C(K, I) = VM*A(K, I)
20     CONTINUE
       CALL INVER(C, B, 6, L)
      DO 50 I = 1, 6
      DO 50 J = 1, 6
      S1(I, J) = 0.
      DO 60 K = 1, 6
 60       S1(I, J) = S1(I, J)+SM(I, K)*A(K, J)*VM
 50       S1(I, J) = S1(I, J)+VF*SF(I, J)
      DO 80 I = 1, 6
      DO 80 J = 1, 6
      S(I, J) = 0.
      DO 90 K = 1, 6
 90   S(I, J) = S(I, J)+S1(I, K)*B(K, J)
 80   CONTINUE
      RETURN
      END
C*************************************************************
         SUBROUTINE ZERO(ML, IL, VF, SF0, SM0, N, X, A1, A2, A5)
         IMPLICIT DOUBLE PRECISION(A-H, O-Z)
         DIMENSION X(N), VF(ML), SF0(ML, 6, 6), SM0(ML, 6, 6),
      &                   SF(3, 3), SM(3, 3), A(3, 3), B(3, 3)
         COMMON /DATA3/WK1(320)
         IF(WK1(16*IL).EQ.0.) THEN
         DO 100 I = 1, N
 100     X(I) = 0.
         DO 10 I = 1, 3
         DO 10 J = 1, 3
         SF(I, J) = SF0(IL, I, J)
 10      SM(I, J) = SM0(IL, I, J)
         CALL BRIGA(VF(IL), SF, SM, A, B, A1, A2, A2)
         X(1) = A(1, 2)
         X(2) = A(1, 3)
         X(3) = A(2, 3)
         DO 20 I = 1, 3
         DO 20 J = 1, 3
         SF(I, J) = SF0(IL, I+3, J+3)
 20      SM(I, J) = SM0(IL, I+3, J+3)
         CALL BRIGA(VF(IL), SF, SM, A, B, A2, A5, A5)
         X(10) = A(1, 2)
         X(14) = A(1, 3)
         X(15) = A(2, 3)
         ELSE
```

```
      DO 30 I = 1, 15
 30   X(I) = WK1((IL-1)*16+I)
      ENDIF
      RETURN
      END
C****************************************************************
      SUBROUTINE FCN(X, F, N)
      IMPLICIT DOUBLE PRECISION (A-H, O-Z)
      DIMENSION X(N), F(N), A(6, 6), B(6, 6), S(6, 6)
      CALL BRIDGE(X, N, A, B, S)
      K = 0
      DO 10 I = 2, 6
      DO 10 J = 1, I-1
      K = K+1
      F(K) = (S(I, J)-S(J, I))*1000.
 10   CONTINUE
      RETURN
      END
C****************************************************************
      SUBROUTINE THERM(ML, IL, VF, SF, SM, ALFF, ALFM, BLFT,
     &                BM, EUF, EUM, LAYER, A_BETA, A_ALFA)
C*** USE Benveniste & Dvorak's formula to calculate thermal stress
      IMPLICIT DOUBLE PRECISION (A-H, O-Z)
      DIMENSION SF(ML, 6, 6), SM(ML, 6, 6), ALFF(ML, 6), ALFM(ML, 6),
     &          BLFT(ML, 6), BM(ML, 3), A_T(3, 3), B_T(3, 3),
     &          S1(3, 3), S2(3, 3), S3(3, 3), EUF(ML, 5), EUM(ML, 5),
     &          BLFT1(ML, 3), SF_T(3, 3), SM_T(3, 3), ALFF_T(3),
     &          ALFM_T(3), EUM1(ML, 5), LAYER(ML)
      SF_T(1, 1) = SF(IL, 1, 1)
      SF_T(1, 2) = SF(IL, 1, 2)
      SF_T(2, 1) = SF(IL, 2, 1)
      SF_T(2, 2) = SF(IL, 2, 2)
      SF_T(1, 3) = SF(IL, 1, 6)
      SF_T(2, 3) = SF(IL, 2, 6)
      SF_T(3, 1) = SF(IL, 6, 1)
      SF_T(3, 2) = SF(IL, 6, 2)
      SF_T(3, 3) = SF(IL, 6, 6)
      SM_T(1, 1) = SM(IL, 1, 1)
      SM_T(1, 2) = SM(IL, 1, 2)
      SM_T(2, 1) = SM(IL, 2, 1)
      SM_T(2, 2) = SM(IL, 2, 2)
      SM_T(1, 3) = SM(IL, 1, 6)
      SM_T(2, 3) = SM(IL, 2, 6)
      SM_T(3, 1) = SM(IL, 6, 1)
      SM_T(3, 2) = SM(IL, 6, 2)
      SM_T(3, 3) = SM(IL, 6, 6)
      ALFF_T(1) = ALFF(IL, 1)
      ALFF_T(2) = ALFF(IL, 2)
      ALFF_T(3) = ALFF(IL, 6)
      ALFM_T(1) = ALFM(IL, 1)
      ALFM_T(2) = ALFM(IL, 2)
```

```
      ALFM_T(3) = ALFM(IL, 6)

      IF(LAYER(IL).LT.0) THEN
       EUM1(IL, 1) = 0.01*EUM(IL, 1)
      EUM1(IL, 3) = 0.01*EUM(IL, 3)
      EUM1(IL, 5) = 0.01*EUM(IL, 5)
      EUM1(IL, 2) = EUM(IL, 2)
      EUM1(IL, 4) = EUM(IL, 4)
      ELSE
       EUM1(IL, 1) = EUM(IL, 1)
      EUM1(IL, 3) = EUM(IL, 3)
      EUM1(IL, 5) = EUM(IL, 5)
      EUM1(IL, 2) = EUM(IL, 2)
      EUM1(IL, 4) = EUM(IL, 4)
      ENDIF

      A1 = EUM1(IL, 1)/EUF(IL, 1)
      A2 = A_BETA+(1.0-A_BETA)*EUM1(IL, 3)/EUF(IL, 3)
      A3 = A_ALFA+(1.0-A_ALFA)*EUM1(IL, 5)/EUF(IL, 5)
      CALL BRIGA(VF, SF_T, SM_T, A_T, B_T, A1, A2, A3)
       VM = 1.-VF
       DO 10 I = 1, 3
       DO 10 J = 1, 3
10     S1(I, J) = SF_T(I, J)-SM_T(I, J)
       CALL INVER(S1, S2, 3, L)
       DO 30 I = 1, 3
      DO 20 J = 1, 3
20    S1(I, J) = 0.
      S1(I, I) = 1.
      DO 30 J = 1, 3
      DO 30 K = 1, 3
30    S1(I, J) = S1(I, J)-A_T(I, K)*B_T(K, J)
      DO 40 I = 1, 3
      DO 40 J = 1, 3
      S3(I, J) = 0.
      DO 40 K = 1, 3
40    S3(I, J) = S3(I, J)+S1(I, K)*S2(K, J)
      DO 50 I = 1, 3
      BM(IL, I) = 0.
      DO 50 J = 1, 3
50    BM(IL, I) = BM(IL, I)+S3(I, J)*(ALFM_T(J)-ALFF_T(J))
      DO 60 I = 1, 3
       BLFT1(IL, I) = VF*ALFF_T(I)+VM*ALFM_T(I)
      DO 60 J = 1, 3
60    BLFT1(IL, I) = BLFT1(IL, I)+VM*(SM_T(I, J)-SF_T(I, J))*BM(IL, J)
       BLFT(IL, 1) = BLFT1(IL, 1)
      BLFT(IL, 2) = BLFT1(IL, 2)
      BLFT(IL, 3) = BLFT1(IL, 2)
       BLFT(IL, 4) = 0.
       BLFT(IL, 5) = 0.
       BLFT(IL, 6) = BLFT1(IL, 3)
```

```
      RETURN
      END
C*****************************************************************
      SUBROUTINE BRIGA(VF, SF_T, SM_T, A_T, B_T, A1, A2, A3)
      IMPLICIT DOUBLE PRECISION (A-H, O-Z)
     DIMENSION A_T(3, 3), B_T(3, 3), SF_T(3, 3), SM_T(3, 3)
     DO 10 I = 1, 3
     DO 10 J = 1, 3
     A_T(I, J) = 0.
10   B_T(I, J) = 0.
      VM = 1.–VF
     A_T(1, 1) = A1
     A_T(2, 2) = A2
     A_T(3, 3) = A3
     IF(SF_T(1, 1).NE.SM_T(1, 1)) A_T(1, 2) = (SF_T(1, 2)
     &-SM_T(1, 2))*(A_T(1, 1)-A_T(2, 2))/(SF_T(1, 1)
     &-SM_T(1, 1))
      D1 = (SM_T(1, 3)–SF_T(1, 3))*(A_T(1, 1)–A_T(3, 3))
      D2 = (SM_T(2, 3)–SF_T(2, 3))*(VF+VM*A_T(1, 1))*(A_T(2, 2)
     &      -A_T(3, 3))+(SM_T(1, 3)-SF_T(1, 3))
     &      *(VF+VM*A_T(3, 3))*A_T(1, 2)
      B11 = SM_T(1, 2)–SF_T(1, 2)
      B12 = SM_T(1, 1)–SF_T(1, 1)
      B22 = (VF+VM*A_T(2, 2))*B11
      B21 = -VM*B11*A_T(1, 2)–(VF+VM*A_T(1, 1))
     &      *(SF_T(2, 2)-SM_T(2, 2))
      IF((B11*B22-B12*B21).NE.0.) THEN
      A_T(1, 3) = (D2*B11–D1*B21)/(B11*B22–B12*B21)
      A_T(2, 3) = (D1*B22–D2*B12)/(B11*B22–B12*B21)
      ENDIF
      C = (VF+VM*A_T(1, 1))*(VF+VM*A_T(2, 2))
     &    *(VF+VM*A_T(3, 3))
      B_T(1, 1) = (VF+VM*A_T(2, 2))*(VF+VM*A_T(3, 3))/C
      B_T(1, 2) = –VM*A_T(1, 2)*(VF+VM*A_T(3, 3))/C
      B_T(1, 3) = (VM**2*A_T(1, 2)*A_T(2, 3)–(VF
     &                +VM*A_T(2, 2))*VM*A_T(1, 3))/C
      B_T(2, 2) = (VF+VM*A_T(1, 1))*(VF+VM*A_T(3, 3))/C
      B_T(2, 3) = –VM*A_T(2, 3)*(VF+VM*A_T(1, 1))/C
      B_T(3, 3) = (VF+VM*A_T(2, 2))*(VF+VM*A_T(1, 1))/C
     RETURN
     END
C*****************************************************************
      SUBROUTINE UPDATE(ML, NL, NL2, SSP, SSF, SSM, A, B, BM,
     &                          LAYER, TS, VF, DTEMP, NINT)
      IMPLICIT DOUBLE PRECISION (A-H, O-Z)
      DIMENSION SSP(ML, 6), SSF(ML, 6), SSM(ML, 6), A(ML, 6, 6), B(ML, 6, 6),
     &             BM(ML, 3), LAYER(ML), TS(ML, 6, 6), VF(ML), S(6),
     &             BM1(ML, 6)
C*** TO UPDATE INTERNAL STRESSES OF EACH PLY.
     IF(NINT.EQ.1) THEN
      DO 100 I1 = 1, NL2
```

```
      IL = 2*I1-1
      DO 10 I = 1, 6
      BM1(IL, I) = 0.
      S(I) = 0.
      DO 10 J = 1, 6
 10   S(I) = S(I)+TS(IL, J, I)*SSP(IL, J)
      BM1(IL, 1) = BM(IL, 1)
      BM1(IL, 2) = BM(IL, 2)
      BM1(IL, 6) = BM(IL, 3)
      DO 20 I = 1, 6
      SSF(IL, I) = SSF(IL, I)-(1.-VF(IL))*BM1(IL, I)*DTEMP*1.E-6/VF(IL)
      SSM(IL, I) = SSM(IL, I)+BM1(IL, I)*DTEMP*1.E-6
      DO 20 J = 1, 6
      SSF(IL, I) = SSF(IL, I)+B(IL, I, J)*S(J)
      DO 20 K = 1, 6
 20   SSM(IL, I) = SSM(IL, I)+A(IL, I, K)*B(IL, K, J)*S(J)
100   CONTINUE
      DO 300 I2 = 1, NL2–1
      IL = 2*I2
      DO 300 I = 1, 6
      SSF(IL, I) = 0.
300   SSM(IL, I) = SSM(IL, I)+SSP(IL, I)

      ELSE
      DO 1100 IL = 1, NL2
      DO 1010 I = 1, 6
      BM1(IL, I) = 0.
      S(I) = 0.
      DO 1010 J = 1, 6
1010  S(I) = S(I)+TS(IL, J, I)*SSP(IL, J)
      BM1(IL, 1) = BM(IL, 1)
      BM1(IL, 2) = BM(IL, 2)
      BM1(IL, 6) = BM(IL, 3)
      DO 1020 I = 1, 6
      SSF(IL, I) = SSF(IL, I)-(1.-VF(IL))*BM1(IL, I)*DTEMP*1.E-6/VF(IL)
      SSM(IL, I) = SSM(IL, I)+BM1(IL, I)*DTEMP*1.E-6
      DO 1020 J = 1, 6
      SSF(IL, I) = SSF(IL, I)+B(IL, I, J)*S(J)
      DO 1020 K = 1, 6
1020  SSM(IL, I) = SSM(IL, I)+A(IL, I, K)*B(IL, K, J)*S(J)
1100  CONTINUE
      ENDIF
      RETURN
      END
C*************************************************************
      SUBROUTINE OVERSTIF(ML, NL, LAYER, TC, SP, CG, ALFT, BTE)
      IMPLICIT DOUBLE PRECISION (A-H, O-Z)
      DIMENSION LAYER(ML), TC(ML, 6, 6), SP(ML, 6, 6), CG(ML, 6, 6),
     &          ALFT(ML, 6), BTE(ML, 6), S(6, 6), C(6, 6), CG1(3, 3), CG2(3, 3)
      DO 100 IL = 1, NL
      DO 10 I = 1, 6
```

```
      DO 10 J = 1, 6
 10   S(I, J) = SP(IL, I, J)
      CALL INVER(S, C, 6, L)
      DO 30 I = 1, 6
      BTE(IL, I) = 0.
      DO 30 J = 1, 6
      CG(IL, I, J) = 0.
      DO 30 K = 1, 6
      BTE(IL, I) = BTE(IL, I)+TC(IL, I, J)*C(J, K)*ALFT(IL, K)
      DO 30 L = 1, 6
      CG(IL, I, J) = CG(IL, I, J)+TC(IL, I, K)*C(K, L)*TC(IL, J, L)
 30   CONTINUE
 100  CONTINUE
      RETURN
      END
C***************************************************************************
      SUBROUTINE UDTAPE(ML, NL, LAYER, VF, EUF, EUM, SF, SM, A, BX,
      &                   SP, ALFF, ALFM, ALF, BM, NL2, A_BETA,
      &                   A_ALFA, NINT)
      IMPLICIT DOUBLE PRECISION (A-H, O-Z)
      DIMENSION LAYER(ML), VF(ML), EUF(ML, 5), EUM(ML, 5),
      &          SF(ML, 6, 6), SM(ML, 6, 6), A(ML, 6, 6), BX(ML, 6, 6),
      &          SP(ML, 6, 6), ALFF(ML, 6),
      &          ALFM(ML, 6), ALF(ML, 6), BM(ML, 6), C(6, 6), D(6, 6)
      IF(NINT.EQ.1) THEN
      DO 100 I = 1, NL2
       IL = 2*I-1
      CALL BRIDGE2(ML, IL, LAYER, EUF, EUM, SF, SM, VF, A, BX,
      &             SP, A_BETA, A_ALFA)
      CALL THERM(ML, IL, VF, SF, SM, ALFF, ALFM, ALF, BM, EUF,
      &             EUM, LAYER, A_BETA, A_ALFA)
100   CONTINUE
      DO 200 I = 1, NL2-1
      IL = 2*I
      DO 200 J1 = 1, 6
      DO 200 J2 = 1, 6
      SP(IL, J1, J2) = SM(IL, J1, J2)
      ALF(IL, J1) = ALFM(IL, J1)
      BM(IL, J1) = 1.
200   CONTINUE
      ELSE
      DO 300 IL = 1, NL2
      CALL BRIDGE2(ML, IL, LAYER, EUF, EUM, SF, SM, VF, A, BX,
      &             SP, A_BETA, A_ALFA)
      CALL THERM(ML, IL, VF, SF, SM, ALFF, ALFM, ALF, BM, EUF,
      &             EUM, LAYER, A_BETA, A_ALFA)
300   CONTINUE
      ENDIF
      RETURN
      END
C*****************************************************************
```

```
      SUBROUTINE BRIDGE2(ML, IL, LAYER, EUF, EUM, SF, SM, VF, A,
     &                  BX, SP, A_BETA, A_ALFA)
      IMPLICIT DOUBLE PRECISION (A-H, O-Z)
      DIMENSION LAYER(ML), EUF(ML, 5), EUM(ML, 5), VF(ML),
     &          SF(ML, 6, 6), SM(ML, 6, 6), A(ML, 6, 6), BX(ML, 6, 6),
     &          SP(ML, 6, 6), C(6, 6), D(6, 6), SP2(6, 6), EUM1(ML, 5)
     IF(LAYER(IL).LT.0) THEN
      EUM1(IL, 1) = 0.01*EUM(IL, 1)
     EUM1(IL, 3) = 0.01*EUM(IL, 3)
     EUM1(IL, 5) = 0.01*EUM(IL, 5)
     EUM1(IL, 2) = EUM(IL, 2)
     EUM1(IL, 4) = EUM(IL, 4)
     ELSE
      DO 5 I = 1, 5
5     EUM1(IL, I) = EUM(IL, I)
     ENDIF
     DO 10 I = 1, 6
     DO 10 J = 1, 6
     BX(IL, I, J) = 0.
10   A(IL, I, J) = 0.
      CALL COUPLE(ML, IL, EUF, EUM1, SF, SM, VF, C, D,
     &                 SP2, A_BETA, A_ALFA)
     DO 20 I = 1, 6
     DO 20 J = 1, 6
     A(IL, I, J) = C(I, J)
     BX(IL, I, J) = D(I, J)
20   SP(IL, I, J) = SP2(I, J)
      RETURN
     END
C*****************************************************************
      SUBROUTINE PROSS(ML, NL2, MUM, MUF, MEM, MEF, NL, NEM,
     &   NEF, MSEG, IPRIT, LFAIL, NSTP, NQ, TL, TU, SS, SA, DSS,
     &   A_BETA, A_ALFA, dis_z, RESIN, FIBER, SSF, SSM, SSP,
     &   ISPLY, VF, LAYER, Z, CG, TS, TC, ALFT, FAIL, SA1, NINT,
     &   JSTP)
      IMPLICIT DOUBLE PRECISION (A-H, O-Z)
      DIMENSION RESIN(2, MUM, MEM), FIBER(MUF, MEF), SSF(ML, 6),
     &           SFAIL(ML, 3), SSM(ML, 6), SSP(ML, 6), EUF(ML, 5),
     &           EUM(ML, 5), SF(ML, 6, 6), SM(ML, 6, 6), SUF(ML, 2),
     &           SUUM(ML, 2), ALFF(ML, 6), ALFM(ML, 6), VF(ML),
     &           LAYER(ML), Z(2, ML+1), SP(ML, 6, 6), CG(ML, 6, 6),
     &           TS(ML, 6, 6), TC(ML, 6, 6), BTE(ML, 6), ALF(ML, 6),
     &           FAIL(ML, 9), A(ML, 6, 6), BM(ML, 6), B(ML, 6, 6),
     &           SS(6), SA(6), DSS(6), SM_Y(ML), CG0(3, 3), DSA(3),
     &           SA1(ML, 3), DSA1(ML, 3), SWRITE(ML, 6), SE1(ML, 2)
C*** CALCULATE COMPLIANCE AND INTERNAL STRESSES OF
C*** EACH LAMINA LAYER AT GIVEN MECHANICAL STRESS,
C*** "DSS", & TEMPERATURE, "DTEMP".
C*** TL - STARTING TEMPERATURE;
C*** TU - ENDING TEMPERATURE; NSTP–TOTAL STEPS
      DTEMP = (TU–TL)/REAL(NSTP)
```

```
      TEMP = TL–DTEMP
     ISTP = 1
     DO 10 IL = 1, NL
10   SM_Y(IL) = 0.
      DO 100 ISP = 1, NSTP
      TEMP = TEMP+DTEMP
      CALL FIBERS(ML, NL, MUF, MEF, NEF, TEMP, FIBER, EUF,
     &               SF, SUF, ALFF)
      CALL MATRIX(ML, NL, MUM, MEM, NEM, MSEG, LAYER,
     &               TEMP, RESIN, SSM, EUM, SM, SUUM, ALFM, SM_Y)
     CALL UDTAPE(ML, NL, LAYER, VF, EUF, EUM, SF, SM, A, B, SP,
     &               ALFF, ALFM, ALF, BM, NL2, A_BETA, A_ALFA, NINT)
     CALL OVERSTIF(ML, NL, LAYER, TC, SP, CG, ALF, BTE)
      CALL STRESS(ML, NL, IDSEC, LAYER, Z, CG, SSP, BTE, DSS, DTEMP,
     &               SS, SA, CG0, DSA, DSA1, SA1, dis_z)
      CALL UPDATE(ML, NL, NL2, SSP, SSF, SSM, A, B, BM, LAYER,
     &               TS, VF, DTEMP, NINT)
      CALL STRTH(ML, NL, NQ, SSF, SSM, SUF, SUUM, FAIL, LAYER,
     &              VF, SS, LFAIL, SWRITE, SE1, ID_F, SA, SA1,
     &              DIS_Z, ID_D, NINT, SFAIL, NFAIL, Z)
      CALL WRTE(ML, NL, IPRIT, LFAIL, ISP, NSTP, NQ, LAYER,
     &             FAIL, SA, ISPLY, SA1, CG0, ALF, SF, SM, SSF, SSM,
     &             SS, SP, SWRITE, dis_z, ID_F, ID_D,
     &             SFAIL, JSTP, ISTP, NFAIL)
 50   IF(ID_F.EQ.1.OR.ID_D.EQ.1) THEN
      return
     ENDIF
100   CONTINUE
      RETURN
210   FORMAT('FINAL FAILURE STRESS')
200   FORMAT(F8.2, 2X, F8.2, 2X, F8.2, 2X, F8.2, 2X, F8.2, 2X, F8.2, 2X)
      END
C*****************************************************************
      SUBROUTINE WRTE(ML, NL, IPRIT, LFAIL, ISP, NSTP, NQ, LAYER,
     &               FAIL, SA, ISPLY, SA1, C, ALFT, SF, SM, SSF, SSM,
     &               SS, SP, SWRITE, dis_z, ID_F, ID_D,
     &               SFAIL, JSTP, ISTP, nfail)
      IMPLICIT DOUBLE PRECISION (A-H, O-Z)
      DIMENSION SF(ML, 6, 6), SM(ML, 6, 6), SP(ML, 6, 6), LAYER(ML),
     &             FAIL(ML, 12), SS(6), SA(6), SA1(ML, 3), SSF(ML, 6),
     &             SSM(ML, 6), C(3, 3), S(3, 3), ALFT(ML, 6),
     &             SWRITE(ML, 6), SFAIL(ML, 6)
      IF(IPRIT.EQ.0) RETURN
      IF(IPRIT.EQ.4.AND.(ISP.EQ.NSTP.OR.ID_F.EQ.1.OR.ID_D.EQ.1))
     & THEN
      WRITE(20, 220)
      DO 180 IL = 1, NL
180   WRITE(20, 190) IL, (SSF(IL, I), I = 1, 6)
     WRITE(20, 225)
      DO 200 IL = 1, NL
200   WRITE(20, 190) IL, (SSM(IL, I), I = 1, 6)
```

```
      WRITE(20, 370)
      DO 360 IL = 1, NL
360   WRITE(20, 190) IL, (ALFT(IL, I), I = 1, 6)
       ENDIF
       IF(ISP.EQ.1.AND.IPRIT.LT.0) THEN
       WRITE(20, 110)
      WRITE(15, 120)
      WRITE(25, 125)
       ENDIF
       IF(ABS(IPRIT).GE.3) GOTO 130
       IF(ISP.EQ.2.OR.ISP.EQ.NSTP.OR.ID_F.EQ.1.OR.ID_D.EQ.1) THEN
      IF(ISP.NE.2.AND.ABS(IPRIT).EQ.1) GOTO 130
      WRITE(20, 230) ISP
       DO 100 IL = 1, NL
       WRITE(20, 40) IL
       DO 20 I = 1, 6
20     WRITE(20, 30) (SF(IL, I, J), J = 1, 6)
       WRITE(20, 50)
       DO 60 I = 1, 6
60     WRITE(20, 30) (SM(IL, I, J), J = 1, 6)
       WRITE(20, 70)
       DO 80 I = 1, 6
80     WRITE(20, 30) (SP(IL, I, J), J = 1, 6)
100    CONTINUE
       ENDIF
130    IF(IPRIT.LT.0.AND.ISP.EQ.ISTP) THEN
       WRITE(15, 140) ISP, (sa(i)*100., i = 1, 3), (ss(i), i = 1, 3)
      IF(ISPLY.NE.0) THEN
       WRITE(25, 140) ISP, (sa1(ISPLY, i)*100., i = 1, 2), dis_z*100.,
      &                (ss(i), i = 4, 6)
      ENDIF
      IF(LFAIL.NE.LFAIL0) THEN
      WRITE(15, 395) LFAIL, (–LAYER(IL), IL = 1, NL)
      WRITE(25, 395) LFAIL, (–LAYER(J), J = 1, NL)
      LFAIL0 = LFAIL
      ENDIF
      ISTP = ISTP+JSTP
       ENDIF
       IF(ID_D.NE.1.AND.ID_F.NE.1.AND.ISP.NE.NSTP) RETURN
      IF(ID_F.EQ.1) THEN
       WRITE(20, 150)
       DO 160 IL = 1, NL
160    WRITE(20, 170) IL, -LAYER(IL), (FAIL(IL, J), J = 1, 6)
       WRITE(20, 151)
       DO 161 IL = 1, NL
161    WRITE(20, 170) IL, -LAYER(IL), (FAIL(IL, J), J = 7, 12)
       IF(NFAIL.GT.0) THEN
      WRITE(20, 145)
      WRITE(20, 150)
      DO 162 JL_2 = NL+1, 2*NL
      IF(LAYER(JL_2).LT.0) THEN
```

```
      WRITE(20, 170) JL_2-NL, -LAYER(JL_2), (FAIL(JL_2, J), J = 1, 6)
 162  ENDIF
      WRITE(20, 151)
      DO 163 JL_2 = NL+1, 2*NL
      IF(LAYER(JL_2).LT.0) THEN
      WRITE(20, 170) JL_2-NL, -LAYER(JL_2), (FAIL(JL_2, J), J = 7, 12)
 163  ENDIF
      ENDIF
      ELSEIF(ID_D.EQ.1) THEN
      WRITE(20, 420)
      WRITE(20, 430)
       WRITE(20, 150)
       DO 570 IL = 1, NL
 570   WRITE(20, 170) IL, -LAYER(IL), (FAIL(IL, J), J = 1, 6)
       WRITE(20, 151)
       DO 571 IL = 1, NL
 571   WRITE(20, 170) IL, -LAYER(IL), (FAIL(IL, J), J = 7, 12)
      WRITE(20, 440)
      WRITE(20, 295) (SS(I), I = 1, 6)
      WRITE(20, 152)
      DO 572 IL = 1, NL
 572  WRITE(20, 195) IL, (SFAIL(IL, J), J = 1, 6)
      ELSE
      WRITE(20, 450)
      WRITE(20, 155)
      WRITE(20, 295) (SS(I), I = 1, 6)
      WRITE(20, 156)
      DO 167 IL = 1, NL
 167   WRITE(20, 195) IL, (swrite(il, j), j = 1, 6)
      ENDIF
       RETURN
 145   FORMAT(1X, 'SECOND PLY IN THE SAME LAYER')
 155  FORMAT(9X, 'LAMINATE STRESSES')
 156   FORMAT(1X, 'PLY', 6X, 'FIBER PRINCIPAL',
      &        1X, 'STRESSES', 4X, 'MATRIX PRINCIPAL STRESSES')
 195  FORMAT(I3, F7.1, 1X, F7.1, X, F7.1, 2X, 2(F7.1, 2X, F7.1, 10X))
 295   FORMAT(3x, F7.1, 1X, F7.1, X, F7.1, 2X, 2(F7.1, 2X, F7.1, 10X))
 40    FORMAT(30X, 'PLY = ', I4/
      &        20X, 'FIBER COMPLIANCE & MODULI:')
 30    FORMAT(5X, 3(E11.4, 2X))
 10    FORMAT(2X, 'E11 = ', E11.4, 1X, 'U12 = ', F5.3, 1X, 'E22 = ', E11.4,
      &        1X, 'U23 = ', F5.3, 1X, 'G12 = ', E11.4)
 50    FORMAT(20X, 'MATRIX COMPLIANCE & MODULI:')
 70    FORMAT(10X, 'LAMINA COMPLIANCE & MODULI IN PLY-SYSTEM:')
 110   FORMAT(/24X, 'LAMINATE PROPERTIES')
 120  FORMAT(20X, 'LAMINATE IN-PLANE STRAINS AND STRESSES'/
      &        1X, 'STEP', 2X, 'XX-STRAIN(%)', 1X, 'YY-STRAIN(%)',
      &        1X, 'XY-STRAIN(%)', 3X, 'XX-STRESS',
      &        1X, 'YY-STRESS', 1X, 'XY-STRESS'/)
 125  FORMAT(20X, 'LAMINATE OUT-OF-PLANE STRAINS AND
      &      STRESSES'/1X, 'STEP', 2X, 'yz-STRAIN(%)', 1X, 'xz-STRAIN(%)',
```

```
     &          1X, 'zz-STRAIN(%)', 3X, 'yz-STRESS',
     &          1X, 'xz-STRESS', 1X, 'zz-STRESS'/)
 140 FORMAT(I5, 4X, F9.4, 2(5X, F9.4), 7X, F8.2, 2(3X, F8.2))
 16  FORMAT(5X, 'E11 = ', E11.4, 1X, 'U12 = ', F5.3, 1X, 'E22 = ', E11.4, 1X,
     &          'G12 = ', E11.4)
 150 FORMAT('PLY', 1X, 'F-ORDER', 9X, 'FAILURE STRESSES,
     & (XX, YY, XY, YZ, XZ ZZ)')
 151  FORMAT('PLY', 1X, 'F-ORDER', 4X, 'FIBER PRINCIPAL STRESSES', 4X,
     &          'MATRIX PRINCIPAL STRESSES')
 152  FORMAT('PLY', 1X, 4X, 'FIBER PRINCIPAL STRESSES', 4X,
     &          'MATRIX PRINCIPAL STRESSES')
 170  FORMAT(I2, 3X, I3, 2X, 3(F7.1, 1X, F7.1, 1X, F7.1, 2X))
 220  FORMAT(20X, 'FIBER STRESSES:')
 225 FORMAT(20X, 'MATRIX STRESSES:')
 370 FORMAT(19X, 'C.T.Es OF LAMINAE:')
 190  FORMAT(1X, 'PLY = ', I3, 4X, 3(F8.2, 2X))
 230 FORMAT(/20X, 'RESULTS AT SOLUTION STEP = ', I4)
 250 FORMAT(17X, 'FIBER-STRESSES', 26X, 'MATRIX-STRESSES')
 260 FORMAT(3(F8.2, 2X), 3X, 3(F8.2, 2X))
 280 FORMAT(33x, 'Lamina ply = ', i3)
 320  FORMAT(18X, 'LAMINATE COMPLIANCE MATRIX & MODULI:')
 340  FORMAT(18X, 3(E11.4, 2X))
 350  FORMAT(10X, 'Exx = ', E11.4, 1X, 'Uxy = ', F5.3, 1X, 'Eyy = ', E11.4, 1X,
     &          'Gxy = ', E11.4)
 390  FORMAT(I5, 4(3X, F9.4), 3X, F8.2)
 395 FORMAT(1X, 'Failed Layers = ', I2, 2X, 'Plies:', 20(I2, 1X))
 410  FORMAT(5X, 2(I2, 1X), 3X, 3(F7.1, 1X, F7.1, 1X, F7.1, 4X))
 420 FORMAT(5x, /'ULTIMATE FAILURE OCCURS DUE TO EXTREME LARGE
STRAIN'/)
 430  FORMAT (5x, 'THE PROGRESSIVE FAILURES BEFORE THE LARGER
     &         STRAIN')
 440  FORMAT(1X, 'THE ULTIMATE FAILURE STRESS
     &         (XX, YY, XY, YZ, XZ, ZZ)')
 450  FORMAT(1X, 'ULTIMATE FAILURE HAS NOT OCCURED')
      END
C***********************************************************************
      SUBROUTINE STRTH(ML, NL, NQ, SSF, SSM, SUF, SUUM, FAIL,
     &                    LAYER, VF, SS, LFAIL, SWRITE, SE1, ID_F,
     &                    SA, SA1, DIS_Z, ID_D, NINT,
     &                    SFAIL, NFAIL, Z)
     USE IMSL
      IMPLICIT DOUBLE PRECISION (A-H, O-Z)
      DIMENSION SSF(ML, 6), SSM(ML, 6), SUF(ML, 2), SUUM(ML, 2), Z(2, ML+1),
     &             FAIL(ML, 12), LAYER(ML), VF(ML), SS(6), S(6),
     &             SFAIL(ML, 6), COEF(4), SS2(3, 3), SWRITE(ML, 6),
     &             SE1(ML, 2), SA(6), SA1(ML, 3)
     DOUBLE COMPLEX C(3)
     NFAIL = 0
     ID_F2 = 0
     DO 5 I = 1, 3
5    IF(ABS(SA(I))*100.GE.12.or.ABS(dis_z)*100.GE.12) ID_D = 1
```

```
      DO 6 IL = 1, NL
        DO I = 1, 3
        IF(ABS(SA(I)+SA(I+3)*(Z(1, IL+1)+Z(1, IL))/2.0)*100.GE.12) ID_D = 1
        END DO
6     CONTINUE
      IF(NINT.EQ.1) THEN
      DO 10 IL = 1, NL
      DO 10 JL = 1, NL
      IF(IL.EQ.(2*JL–1)) THEN
      IF(ABS(SA1(IL, 1))*100.GE.12.or.ABS(SA1(IL, 2))*100.GE.12) ID_D = 1
      ENDIF
10    CONTINUE
      ELSE
      DO 510 IL = 1, NL
      IF(ABS(SA1(IL, 1))*100.GE.12.or.ABS(SA1(IL, 2))*100.GE.12) ID_D = 1
510    CONTINUE
      ENDIF

      DO 100 IL = 1, NL
      ID_F = 0
      ID = 0
      SS2(1, 1) = SSF(IL, 1)
      SS2(2, 2) = SSF(IL, 2)
      SS2(3, 3) = SSF(IL, 3)
      SS2(1, 2) = SSF(IL, 6)
      SS2(1, 3) = SSF(IL, 5)
      SS2(2, 1) = SSF(IL, 6)
      SS2(3, 1) = SSF(IL, 5)
      SS2(2, 3) = SSF(IL, 4)
      SS2(3, 2) = SSF(IL, 4)
      COEF(1) = -(SS2(1, 1)*SS2(2, 2)*SS2(3, 3)+SS2(1, 2)*SS2(2, 3)*SS2(3, 1)+
     &   SS2(1, 3)*SS2(2, 1)*SS2(3, 2)–SS2(3, 1)*SS2(2, 2)*SS2(1, 3)–
     &   SS2(1, 1)*SS2(2, 3)*SS2(3, 2)–SS2(1, 2)*SS2(2, 1)*SS2(3, 3))
      COEF(2) = SSF(IL, 1)*SSF(IL, 2)+SSF(IL, 1)*SSF(IL, 3)+SSF(IL, 2)
     &          *SSF(IL, 3)-SSF(IL, 4)*SSF(IL, 4)–SSF(IL, 5)*SSF(IL, 5)
     &          –SSF(IL, 6)*SSF(IL, 6)
      COEF(3) = –(SSF(IL, 1)+SSF(IL, 2)+SSF(IL, 3))
      COEF(4) = 1
      CALL DZPLRC(3, COEF, C)
      S1 = C(1)
      S2 = C(2)
      S3 = C(3)
      IF(S1.LT.S2) THEN
      SHUAN = S2
      S2 = S1
      S1 = SHUAN
      ENDIF
      IF(S1.LT.S3) THEN
      SHUAN = S3
      S3 = S1
      S1 = SHUAN
```

```
      ENDIF
      IF(S3.GT.S2) THEN
      SHUAN = S3
      S3 = S2
      S2 = SHUAN
      ENDIF
      SE = S1
      SWRITE(IL, 1) = S1
      SWRITE(IL, 2) = S2
      SWRITE(IL, 3) = S3
       QN = 1./REAL(NQ)
       IF(S3.GE.(-0.000001)) SE = (S1**NQ+S2**NQ+S3**NQ)**QN
      SE1(IL, 1) = SE
      IF(ID_D.EQ.1) THEN
      SFAIL(IL, 1) = S1
      SFAIL(IL, 2) = S2
      SFAIL(IL, 3) = S3
      GOTO 5000
      ENDIF
!**************************************************************
!     modified
      IF(S1.le.(-0.000001)) s3 = s3-s1
       IF(SE.GE.SUF(IL, 1)) ID_F = 1
       IF(S3.LE.(-SUF(IL, 2))) ID_F = 1
!**************************************************************
5000   CONTINUE
       SS2(1, 1) = SSM(IL, 1)
      SS2(2, 2) = SSM(IL, 2)
      SS2(3, 3) = SSM(IL, 3)
       SS2(1, 2) = SSM(IL, 6)
      SS2(1, 3) = SSM(IL, 5)
      SS2(2, 1) = SSM(IL, 6)
      SS2(3, 1) = SSM(IL, 5)
       SS2(2, 3) = SSM(IL, 4)
      SS2(3, 2) = SSM(IL, 4)
       COEF(1) = -(SS2(1, 1)*SS2(2, 2)*SS2(3, 3)+SS2(1, 2)*SS2(2, 3)*SS2(3, 1)+
      &   SS2(1, 3)*SS2(2, 1)*SS2(3, 2)-SS2(3, 1)*SS2(2, 2)*SS2(1, 3)-
      &   SS2(1, 1)*SS2(2, 3)*SS2(3, 2)-SS2(1, 2)*SS2(2, 1)*SS2(3, 3))
      COEF(2) = SSM(IL, 1)*SSM(IL, 2)+SSM(IL, 1)*SSM(IL, 3)+SSM(IL, 2)
      &          *SSM(IL, 3)-SSM(IL, 4)*SSM(IL, 4)-SSM(IL, 5)*SSM(IL, 5)
      &          -SSM(IL, 6)*SSM(IL, 6)
      COEF(3) = -(SSM(IL, 1)+SSM(IL, 2)+SSM(IL, 3))
      COEF(4) = 1
      CALL DZPLRC(3, COEF, C)
       S1M = C(1)
      S2M = C(2)
      S3M = C(3)
      IF(S1M.LT.S2M) THEN
      SHUAN = S2M
      S2M = S1M
      S1M = SHUAN
```

```
      ENDIF
      IF(S1M.LT.S3M) THEN
      SHUAN = S3M
      S3M = S1M
      S1M = SHUAN
      ENDIF
      IF(S3M.GT.S2M) THEN
      SHUAN = S3M
      S3M = S2M
      S2M = SHUAN
      ENDIF
      SWRITE(IL, 4) = S1M
      SWRITE(IL, 5) = S2M
      SWRITE(IL, 6) = S3M
      IF(ID_D.EQ.1) THEN
      SFAIL(IL, 4) = S1M
      SFAIL(IL, 5) = S2M
      SFAIL(IL, 6) = S3M
      GOTO 100
      ENDIF
      SE = S1M
        IF(S1M.LT.0) THEN
      S2M = S2M–S1M
      S3M = S3M–S1M
      S1M = 0.
      ENDIF
        IF(S3M.GT.(-0.000001)) SE = (S1M**NQ+S2M**NQ+S3M**NQ)**QN
      SE1(IL, 2) = SE
        IF(LAYER(IL).LT.0) GOTO 20
        IF(SE.GE.SUUM(IL, 1)) ID = ID+1
20      CONTINUE
        IF(S3M.LE.( –SUUM(IL, 2))) THEN
      IF(NINT.EQ.1) THEN
      DO 30 JL_1 = 1, NL
      IF(IL.EQ.(2*JL_1–1)) ID_F = 1
30    CONTINUE
        ID = ID+1
      ELSE
      ID_F = 1
      ENDIF
      ENDIF
40      IF(ID.GT.0.OR.ID_F.EQ.1) THEN
      JL = IL
        IF(LAYER(IL).LT.0.AND.ID_F.EQ.0) GOTO 120
      IF(LAYER(IL).LT.0.AND.ID_F.EQ.1) THEN
      JL = NL+IL
      NFAIL = NFAIL+1
      ENDIF
        LFAIL = LFAIL+1
        LAYER(JL) = –LFAIL
        DO 50 I = 1, 6
```

```
50      FAIL(JL, I) = SS(I)
        FAIL(JL, 7) = S1
        FAIL(JL, 8) = S2
       FAIL(JL, 9) = S3
        FAIL(JL, 10) = S1M
        FAIL(JL, 11) = S2M
        FAIL(JL, 12) = S3M
120    CONTINUE
       IF(NL.EQ.1.AND.ID.GT.0) ID_F = 1
       IF(ID_F.EQ.1) ID_F2 = 1
        ENDIF
 100    CONTINUE
        ID_F = ID_F2
        RETURN
        END
```

6.5 Examples

Four examples containing either UD lamina or multi-directional laminate are analyzed using the computer routine given above. Illustrations with these examples are given below by showing the input data and the output results. The simulated results include an ultimate failure strength, progressive failure information, internal stresses in the fiber and matrix and stress/strain curves of the laminate, depending on different requirements. In the following, for easy understanding, the words and sentences underlined are explanative, whereas those without underlining are input or output data.

6.5.1 Example 6-1

The first example is a UD lamina made of E-glass and MY750/HY917/DY063 epoxy matrix, which is one of the material systems used in the WWFE-I exercise. The constituent properties, the fiber volume fraction and the bridging parameters β and α used for this composite are given in Section 5.14, see Tables 5.15 – 5.19 for details. In this example, the UD lamina is subjected to a longitudinal tensile load until a failure is attained. The objective is to obtain the longitudinal tensile strength of this lamina as well as the internal stress states in the fiber and matrix when the failure occurs. The minimum longitudinal load is set to 0, whereas the maximum load is assigned to be 3,000 MPa. The load step number of 6,000 is chosen for the analysis of this problem.

The input data for this example are listed as follows:

```
3,          !NQ
```

```
0.45,0.35,    !BETA,ALFA
1,6000,4,1,0,0,0,0,0,0, !NL,NSTP,JPRIT,IDSOL,JSTP,IDRES,IDPLY,IDSEC,ISPLY,NINT
-1.0,1.0,    !Z(1,I)
1.0,             !Z(2,1)
0.,            !ANGLE(I)
120.0,25.0,25.0,25.0,
100,
0.0,0.0,0.0,0.0,0.0,0.0,3000.,0.,0.,0.0,0.0,0.,   !SSMIN,SSMAX
2,                               !NEF
74000.0,0.2,74000.0,0.2,30800.0,             !FIBER PROPERTIES
25.,120.,
74000.0,74000.0,
2092.8,2092.8,
1311.8,1311.8,
4.9,4.9,
4.9,4.9,
8,2,                                                  !MATRIX PROPERTIES
25.,120.,
58.,58.,
0.35,0.35,
60.8,60.8,
74.8,74.8,
32.6,32.6,
39.9,39.9,
46.8,46.8,
52.,52.,
55.6,55.6,
58.,58.,
60.1,60.1,
62.0,62.0,
3350.0,3350.0,
1698.0,1698.0,
1387.0,1387.0,
918.0,918.0,
542.0,542.0,
317.0,317.0,
244.0,244.0,
186.0,186.0,
32.6,32.6,
39.9,39.9,
46.8,46.8,
52.,52.,
55.6,55.6,
58.,58.,
60.1,60.1,
62.0,62.0,
3350.0,3350.0,
1698.0,1698.0,
1387.0,1387.0,
918.0,918.0,
542.0,542.0,
```

```
317.0,317.0,
244.0,244.0,
186.0,186.0,
0.60,                                   ! VF
```

In the input file, the words and phrases after the symbol "!" are for comment only and will not be read by the computer routine. After running the computer routine, the main results can be found in the data file "moduli_output.dat". The results for this problem are as follows.

```
INDEX IN MODIFIED MAXIMUM NORMAL STRESS CRITERION=   3
 Beta(in defining bridging elements a2) = 0.450   Alfa(in defining a3)=0.350

FIBER MODULI(E1,U12,E2,U23,G12) = 0.7400E+05 0.200 0.7400E+05 0.200 0.3080E+05

 TEMPERATURE POINTS (Nef) =    2   ON WHICH FIBER PROPERITES ARE
CHANGED
   T(1-Nef) =     25.0     120.0
  E1(1-Nef) = 74000.   74000.
  Su(1-Nef) = 2092.8   2092.8
Suc(1-Nef) = 1311.8   1311.8
Af1(1-Nef) =   4.900     4.900
Af2(1-Nef) =   4.900     4.900

  SEGMENTS (Mseg) OF THE MATRIX STRESS-STRAIN CURVE =     8

                    MATRIX PROPERTIES UNDER TENSION:
    TEMPERATURE POINTS (ON WHICH MATRIX PROPERTIES ARE VARIED) =    2
  TEMP.    Alfm     Poissons Ratio  Tensile Strength  Com. Strength
   25.0   58.00      0.350              60.8               74.8
  120.0   58.00      0.350              60.8               74.8
  TEMP.                       YILED  STRENGTH,  Ys(1--Nem)*check  YILED  and  not
YIELD
   25.0      32.6     39.9     46.8     52.0     55.6     58.0    60.1    62.0
   120.0     32.6     39.9     46.8     52.0     55.6     58.0    60.1    62.0
  TEMP.                    TANGENTENT MODULI, ET(1--Nem)
   25.0   3350.0   1698.0    1387.0    918.0   542.0   317.0   244.0   186.0
  120.0     3350.0    1698.0    1387.0     918.0      542.0     317.0     244.0     186.0

                  MATRIX PROPERTIES UNDER COMPRESSION:
    TEMPERATURE POINTS (ON WHICH MATRIX PROPERTIES ARE VARIED) =    2
  TEMP.    Alfm     Poissons Ratio  Tensile Strength  Com. Strength
   25.0   58.00      0.350              60.8               74.8
  120.0   58.00      0.350              60.8               74.8
  TEMP.                     YIELD STRENGTH, Ys(1--Nem)
   25.0        32.6      39.9      46.8      52.0      55.6      58.0      60.1      62.0
  120.0        32.6      39.9      46.8      52.0      55.6      58.0      60.1      62.0
  TEMP.                    TANGENTENT MODULI, ET(1--Nem)
   25.0    3350.0   1698.0   1387.0    918.0    542.0    317.0    244.0    186.0
  120.0    3350.0   1698.0   1387.0    918.0    542.0    317.0    244.0    186.0
```

```
PLY   PLY-ANGLE      Vf        Z(i)        Z(i+1)
 1      0.00       0.600     -1.000        1.000
```

In the first part of the results file, the basic information for the laminate is output. The user can check it and make necessary changes accordingly.

```
        THERMAL RESIDUAL STRESSES COOLING FROM     120.00     TO
25.00

                    FIBER STRESSES:
PLY=  1       -13.55      -7.32       0.00
  0.00      0.00       0.00
                    MATRIX STRESSES:
PLY=  1        20.33      10.98       0.00
  0.00      0.00       0.00
                    C.T.Es OF LAMINAE:
PLY=  1         6.46      21.67      21.67
  0.00      0.00       0.00
ULTIMATE FAILURE HAS NOT OCCURED
            LAMINATE STRESSES
      0.0      0.0      0.0      0.0      0.0        0.0
PLY      FIBER PRINCIPAL STRESSES    MATRIX PRINCIPAL STRESSES
 1    0.0    -7.3   -13.6     20.3     11.0          0.0
```

The above outputs are the thermal residual stresses with six components in the fiber and matrix if the stress free temperature TL is not equal to the initial working temperature T0.

```
                    FIBER STRESSES:
PLY=  1      2075.38      -7.32       0.00
  0.00      0.00       0.00
                    MATRIX STRESSES:
PLY=  1        60.68      10.98       0.00
  0.00      0.00       0.00
                    C.T.Es OF LAMINAE:
PLY=  1         5.05      21.47      21.47
  0.00      0.00       0.00
PLY F-ORDER            FAILURE STRESSES (XX,YY,ZZ,YZ,XZ,XY)
 1     1   1269.5     0.0     0.0      0.0     0.0     0.0
PLY F-ORDER    FIBER PRINCIPAL STRESSES     MATRIX PRINCIPAL STRESSES
 1     1   2075.4     0.0    -7.3     60.7    11.0     0.0
```

The above outputs are the ultimate failure stresses on the laminate and the internal stresses and the principal stresses in the fiber and matrix materials when the failure occurs.

6.5.2 Example 6-2

In this example, the same UD lamina as in Example 6-1 is used. However, a transverse compressive load is applied to the lamina. The objective of this example is to obtain the stress/strain curves of the lamina up to failure. Thus, the controlling Parameter JPRIT is set to –3 and JSTP is set to 10. The input data can be changed accordingly, which are given below.

```
3,              !NQ
0.45,0.35,   !BETA,ALFA
1,6000,-3,1,10,0,0,0,0,0, !NL,NSTP,JPRIT,IDSOL,JSTP,IDRES,IDPLY,IDSEC,ISPLY,NINT
-1.0,1.0,     !Z(1,I)
1.0,            !Z(2,1)
0.,            !ANGLE(I)
120.0,25.0,25.0,25.0,
100,
0.0,0.0,0.0,0.0,0.0,0.0,0.,-300.,0.,0.0,0.0,0.,   !SSMIN,SSMAX
2,                                  !NEF
74000.0,0.2,74000.0,0.2,30800.0,             !FIBER PROPERTIES
25.,120.,
74000.0,74000.0,
2092.8,2092.8,
1311.8,1311.8,
4.9,4.9,
4.9,4.9,
8,2,                                                !MATRIX PROPERTIES
25.,120.,
58.,58.,
0.35,0.35,
60.8,60.8,
74.8,74.8,
32.6,32.6,
39.9,39.9,
46.8,46.8,
52.,52.,
55.6,55.6,
58.,58.,
60.1,60.1,
62.0,62.0,
3350.0,3350.0,
1698.0,1698.0,
1387.0,1387.0,
918.0,918.0,
542.0,542.0,
317.0,317.0,
244.0,244.0,
186.0,186.0,
```

```
32.6,32.6,
39.9,39.9,
46.8,46.8,
52.,52.,
55.6,55.6,
58.,58.,
60.1,60.1,
62.0,62.0,
3350.0,3350.0,
1698.0,1698.0,
1387.0,1387.0,
918.0,918.0,
542.0,542.0,
317.0,317.0,
244.0,244.0,
186.0,186.0,
0.60,                                   ! VF
```

The output results in the data file "deflec1.dat" are shown as in the following formats:

```
                    LAMINATE IN-PLANE STRAINS AND STRESSES
 STEP   XX-STRAIN(%)  YY-STRAIN(%)  XY-STRAIN(%)     XX-STRESS  YY-STRESS
XY-STRESS

    1      0.0000       -0.0004       0.0000          0.00       -0.05       0.00
   11      0.0003       -0.0041       0.0000          0.00       -0.55       0.00
   21      0.0006       -0.0078       0.0000          0.00       -1.05       0.00
   31      0.0009       -0.0115       0.0000          0.00       -1.55       0.00
......
......
 2871      0.0802       -2.7834       0.0000          0.00     -143.55       0.00
 2881      0.0803       -2.8306       0.0000          0.00     -144.05       0.00
 2891      0.0805       -2.8780       0.0000          0.00     -144.55       0.00
```

Based on these data, the transverse compressive stress-strain curve can be plotted in Fig. 6.4.

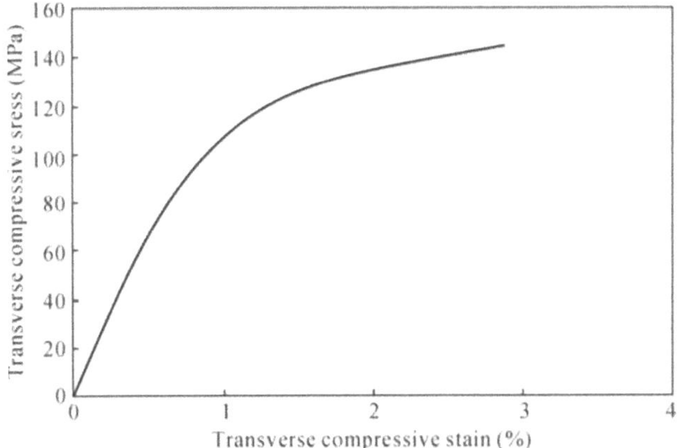

Fig. 6.4 Transverse compressive stress-strain curve for Example 6-2

6.5.3 *Example 6-3*

This is a multi-directional laminate made of $[55°/-55°]_s$ lay ups and having the same material system as used in Example 6-1. A uniaxial load is applied along the global *x*-direction. The ultimate failure strength and the progressive failure information are the main concerns. Pure matrix inter-layers are not introduced for this laminate. The data input file is shown as follows:

```
3,              !NQ
0.45,0.35,   !BETA,ALFA
4,6000,4,1,10,0,0,0,0,0, !NL,NSTP,JPRIT,IDSOL,JSTP,IDRES,IDPLY,IDSEC,ISPLY,NINT
-2.0,-1.0,0.,1.0,2.0,    !Z(1,I)
1.0,            !Z(2,1)
55.,-55.,-55.,55.,       !ANGLE(I)
120.0,25.0,25.0,25.0,
100,
0.0,0.0,0.0,0.0,0.0,0.0,2000.,0.,0.,0.0,0.0,0.,  !SSMIN,SSMAX
2,                               !NEF
74000.0,0.2,74000.0,0.2,30800.0,          !FIBER PROPERTIES
25.,120.,
74000.0,74000.0,
2092.8,2092.8,
1311.8,1311.8,
4.9,4.9,
4.9,4.9,
8,2,                                          !MATRIX PROPERTIES
25.,120.,
58.,58.,
0.35,0.35,
60.8,60.8,
```

```
74.8,74.8,
32.6,32.6,
39.9,39.9,
46.8,46.8,
52.,52.,
55.6,55.6,
58.,58.,
60.1,60.1,
62.0,62.0,
3350.0,3350.0,
1698.0,1698.0,
1387.0,1387.0,
918.0,918.0,
542.0,542.0,
317.0,317.0,
244.0,244.0,
186.0,186.0,
32.6,32.6,
39.9,39.9,
46.8,46.8,
52.,52.,
55.6,55.6,
58.,58.,
60.1,60.1,
62.0,62.0,
3350.0,3350.0,
1698.0,1698.0,
1387.0,1387.0,
918.0,918.0,
542.0,542.0,
317.0,317.0,
244.0,244.0,
186.0,186.0,
0.60,
```

The output file “moduli_output.dat” contains the following results:

```
  T(1-Nef)=    25.0     120.0
 E1(1-Nef)= 74000.   74000.
 Su(1-Nef)= 2092.8   2092.8
Suc(1-Nef)= 1311.8   1311.8
Af1(1-Nef)=   4.900    4.900
Af2(1-Nef)=   4.900    4.900

  SEGMENTS (Mseg) OF THE MATRIX STRESS-STRAIN CURVE =    8

                    MATRIX PROPERTIES UNDER TENSION:
    TEMPERATURE POINTS (ON WHICH MATRIX PROPERTIES ARE VARIED) =    2
  TEMP.    Alfm    Poissons Ratio   Tensile Strength   Com. Strength
   25.0    58.00      0.350              60.8               74.8
  120.0    58.00      0.350              60.8               74.8
```

TEMP.				YIELD STRENGTH, Ys(1--Nem)				
25.0	32.6	39.9	46.8	52.0	55.6	58.0	60.1	62.0
120.0	32.6	39.9	46.8	52.0	55.6	58.0	60.1	62.0
TEMP.				TANGENTENT MODULI, ET(1--Nem)				
25.0	3350.0	1698.0	1387.0	918.0	542.0	317.0	244.0	186.0
120.0	3350.0	1698.0	1387.0	918.0	542.0	317.0	244.0	186.0

MATRIX PROPERTIES UNDER COMPRESSION:

TEMPERATURE POINTS (ON WHICH MATRIX PROPERTIES ARE VARIED) = 2

TEMP.	Alfm	Poissons Ratio	Tensile Strength	Com. Strength
25.0	58.00	0.350	60.8	74.8
120.0	58.00	0.350	60.8	74.8

TEMP.				YIELD STRENGTH, Ys(1--Nem)				
25.0	32.6	39.9	46.8	52.0	55.6	58.0	60.1	62.0
120.0	32.6	39.9	46.8	52.0	55.6	58.0	60.1	62.0
TEMP.				TANGENTENT MODULI, ET(1--Nem)				
25.0	3350.0	1698.0	1387.0	918.0	542.0	317.0	244.0	186.0
120.0	3350.0	1698.0	1387.0	918.0	542.0	317.0	244.0	186.0

PLY	PLY-ANGLE	Vf	Z(i)	Z(i+1)
1	55.00	0.600	–2.000	–1.000
2	–55.00	0.600	–1.000	0.000
3	–55.00	0.600	0.000	1.000
4	55.00	0.600	1.000	2.000

The above results are the definition for the laminate to be analyzed, concerning constituent material properties and laminate geometric parameters.

THERMAL RESIDUAL STRESSES COOLING FROM 120.00 TO 25.00

FIBER STRESSES:

PLY = 1	–32.58	6.50	0.00
0.00	0.00	5.29	
PLY = 2	–32.58	6.50	0.00
0.00	0.00	–5.29	
PLY = 3	–32.58	6.50	0.00
0.00	0.00	–5.29	
PLY = 4	–32.58	6.50	0.00
0.00	0.00	5.29	

MATRIX STRESSES:

PLY = 1	21.58	17.54	0.00
0.00	0.00	1.99	
PLY = 2	21.58	17.54	0.00
0.00	0.00	–1.99	
PLY = 3	21.58	17.54	0.00
0.00	0.00	–1.99	
PLY = 4	21.58	17.54	0.00
0.00	0.00	1.99	

C.T.Es OF LAMINAE:

PLY = 1	6.46	21.67	21.67
0.00	0.00	0.00	
PLY = 2	6.46	21.67	21.67

```
      0.00          0.00          0.00
  PLY =  3               6.46          21.67          21.67
      0.00          0.00          0.00
  PLY =  4               6.46          21.67          21.67
      0.00          0.00          0.00
  ULTIMATE FAILURE HAS NOT OCCURRED
              LAMINATE STRESSES
          0.0        0.0        0.0        0.0        0.0             0.0
  PLY         FIBER PRINCIPAL STRESSES      MATRIX PRINCIPAL STRESSES
   1      7.2        0.0     -33.3        22.4       16.7              0.0
   2      7.2        0.0     -33.3        22.4       16.7              0.0
   3      7.2        0.0     -33.3        22.4       16.7              0.0
   4      7.2        0.0     -33.3        22.4       16.7              0.0
```

The above results are the thermal residual stresses in the fiber and matrix materials together with the thermal expansion coefficients of each ply in the laminate.

```
FIBER STRESSES:
  PLY =  1              74.12          54.25           1.59
      0.00          0.00        -74.17
  PLY =  2              74.12          54.25           1.59
      0.00          0.00         74.17
  PLY =  3              74.12          54.25           1.59
      0.00          0.00         74.17
  PLY =  4              74.12          54.25           1.59
      0.00          0.00        -74.17
                            MATRIX STRESSES:
  PLY =  1              21.54          42.58          -2.38
      0.00          0.00        -26.91
  PLY =  2              21.54          42.58          -2.38
      0.00          0.00         26.91
  PLY =  3              21.54          42.58          -2.38
      0.00          0.00         26.91
  PLY =  4              21.54          42.58          -2.38
      0.00          0.00        -26.91
                            C.T.Es OF LAMINAE:
  PLY =  1               4.90          18.47          18.47
      0.00          0.00          0.80
  PLY =  2               4.90          18.47          18.47
      0.00          0.00         -0.80
  PLY =  3               4.90          18.47          18.47
      0.00          0.00         -0.80
  PLY =  4               4.90          18.47          18.47
      0.00          0.00          0.80

ULTIMATE FAILURE OCCURS DUE TO EXTREMELY LARGE STRAIN
      PROGRESSIVE FAILURES BEFORE THE LARGER STRAIN
PLY F-ORDER              FAILURE STRESSES (xx,yy,xy,yz,xz,zz)
 1      1      101.7      0.0        0.0          0.0        0.0        0.0
```

```
 2    2   101.7     0.0      0.0      0.0      0.0      0.0
 3    3   101.7     0.0      0.0      0.0      0.0      0.0
 4    4   101.7     0.0      0.0      0.0      0.0      0.0
PLY F-ORDER    FIBER PRINCIPAL STRESSES    MATRIX PRINCIPAL STRESSES
 1    1   136.4     1.5    –10.2     60.8      3.9     –2.3
 2    2   136.4     1.5    –10.2     60.8      3.9     –2.3
 3    3   136.4     1.5    –10.2     60.8      3.9     –2.3
 4    4   136.4     1.5    –10.2     60.8      3.9     –2.3
 THE ULTIMATE FAILURE STRESS (XX,YY,ZZ,YZ,XZ,XY)
       102.7     0.0      0.0      0.0      0.0            0.0
PLY      FIBER PRINCIPAL STRESSES    MATRIX PRINCIPAL STRESSES
  1  139.0     1.6   –10.6     60.9      3.2            –2.4
  2  139.0     1.6   –10.6     60.9      3.2            –2.4
  3  139.0     1.6   –10.6     60.9      3.2            –2.4
  4  139.0     1.6   –10.6     60.9      3.2            –2.4
```

In the above outputs, the failure information is shown. It is seen that the ultimate failure of this laminate is caused by an extremely large strain. Before that, four plies fail simultaneously at 101.7 MPa due to a matrix tensile failure. Then the ultimate failure occurs at 102.7 MPa. The fiber and matrix principal stresses at each failure are provided.

6.5.4 Example 64

This example is the same as Example 6-3. The only difference is that pure matrix inter-layers are incorporated into the analysis. No other change for the input data listed in the above example 6–3 is made except that the controlling parameter NINT (in the third line of the input data file) is set to 1. Thus the full input file will not be repeated here, and only some different output results are shown for illustration.

In the first part of the output file, the laminate lay-up information is modified to:

```
  PLY   PLY-ANGLE      Vf        Z(i)       Z(i+1)
   1      55.00      0.615     –2.000      –1.025
   2       0.00      0.000     –1.025      –0.975
   3     –55.00      0.632     –0.975      –0.025
   4       0.00      0.000     –0.025       0.025
   5     –55.00      0.632      0.025       0.975
   6       0.00      0.000      0.975       1.025
   7      55.00      0.615      1.025       2.000
```

The later analysis is then based on the modified "7" layers of the laminate and failure information is more complicated than that for Example 6-3.

```
ULTIMATE FAILURE OCCURS DUE TO EXTREMELY LARGE STRAIN
       PROGRESSIVE FAILURES BEFORE THE LARGER STRAIN
```

PLY	F-ORDER	FAILURE STRESSES (xx,yy,xy,yz,xz,zz)					
1	3	101.3	0.0	0.0	0.0	0.0	0.0
2	5	101.7	0.0	0.0	0.0	0.0	0.0
3	1	100.7	0.0	0.0	0.0	0.0	0.0
4	6	101.7	0.0	0.0	0.0	0.0	0.0
5	2	100.7	0.0	0.0	0.0	0.0	0.0
6	7	101.7	0.0	0.0	0.0	0.0	0.0
7	4	101.3	0.0	0.0	0.0	0.0	0.0

PLY	F-ORDER	FIBER PRINCIPAL STRESSES			MATRIX PRINCIPAL STRESSES		
1	3	135.4	1.4	−11.2	61.0	3.9	−2.3
2	5	0.0	0.0	0.0	60.8	9.3	0.0
3	1	134.7	1.3	−10.4	60.9	4.3	−2.3
4	6	0.0	0.0	0.0	60.8	9.3	0.0
5	2	134.7	1.3	−10.4	60.9	4.3	−2.3
6	7	0.0	0.0	0.0	60.8	9.3	0.0
7	4	135.4	1.4	−11.2	61.0	3.9	−2.3

THE ULTIMATE FAILURE STRESS (XX,YY,ZZ,YZ,XZ,XY)

102.7	0.0	0.0	0.0	0.0	0.0

PLY	FIBER PRINCIPAL STRESSES			MATRIX PRINCIPAL STRESSES		
1	138.4	1.5	−11.4	61.1	3.1	−2.4
2	0.0	0.0	0.0	60.9	8.8	0.0
3	138.8	1.4	−10.0	61.0	3.4	−2.4
4	0.0	0.0	0.0	60.9	8.8	0.0
5	138.8	1.4	−10.0	61.0	3.4	−2.4
6	0.0	0.0	0.0	60.9	8.8	0.0
7	138.4	1.5	−11.4	61.1	3.1	−2.4

It is seen that the mid [−55°] layers fail first at σ_{xx} = 100.7 MPa, followed by the failures of the surface [55°] layers at σ_{xx} = 101.3 MPa. The pure matrix inter-layer failures occur at σ_{xx} = 101.7 MPa. After all of these progressive failures, the ultimate failure of the laminate occurs due to an extremely large strain at σ_{xx} = 102.7 MPa, the same as that in Example 6-3. This means that the introduction of the pure matrix inter-layers does not have any effect on the ultimate failure strength prediction for this problem.

Index

T

U

V

W

Y

Zeitfracht Medien GmbH
Ferdinand-Jühlke-Straße 7
99095 Erfurt, Deutschland
produktsicherheit@kolibri360.de